The Facts On File

DICTIONARY
of
COMPUTER SCIENCE

Revised Edition

The Facts On File

DICTIONARY
of
COMPUTER SCIENCE

Revised Edition

Edited by
John Daintith
Edmund Wright

Facts On File
An imprint of Infobase Publishing

The Facts On File Dictionary of Computer Science
Revised Edition

Facts On File, Inc.
An imprint of Infobase Publishing
132 West 31st Street
New York NY 10001

Library of Congress Cataloging-in-Publication Data
The Facts on File dictionary of computer science. / edited by John Daintith, Edmund Wright. — Rev. ed.
    p.   cm.
Includes bibliographical references
ISBN 0-8160-5999-3
1. Computer science—Dictionaries.   I. Daintith, John.   II Wright, Edmund (Thomas Edmund Farnsworth).   III. Facts on File, Inc.
QA76.15.F345 2006
004.03—dc22

2006042004

Compiled and typeset by Market House Books Ltd, Aylesbury, UK

Printed in the United States of America

  MP   10 9 8 7 6 5 4 3 2 1

This book is printed on acid-free paper

PREFACE

This dictionary is one of a series designed for use in schools. It is intended for students of computer science, but we hope that it will also be helpful to other science students and to anyone interested in computers. Facts On File also publishes dictionaries in a variety of disciplines, including astronomy, chemistry, biology, earth science, physics, mathematics, forensic science, weather and climate, and marine science.

The first edition of this book was published in 2001. This second edition has been extensively revised and extended and now contains over 2,450 headwords covering the terminology of computing. A totally new feature of this edition is the inclusion of over 500 pronunciations for terms that are not in everyday use. A number of appendixes have been included covering mathematical symbols and notation, file extensions, domain names, and numerical conversions. There is also a list of Web sites and a bibliography. A guide to using the dictionary has also been added to this latest version of the book.

We would like to thank all the people who have cooperated in producing this book. A list of contributors is given on the acknowledgments page. We are also grateful to the many people who have given additional help and advice.

ACKNOWLEDGMENTS

Editors (previous editions)

Valerie Illingworth M.Phil.
John O. E. Clark B.Sc.

Contributors

John Illingworth B.Sc., M.Sc.
Martin Tolley B.Sc.
Margaret Tuthill B.Sc.

Pronunciations

William Gould B.A.

CONTENTS

GUIDE TO USING THE DICTIONARY

The main features of dictionary entries are as follows.

Headwords
The main term being defined is in bold type:

> **access** In general, to obtain or write data; for example, to access a file or to access a Web site.

Variants
Sometimes a word has a synonym or alternative spelling. This is placed in brackets after the headword, and is also in bold type:

> **flip-flop** (**bistable**) An electronic circuit that has two stable states, either of which can be maintained until the circuit is made to switch, or 'flip', to the other state.

Here, 'bistable' is another term for 'flip-flop'. Generally, the entry for the synonym consists of a simple cross-reference:

> **bistable** ... *See* flip-flop.

Abbreviations
Abbreviations for terms are treated in the same way as variants:

> **storage area network** (SAN) A high-speed device connected to a network's servers, on which their data is stored rather than on local hard disks.

The entry for the abbreviation consists of a reference:

> **SAN** *See* storage area network.

Often, the main entry is given at the abbreviation:

> **DLL** (**dynamic linked library**) A Microsoft Windows standard for sharing code between programs.

In these cases, the entry for the full form consists of a reference:

> **dynamic link library** *See* DLL.

Multiple definitions

Some terms have two or more distinct senses. These are numbered in bold type

> **cell** **1.** A LOCATION in memory or a REG-
> ISTER... .
> **2.** A box on the screen in a spreadsheet
> program or in a spreadsheet-like grid ... .

Cross-references

These are references within an entry to other entries that may give additional useful information. Cross-references are indicated in two ways. When the word appears in the definition, it is printed in small capitals:

> **case statement** A conditional CONTROL
> STRUCTURE that appears in many program-
> ming languages and allows a selection to be
> made between several choices... .

In this case the cross-reference is to the entry for 'control structure'.

Alternatively, a cross-reference may be indicated by 'See', 'See also', or 'Compare', usually at the end of an entry:

> **scalable font** Any font that can be
> scaled to produce characters of different
> sizes without any loss in quality. *See also*
> outline font.

Hidden entries

Sometimes it is convenient to define one term within the entry for another term:

> **scripting language** A language that al-
> lows the user to execute a sequence of op-
> erations automatically ... by specifying
> them in a text file called a *script*.

Here, 'script' is a hidden entry under 'scripting language', and is indicated by italic type.
The individual entries consists of a simple cross-reference:

> **script** *See* scripting language.

Pronunciations

Where appropriate pronunciations are indicated immediately after the headword, enclosed in forward slashes:

> **biometrics** /bў-ŏ-**met**-riks/ The study,
> measurement, and analysis of human bio-
> logical characteristics.

Note that simple words in everyday language are not given pronunciations. Also headwords that are two-word phrases do not have pronunciations if the component words are
pronounced elsewhere in the dictionary.

Pronunciation Key

A consonant is sometimes doubled to prevent accidental mispronunciation of a syllable resembling a familiar word; for example, /**ass**-id/ /acid/,rather than /**as**-id/ and /ul-tră-/**sonn**-iks// /ultrasonics/, rather than /ul-tră-**son**-iks/. An apostrophe is used: (a) between two consonants forming a syllable, as in /**den**-t'l/ /dental/,and (b) between two letters when the syllable might otherwise be mispronounced through resembling a familiar word, as in /**th'e**-ră-pee/ /therapy/ and /tal'k/ /talc/. The symbols used are:

/a/ *as in* back /bak/, active /**ak**-tiv/
/ă/ *as in* abduct /ăb-**dukt**/, gamma /**gam**-ă/
/ah/ *as in* palm /pahm/, father /**fah**-*th*er/
/air/ *as in* care /kair/, aerospace /**air**-ŏ-spays/
/ar/ *as in* tar /tar/, starfish /**star**-fish/, heart /hart/
/aw/ *as in* jaw /jaw/, gall /gawl/, taut /tawt/
/ay/ *as in* mania /**may**-niă/ ,grey /gray/
/b/ *as in* bed /bed/
/ch/ *as in* chin /chin/
/d/ *as in* day /day/
/e/ *as in* red /red/
/ĕ/ *as in* bowel /**bow**-ĕl/
/ee/ *as in* see /see/, haem /heem/, caffeine //**kaf**-een/,/ baby /**bay**-bee/
/eer/ *as in* fear /feer/, serum /**seer**-ŭm/
/er/ *as in* dermal /**der**-măl/, labour /**lay**-ber/
/ew/ *as in* dew /dew/, nucleus /**new**-klee-ŭs/
/ewr/ *as in* epidural /ep-i-**dewr**-ăl/
/f/ *as in* fat /fat/, phobia /**foh**-biă/, rough /ruf/
/g/ *as in* gag /gag/
/h/ *as in* hip /hip/
/i/ *as in* fit /fit/, reduction /ri-**duk**-shăn/
/j/ *as in* jaw /jaw/, gene /jeen/, ridge /rij/
/k/ *as in* kidney /**kid**-nee/, chlorine /**klor**-een/, crisis /**krÿ**-sis/
/ks/ *as in* toxic /**toks**-ik/
/kw/ *as in* quadrate /**kwod**-rayt/
/l/ *as in* liver /**liv**-er/, seal /seel/
/m/ *as in* milk /milk/
/n/ *as in* nit /nit/
/ng/ *as in* sing /sing/

/nk/ *as in* rank /rank/, bronchus /**bronk**-ŭs/
/o/ *as in* pot /pot/
/ô/ *as in* dog /dôg/
/o/ *as in* buttock /**but**-ŏk/
/oh/ *as in* home /hohm/, post /pohst/
/oi/ *as in* boil /boil/
/oo/ *as in* food /food/, croup /kroop/, fluke /flook/
/oor/ *as in* pruritus /proor-ÿ-tis/
/or/ *as in* organ /**or**-găn/, wart /wort/
/ow/ *as in* powder /**pow**-der/, pouch /powch/
/p/ *as in* pill /pil/
/r/ *as in* rib /rib/
/s/ *as in* skin /skin/, cell /sel/
/sh/ *as in* shock /shok/, action /**ak**-shŏn/
/t/ *as in* tone /tohn/
/th/ *as in* thin /thin/, stealth /stelth/
/*th*/ as in then /*th*en/, bathe /bay*th*/
/u/ *as in* pulp /pulp/, blood /blud/
/ŭ/ *as in* typhus /**tÿ**-fŭs/
/û/ *as in* pull /pûl/, hook /hûk/
/v/ *as in* vein /vayn/
/w/ *as in* wind /wind/
/y/ *as in* yeast /yeest/
/ÿ/ *as in* bite /bÿt/, high /hÿ/, hyperfine /**hÿ**-per-fÿn/
/yoo/ *as in* unit /**yoo**-nit/, formula /**form**-yoo-lă/
/yoor/ *as in* pure /pyoor/, ureter /yoor-**ee**-ter/
/ÿr/ *as in* fire /fÿr/
/z/ *as in* zinc /zink/, glucose /**gloo**-kohz/
/zh/ *as in* vision /**vizh**-ŏn/

abend /ă-**bend**/   The premature stopping of a program caused by a program error, system failure, power failure, etc. The term is a contraction of 'abnormal end'. *See also* abort.

abort   To halt a processing activity in a computer before some planned conclusion has been reached. The activity is halted, by the OPERATING SYSTEM or by human intervention, because a point has been reached beyond which processing cannot continue. The process may have tried to obey an undefined instruction or may have failed to obey some operating condition of the computer. It is also possible for the processing activity to bring itself to a halt when it becomes apparent that it cannot reach a successful conclusion. A person monitoring the progress of a program may terminate it because it has gone on too long, or is producing meaningless or incorrect output.

ABS /abz/   *See* Basic.

absolute address   **1.** An ADDRESS that identifies a storage location (or a device) without the use of any intermediate reference. *See also* addressing mode.
2. An address permanently assigned by the machine designer to a storage location.

absolute addressing   *See* direct addressing.

accelerated graphics port   *See* AGP.

accelerator board (accelerator card)   A printed circuit board that replaces or augments the processor to give enhanced performance.

accelerator key (shortcut key)   A key or combination of keys that can be used to carry out a specified function. For example, ctrl+V (pressing the control key and the letter V) is a method of pasting data in many applications.

access   In general, to obtain or write data; for example, to access a file or to access a Web site.

Access (Microsoft Access)   A DATABASE MANAGEMENT SYSTEM for RELATIONAL DATABASES developed by Microsoft. First released in 1992, it has been upgraded several times and the current version is Access 2003. Access forms part of some versions of Microsoft OFFICE.

access methods   *See* file; database.

access privilege   The status given to a particular user or group of users on a single computer or on a network, specifying what data can be viewed and which system operations are allowable. Payroll information or medical records, for example, would generally be restricted in availability.

access provider   An organization, usually a commercial company, that provides users with access to the Internet. *See* ISP.

access time   The time taken to retrieve a particular item of data from a storage location. This is equivalent to the time interval between an item of data being called from the location and being ready to use. The access time depends on the method by which the data is accessed and on the type of storage.
   There is RANDOM ACCESS to data in MAIN STORE, and the data is available for use al-

most immediately. There is also random access to data on magnetic disk BACKING STORE, but the access time is considerably longer than for main store. The access time for disk storage has two components: the time taken for the read/write head to get to a particular TRACK is called the *seek time*; the delay allowing an item of data in that track to become available under the read/write head is known as the *latency*, and on average is equal to half the time for a complete revolution of the disk. The time required to transfer data between disk and main store also forms part of the access time.

In the case of MAGNETIC TAPE, data is retrieved by SERIAL ACCESS and it may take several minutes to reach a particular storage location.

account   A method of identifying a particular user on a computer or a network, usually by a unique name. An account can be used for the allocation of access and facilities such as E-MAIL. On a commercial network accounts can be used to keep track of billing information. An account is usually protected by a password, which is known only to the user and can be changed only by the user or the system administrator.

accounting package   A software application used to automate the accounting operations of a business.

accumulator   An electronic device – a REGISTER – that acts as a temporary store in the arithmetic and logic unit (ALU) and in which the result of an arithmetic or logic operation is formed. For example, a number (held in main store) can be added to the existing contents of the accumulator and the total would then appear in the accumulator; the addition is performed by the ALU. If, say, the number 5 is held in the accumulator and the number 12 is to be added, then the contents will change to 17. The contents of the accumulator are in binary form and the number of bits stored is normally equal to the WORD LENGTH. The above example would cause the contents of an 8-bit accumulator to change from

00000101

to

00010001

In some processors there is only one accumulator. In others there are a number of general-purpose registers, any one of which can act as an accumulator.

ACK /ak/   A message ('acknowledge'), such as ASCII character 6 transmitted as a control code, sent from a receiving device to a sending device to show that a transmission was successfully received and, if appropriate, that the receiving device is ready for the next transmission. *See also* acknowledgment.

acknowledgment   A means by which a device receiving transmitted data can indicate to the sending device whether the data has been correctly received or not. The receiving device returns a confirming message, known as a *positive acknowledgment*, for each block of data received correctly. When an error is detected in a block of data, the receiver returns a *negative acknowledgment* to indicate that this block should be retransmitted by the sending device. *See also* ACK; NAK.

A failure on the part of the receiver to send an acknowledgment of either kind within a set time, or the failure of the sender to respond to an acknowledgment by sending another block in a set time, is called a TIMEOUT, and normally results in retransmission of the last block or another acknowledgment.

acoustic coupler   A device formerly used to convert binary data into a form that could be transmitted along a telephone line. An ordinary telephone handset was placed in a depression in the acoustic coupler to make the connection between coupler and telephone system. The resulting telephone signal was a sequence of low- and high-pitched sounds representing the binary 0s and 1s making up the data. An acoustic coupler could also accept such a signal from a telephone line, converting it into the corresponding data that could then be fed to a computer. The binary data was fed into and out of the acoustic coupler by cable. Acoustic couplers have now been superseded by MODEMS.

Acrobat *Trademark* A collection of programs developed by Adobe Systems, Inc. for reproduction of printed documents in their original form and for electronic publications. The system involves conversion of a formatted file (from a word processor, DTP application, etc.) into a portable document format (PDF) file. This file format can be read by the application *Acrobat Reader*, which is freely available. The PDF file contains all the formatting and fonts of the original document and is particularly useful for electronic transmission or publication of documents in which layout is important (e.g. official publications, legal documents, etc.). PDF files can be generated directly from many applications. They can also be produced from PostScript files using the application *Acrobat Distiller*. PDF files can also have bookmarks and hyperlinks embedded and can be indexed. *See also* PostScript.

active Operating or being used. For example, the *active cell* in a spreadsheet is the one selected at a particular time; the *active window* in an application is the one affected by input from the keyboard or cursor.

Active Directory *Trademark* A method of organizing a Windows network. Every object on the network – users, groups of users, printers, disk DIRECTORIES, network services (e.g. e-mail) – can be arranged in a logical structure that controls which objects can interact with which other objects and in what way: for example which users can access the contents of a given disk directory, which users can manage which other users, etc. This logical structure need not match the physical structure of the network. If the sales department of a large company is spread across multiple sites and so multiple LANs, Active Directory can be configured to make it appear as a single unit to the rest of the network.

active matrix display (**thin film transis-**tor display; **TFT display**) A high-quality color display using LCD technology. Formerly used only in laptop and notepad computers, active matrix displays are now found in flat panel displays for desktop computers.

Active Server Pages *See* ASP.

ActiveX /ak-tă-veks/ *Trademark* A development by Microsoft of a loosely defined collection of technologies based on COM used to enable different software components to interact. The most common example is *ActiveX controls*, which are separately developed components that add functionality to programs or Web pages. Examples range from reusable program components, such as calendars, clocks, and various types of data entry/display tools, to complex special effects for Web pages.

Ada /ay-dă/ *Trademark* A programming language named after Ada, Lady Lovelace, the mathematician and sometime assistant to Charles Babbage. Ada was developed as a result of a competition for a new programming language organized by the US Department of Defense in the late 1970s. It was designed to be used in EMBEDDED systems, i.e. in situations where a computer forms part of a specialized system such as a cruise missile or a manufacturing plant. Ada is a large high-level language embodying concepts developed in earlier languages such as PASCAL, but carrying them much further. There is now an international standard for Ada.

adapter A printed circuit board inserted into a computer in order to enable it to interface with a peripheral device, such as a modem or sound card.

Adaptive Server Enterprise *See* SQL Server.

ADC *See* A/D converter.

A/D converter (**analog to digital converter; ADC**) A device for sampling an ANALOG SIGNAL and converting it into an equivalent DIGITAL SIGNAL, i.e. for converting a continuously varying signal, such as a voltage, into a series of binary values that are suitable for use by a computer. *See also* D/A converter.

A	0110
B	0100
S	1010
C	0100

Binary addition

adder   An electronic device – a LOGIC CIRCUIT – that adds together two numbers. It can also subtract two numbers using two's complement arithmetic (*see* complement). The numbers are represented as binary signals that are fed into the device (*see* digital signal). The binary output represents the sum of the numbers.

When two binary (or decimal) numbers are added, each step of the addition generates a *sum digit*, which forms part of the answer, and a *carry digit*, which is passed to the next step. In the table, S is the sum and C the carry resulting from the addition of two binary numbers A and B, equal to 0110 (i.e. 6) and 0100 (i.e. 4). This is a type of TRUTH TABLE.

Within the adder, addition is performed electronically on corresponding pairs of bits (the two least significant bits, the two next least significant bits, etc.). The circuitry involved in this must be able not only to add a particular pair of bits to produce a sum bit but also to add the carry bit from the previous step in the addition. The sum bits together form the final answer that is output from the device. The final carry bit is discarded or is used to detect OVERFLOW.

add-in card   *See* expansion card.

add-on   1. A device that is external to a computer and plugged into it in order to increase its capabilities.
2. A small program that enhances a larger one.

address   1. A number used to specify a LOCATION within computer memory, usually to allow storage or retrieval of an item of information at this location. The words 'address' and 'location' are actually used as synonyms, as in ordinary speech. In backing store an address identifies a specific SECTOR on a magnetic disk or a specific BLOCK on a magnetic tape. In main store an address identifies a particular location holding a WORD or a BYTE. When a location is specified it is said to be *addressed*. In some computers, processor REGISTERS and/or INPUT/OUTPUT DEVICES can also be addressed.

The reference to an address appears in a MACHINE INSTRUCTION. An instruction is normally made up of a combination of an OPERATION CODE and some way of specifying the OPERAND (or operands) upon which the operation is to be performed. This is commonly done by specifying the MACHINE ADDRESS of the operand. The process is known as DIRECT ADDRESSING.

Direct addressing is one of a variety of processes by which the storage location to be accessed is determined from the machine instruction. Alternative ADDRESSING MODES are used in cases where the machine address is too large to be comfortably included in a machine instruction, or where it is not necessary or not possible to assign an explicit address.
2. The means used to specify the location of a participant in a computer NETWORK; in particular, the sequence of characters used to specify the destination of an e-mail.

address bus   *See* bus.

addressing mode   Any of various methods by which an ADDRESS can be specified in the address part of a MACHINE INSTRUCTION. The methods include DIRECT ADDRESSING, IMMEDIATE ADDRESSING, INDIRECT ADDRESSING, INDEXED ADDRESSING, and RELATIVE ADDRESSING, all of which are used in computer systems. Whatever method is used, the computer's hardware applies appropriate rules to derive an ABSOLUTE ADDRESS. See table overleaf.

address part   A part of a MACHINE IN-

ADDRESSING MODES	
Instruction	*Source of value*
move 1,1	the contents of location 1
move immediate 1,1	the value 1, given in the instruction
move indirect 1,1	the value found at the address stored at location 1
move indexed 1,1 (2)	the value found at the address formed by adding 1 to contents of register 2
move 1, TABLE + 3	the value found at the address formed by adding 3 to the value of the symbol 'TABLE'

STRUCTION that usually contains only an ADDRESS, or information on where to find or how to construct an address. There may be more than one address part in an instruction. In a JUMP INSTRUCTION one of the addresses may specify the location of the next instruction to be executed.

address register A REGISTER (i.e. a temporary location) in which an ADDRESS is stored. *See also* instruction address register; memory address register.

address resolution The specification of a computer's address in a form that enables it to communicate effectively with the Internet.

address resolution protocol (ARP) A standard used to specify the address of a computer or other item of hardware on a local area network (LAN) that is connected to the Internet.

address space *See* machine address.

administrator (**superuser**; **supervisor**) The controlling ACCOUNT on a computer or network. People who log on to this account have few, if any, restrictions on their actions: they can access all data, issue any command supported by the operating system or application programs, and create, manage, and delete other accounts. Such accounts are necessary for effective system management, but are security risks. It is therefore important that their passwords are known only to authorized people.

Adobe Type Manager /ă-**doh**-bee/ (**ATM**) Software produced by Adobe Systems, Inc. used to control PostScript FONTS.

ADSL (**asymmetric digital subscriber line**) A standard for high-speed digital information transfer, including video, through ordinary copper telephone lines using a variant of DSL. It transfers data much faster in one direction than in the other and can share a wire with a telephone; this makes it suitable for Internet users, who typically download much more data than they upload. Maximum speeds are 8Mbps downloading and 800 Kbps uploading. The actual speed declines with the length of the wires and operational speeds are slower: 1Mbps downloading and 256Kbs uploading. ADSL was developed to compete with cable companies using fiber-optic transmission lines. *Compare* SCSL. *See also* VDSL.

ADSL transceiver A peripheral device that translates computer BINARY DATA into the format required for ADSL transmission, and vice versa. Because of the similarity of its functions to those of a MODEM, it is often incorrectly called an *ADSL modem*.

Advanced Research Project Agency Network *See* ARPANET.

Advanced Technology Attachment
See ATA.

adware 1. A piece of software that displays unwanted advertisements. It is usually concealed in Web pages or software downloaded from the Internet. *See also* spyware; malware.
2. Applications that display advertisements. The revenue from these allows the program to be cheap or even free of charge to the user.

agent A program that runs in the background but informs the user when specified conditions are met. A common type of agent searches through databases for information on a particular user-defined topic. Such programs are useful for searching the Internet, usually across one type of information resource only. They can be set to perform searches at regular intervals, reporting back with any information, or its location, when found.

AGP (accelerated graphics port) A specification for a high-performance BUS running directly between a graphics controller and a computer's memory. With AGP, video data does not have to be transferred through the primary (PCI) bus and the consequent increase in speed gives smoother and faster image transfer and display.

AI *See* artificial intelligence.

alert box A panel displayed on-screen to give a user a warning or important message.

Algol /al-gol/ A high-level PROGRAMMING LANGUAGE whose name is derived from **algo**rithmic language. It was the result of the deliberations of an international committee in the late 1950s, which culminated in the Algol-60 Report in 1960. Although never widely used and now defunct, Algol-60 was a turning point in language development since it was designed to reflect mathematical ALGORITHMS rather than the architecture of computers. It strongly influenced many succeeding languages, such as PASCAL, C, and ADA.

algorithm /al-gŏ-ri*th*-'m/ A set of simple and clearly defined instructions or steps that when followed precisely enables some operation to be performed. No personal judgments have to be made. The operation could involve, say, baking a cake, constructing a cupboard, building a house, producing business accounts, and so on. It may involve a mathematical calculation – e.g. addition, multiplication, taking a square root – or a selection process. Algorithms can be devised for a great variety of operations, ranging widely in complexity. As the complexity increases, the number of steps to be taken increases. The same operation can often be performed by different algorithms.
  Algorithms are used for the solution of problems. Since they involve no personal judgment, they are suitable subjects for computers to handle. When a computer is to perform some specific task, an appropriate algorithm is selected or is specially written. It is expressed in a formal notation – a PROGRAMMING LANGUAGE – and as such forms a major part of a computer PROGRAM. Much that is said about programs applies to algorithms, and vice versa.

algorithmic language /al-gŏ-**ri***th*-mik/ A programming language, for example C++ or Basic, that uses algorithms for processing.

alias A name for a person or group using e-mail that is simpler or easier to remember than the full name.

aliasing The production of visibly jagged edges on curves or diagonals, caused when the resolution of a screen or printer is too low.

align To space text so that it neatly touches a reference line, especially the left or right margin. *See also* justify.

allocation *See* resource allocation; storage allocation.

alpha /al-fă/ Describing the initial version of a new piece of software, tested by a small number of users. *See* beta.

alphanumeric character   Any of the 26 letters of the alphabet, capital or lower case (i.e. A–Z, a–z), or any of the digits 0–9. An *alphanumeric character set* is a CHARACTER SET that contains the alphanumeric characters and may in addition include some SPECIAL CHARACTERS and the SPACE CHARACTER.

AltaVista /al-tă-**viss**-tă/   A highly popular Internet SEARCH ENGINE.

alt key /awlt/   A key found on some keyboards that is similar to the control key and is used in conjunction with other keys. In Windows, for example, pressing the alt key and the tab key cycles the focus through the active applications. An alt key is a standard feature on PC keyboards.

ALU   *See* arithmetic and logic unit.

AMD   *Trademark* The world's second-largest manufacturer of MICROPROCESSORS, Advanced Micro Devices, Inc., founded in 1969. Its chips are compatible with those produced by INTEL.

America Online   *See* AOL.

amplitude   *See* signal.

analog computer   A computer that accepts and processes data in the form of a continuously variable quantity whose magnitude is made proportional to the value of the data. An analog computer solves problems by physical analogy, usually electrical. Each quantity (e.g. temperature, speed, displacement) that occurs in the problem is represented in the computer by a continuously variable quantity such as voltage. Circuit elements in the computer are interconnected in such a way that the voltages fed to them interact in the same way as the quantities in the real-life problem interact. The output voltage then represents the numerical solution of the problem.

Analog computers are used in situations where a continuous representation of data is more appropriate and so both more accurate and much faster than the discrete representation that is required by a DIGITAL COMPUTER. Their major application is in SIMULATION, i.e. in imitating the behavior of some existing or intended system, or some aspect of that behavior, for purposes of design, research, etc. Another important application is PROCESS CONTROL in industry or manufacturing, where data has to be continuously monitored.

An analog computer can be used in conjunction with a digital computer, forming what is known as a *hybrid computer*. A hybrid computer, which can process both analog and discrete data, often has advantages over an analog computer in, say, process control.

analog signal   A continuously varying value of voltage or current. *Compare* digital signal.

analog to digital converter   *See* A/D converter.

Analytical Engine   A calculating machine planned by Charles Babbage in 1834, but never constructed. The machine was to follow his DIFFERENCE ENGINE but to be of more general application and to be programmed by punched cards. Babbage was supported in this work by Ada Lovelace (*see* Ada). A full-scale working version of the machine was constructed at the London Science Museum to celebrate Babbage's bicentenary in 1992.

AND gate   A LOGIC GATE whose output is high only when all (two or more) inputs are high; otherwise the output is low. It thus performs the AND OPERATION on its inputs and has the same TRUTH TABLE. The truth table for a gate with two inputs (A, B) is shown overleaf, where 0 represents a low signal level and 1 a high level.

AND operation   A LOGIC OPERATION combining two statements or formulae, A and B, in such a way that the outcome is true only if A and B are both true; otherwise the outcome is false. The TRUTH TABLE for the AND operation is shown overleaf, where the two truth values are represented by 0 (false) a

A	B	output/outcome
0	0	0
0	1	0
1	0	0
1	1	1

Truth table for AND gate and AND operation

nd 1 (true). The operation can be written in several ways, including

A AND B   A & B   A.B

A and B are known as the OPERANDS and the symbol between them (AND, &, .) is a LOGIC OPERATOR for the AND operation.

In a computer the AND operation is used in HIGH-LEVEL LANGUAGES to combine two logical expressions according to the rules stated above. It is used in both high- and LOW-LEVEL LANGUAGES to combine two BYTES or WORDS, producing a result by performing the AND operation on each corresponding pair of bits in turn. If 8-bit words are used, then for example with operands

01010001
11110000

the outcome of the AND operation is

01010000

See also AND gate.

animation   The creation of moving pictures. The technique is to display a succession of static pictures, each slightly different. Software also exists for automatically producing intermediate images between a starting image and a finishing image. Sophisticated computer animation is now an established technique in the film industry.

annotation   Explanation added to a program to assist the reader, usually in the form of COMMENTS included in the program text.

anonymous FTP (**anonymous file transfer protocol**)   A system that allows a user to access an FTP site on a network and download files without having an account by using the special user 'anonymous'. This gives only limited privileges on the site (e.g. access only to certain directories).

ANSI /**an**-see/   The American National Standards Institute, the organization that determines US industrial standards, including some used in computing, and ensures that these correspond to those set by the International Organization for Standardization, ISO.

AOL (**America Online**)   A large commercial Internet service provider.

API (**application programming interface**)   A tightly defined set of FUNCTIONS and PROCEDURES that an application program or the operating system makes available for utilization by other application programs.

APL   A high-level PROGRAMMING LANGUAGE whose name is derived from the words **a** programming language. It was developed by Kenneth Iverson in the early 1960s for interactive computing, which was a new concept at the time. Its main feature is its unusual SYNTAX, in which single character operators perform complex tasks. Unfortunately the APL CHARACTER SET is quite different to other programming languages.

Apple   *Trademark*   A manufacturer of microcomputers and other computer-related products founded in 1976 by Steve Jobs and Steve Wozniak. It is noted for its MACINTOSH range of personal computers.

applet /**ap**-let/   A small program usually written in JAVA that can be downloaded onto a user's computer along with a Web page and run by the browser. Applets are commonly used to add interest to Web pages, for example, animation.

AppleTalk /**ap**-'l-tawk/   A set of network PROTOCOLS designed by APPLE and incorporated in MACINTOSH computers. It has now been superseded by TCP/IP but is still supported for BACKWARD COMPATIBILITY. *See also* LocalTalk.

application   A computer program that has a specific use; for example, a word-processing program, a graphics program, or a database.

application package (software package) A set of programs directed at some application in general, such as COMPUTER GRAPHICS, WORD PROCESSING, CAD (computer-aided design), or statistics.

application programming interface *See* API.

applications-orientated language *See* high-level language.

Archie /ar-chee/ A utility for searching for files in FTP sites on the Internet.

architecture The description of a computer system at a somewhat general level, including details of the instruction set and user interface, input/output operation and control, the organization of memory and the available addressing modes, and the interconnection of the major units. A particular architecture may be implemented by a computer manufacturer in the form of several different machines with different performance and cost.

archive *See* file archive.

argument (actual parameter) A value or address that is passed to a PROCEDURE, SUBROUTINE, or FUNCTION at the time it is called. For example, in the Basic statement
$$Y = SQR(X)$$
X is the argument of the SQR (square root) function. *See also* parameter.

arithmetic and logic unit /ă-rith-mĕ-tik/ (ALU; arithmetic unit) The part of the processing unit of a computer in which ARITHMETIC OPERATIONS, LOGIC OPERATIONS, and related operations are performed. The ALU is thus able, for example, to add or subtract two numbers, negate a number, compare two numbers, or do AND, OR, or NOT operations. The choice of operations that can be carried out by an ALU depends on the type (and cost) of the computer. The ALU is wholly electronic.

The operations are performed on items of data transferred from main store into registers – temporary stores – within the ALU. In some processors all arithmetic and logic OPERANDS have to be placed in a special register called the ACCUMULATOR, while in other processors any register may be used. The result of the operation is subsequently transferred back to the main store. The movement of data between main store and ALU is under the direction of the CONTROL UNIT. The arithmetic or logic operations to be performed in the ALU are specified in the operation part of MACHINE INSTRUCTIONS. The control unit interprets each instruction as it is fetched from main store and directs the ALU as to which operation (if any) is required.

arithmetic instruction A machine instruction specifying an ARITHMETIC OPERATION and the OPERAND or operands on which the arithmetic operation is to be performed. An example, expressed in ASSEMBLY LANGUAGE, might be
ADDI 3 4
This is an instruction to add 4 to contents of register 3, placing the result in register 3 and setting the carry bit if the result is too big to fit.
See also logic instruction.

arithmetic/logic unit *See* arithmetic and logic unit.

arithmetic operation An operation that follows the rules of arithmetic, the most commonly occurring examples being addition, subtraction, multiplication, and division. In computing, arithmetic operations may be carried out on signed or unsigned INTEGERS or REAL NUMBERS. They are normally performed in the ARITHMETIC AND LOGIC UNIT of a computer. *See also* arithmetic operator; operand.

arithmetic operator A symbol representing a simple arithmetic operation (e.g. addition or multiplication) that is to be performed on numerical data, quantities, etc. The operators used in a particular programming language may differ from those in general use, as shown in the table. The operations
7 multiplied by 2
6 divided by 3

ARITHMETIC OPERATORS		
Operation	*Operators*	
	In general use	*In Basic*
addition	+	+
subtraction	–	–
multiplication	× or .	*
division	÷ or /	/
exponentiation	5^2	$5 \uparrow 2$

would thus be written as 7*2 and 6/3 in most high-level languages. Some languages have separate operators for integer division and remaindering (*see* integer arithmetic). Some languages do not have an operator for EXPONENTIATION.

An example of how arithmetic operators are used, in Basic, is as follows:

D = SQR((X(1) – X(0)) $\uparrow$ 2 + X(2)*3)

This is equivalent to

$$D = \sqrt{((X_1 - X_0)^2 + 3X_2)}$$

arithmetic shift *See* shift.

arithmetic unit *See* arithmetic and logic unit.

ARP *See* address resolution protocol.

ARPANET /ar-pă-net/ (**Advanced Research Project Agency Network**) A network introduced in 1969 connecting a small number of research institutions, funded by the US Department of Defense. ARPANET used PACKET SWITCHING and pioneered many of the protocols used in network operation. It was the forerunner of the INTERNET.

array One form in which a collection of related data items can be stored in computer memory. The data items in an array are arranged in a particular order or pattern and are all of the same type, for example all integers or all real numbers. This collection of data items is referred to as an array. More usually, however, the word array refers to the set of storage LOCATIONS in which the data items are placed, keeping their original arrangement.

The set of locations forming an array is referenced by a single IDENTIFIER, chosen by the programmer. Each element in an array (i.e. a location or its contents) can be specified by combining one or more *subscript* values with the identifier. Subscripts are usually integers and are generally placed in brackets after the identifier. The number of subscripts required to specify an element gives the *dimension* of the array.

The simplest array is a single sequence of elements. This is a *one-dimensional array*, only one subscript being necessary to select a particular element. For example, a list of people's ages could form an array named AGE; if the subscript of the first element is 1, the age of the eighth person in the list is found by specifying AGE(8). The subscript may be a VARIABLE or an EXPRESSION. A one-dimensional array is also known as a *vector*.

In a *two-dimensional array* (also called a *matrix*), the elements are arranged in the form of a table with a fixed number of *rows* and a fixed number of *columns*. Each element is distinguished by a pair of subscripts; the first subscript gives the row number, the second gives the column number. For example, A(3,7) refers to the element in row three and column seven of the array A. Again, the subscripts may be variables or expressions.

The values of a subscript range from a lower limit (usually 1 or 0 unless otherwise specified) to an upper limit. These limits specify the total number of elements in an array, and are called *bounds*. The bounds of an array can be declared in various ways, depending on the programming language. In Basic, for example, a DIMENSION statement is used:

DIM X(4,10)

This is a declaration of a two-dimensional array, X, with 5 rows and 11 columns, since the default lower bound in Basic is zero.

arrow keys Four keys on a KEYBOARD that are labeled with up, down, left, and right arrow symbols. The arrow keys can be used for control of the CURSOR on a display screen.

artificial intelligence (AI) The branch of computer science concerned with programs that carry out tasks requiring intelligence when done by humans. Many of these tasks involve a lot more computation than is immediately apparent because much of the computation is unconscious in humans, making it hard to simulate. Programs now exist that play chess and other games at the highest level, take decisions based on available evidence, prove theorems in certain branches of mathematics, recognize connected speech using a limited vocabulary, and use television cameras to recognize objects. Although these examples sound impressive, the programs have limited ability, no creativity, and each can only carry out a limited range of tasks. There is still a lot more research to be done before the ultimate goal of artificial intelligence is achieved, which is to understand intelligence well enough to make computers more intelligent than people. In fact there is considerable controversy about the whole subject, with many people arguing that the human thought process is different in kind to the computational operation of computer processes and so cannot be reproduced using computers. *See also* expert system; robot.

artificial life A branch of ARTIFICIAL-INTELLIGENCE research concerned with investigations into the behavior of living organisms. Computer simulations are used to show how self-replicating organisms might behave.

ascending order The arrangement of a set of data in order starting with the lowest character value, as in A to Z or 1 to 50.

ASCII /*as*-kee/ American National Standard Code for Information Interchange, a standard code for the interchange of information between computer systems, data communication systems, and associated equipment. Since it is a standard code (rather than one developed by a particular manufacturer), it allows equipment of different manufacturers to exchange information. It is thus widely used. ASCII encoding produces coded characters of 7 bits, and hence provides 2^7, i.e. 128, distinct bit patterns. These 128 characters make up the ASCII CHARACTER SET: they consist of ALPHANUMERIC CHARACTERS, the SPACE CHARACTER, SPECIAL CHARACTERS, and CONTROL CHARACTERS. The character set and the control characters are explained in the table overleaf. The binary encodings shown in the table are for the character

$$b_7b_6b_5b_4b_3b_2b_1$$

Thus the encoding for G is 1000111 and the encoding for g is 1100111. ASCII became the de facto standard character set for most computers and has been incorporated in such larger character sets as UNICODE. The coding from values 129 to 256 is sometimes called *extended ASCII*. *See also* ISO-7.

ASP (Active Server Pages) A SCRIPTING LANGUAGE developed by Microsoft for SERVER-SIDE SCRIPTING. ASP was first released in 1996 and the latest version, ASP.NET, is part of the .NET software-development platform.

aspect ratio The ratio of width to height for a rectangular shape, such as an illustration or window.

assembler A program that takes as input a program written in ASSEMBLY LANGUAGE and translates it into MACHINE CODE. Each instruction in assembly language is

ASCII CHARACTER SET

b4 b3 b2 b1	b7 b6 b5 = 0 0 0	0 0 1	0 1 0	0 1 1	1 0 0	1 0 1	1 1 0	1 1 1
0 0 0 0	NUL	DLE	space	0	@	P	`	p
0 0 0 1	SOH	DC1	!	1	A	Q	a	q
0 0 1 0	STX	DC2	"	2	B	R	b	r
0 0 1 1	ETX	DC3	#	3	C	S	c	s
0 1 0 0	EOT	DC4	$	4	D	T	d	t
0 1 0 1	ENQ	NAK	%	5	E	U	e	u
0 1 1 0	ACK	SYN	&	6	F	V	f	v
0 1 1 1	BEL	ETB	'	7	G	W	g	w
1 0 0 0	BS	CAN	(	8	H	X	h	x
1 0 0 1	HT	EM	)	9	I	Y	i	y
1 0 1 0	LF	SUB	*	:	J	Z	j	z
1 0 1 1	VT	ESC	+	;	K	[	k	{
1 1 0 0	FF	FS	,	<	L	\	l	|
1 1 0 1	CR	GS	–	=	M	]	m	}
1 1 1 0	SO	RS	.	>	N	^	n	~
1 1 1 1	SI	US	/	?	O	—	o	DEL

ASCII CONTROL CHARACTERS

NUL	null character	DLE	data link escape
SOH	start of header	DC1	device control 1
STX	start of text	DC2	device control 2
ETX	end of text	DC3	device control 3
EOT	end of transmission	DC4	device control 4
ENQ	enquiry	NAK	negative acknowledge
ACK	acknowledge	SYN	synchronous idle
BEL	bell	ETB	end of transmission block
BS	backspace	CAN	cancel
HT	horizontal tabulation	EM	end of medium
LF	line feed	SUB	substitute
VT	vertical tabulation	ESC	escape
FF	form feed	FS	file separator
CR	carriage return	GS	group separator
SO	shift out	RS	record separator
SI	shift in	US	unit separator
		DEL	delete

usually converted into one machine instruction. The input to the assembler is called the *source program*; the output is called an OBJECT MODULE. The translation process is known as *assembly* and the program that is translated is said to have been *assembled*. The entire program must be assembled before it can be executed. *See also* compiler; interpreter.

assembly language A type of PROGRAMMING LANGUAGE that is a readable and convenient notation (in human terms) for representing programs in MACHINE CODE. Assembly language was originally devised to alleviate the tedious and time-consuming task of writing programs actually in machine code. It may be used nowadays, in preference to a HIGH-LEVEL LANGUAGE, for reasons of speed or compactness.

Assembly language is the most commonly used LOW-LEVEL LANGUAGE. Different forms are developed for different computers, usually by the computer manufacturers. The features of a particular assembly language thus reflect the facilities of the machine on which it is employed.

Each MACHINE INSTRUCTION has a corresponding assembly language instruction. The programmer can use alphabetic OPERATION CODES with mnemonic significance (e.g. LDA is commonly used for 'load accumulator'), and can use symbolic names, personally chosen, for ADDRESSES of storage locations or registers (*see* symbolic addressing). Addressing modes can also be expressed in a convenient way. In addition assembly language allows the use of various number systems (e.g. DECIMAL, OCTAL, HEXADECIMAL NOTATION) for numerical constants, and allows the programmer to attach LABELS to lines of the program.

A program known as an ASSEMBLER translates the assembly language program into machine code. It can then be executed by the computer.

assignment statement A program STATEMENT that assigns a new value to a VARIABLE. Each variable is associated with a particular LOCATION or group of locations in memory. An assignment statement will thus cause a new value to be placed at the appropriate storage location(s). The assignment statement is fundamental to most programming languages. It is indicated by a special symbol, such as = or :=. This is called the *assignment operator*. The symbol used depends on the language.

The expression on the right of the assignment operator is given to the variable whose name appears on the left. If, say, the value of a variable C is to be increased by one, this is expressed in Basic, Fortran, C, and Java as

C = C + 1

and in Pascal and Algol as

C := C + 1;

and is read as C becomes C + 1. In a program FLOWCHART the symbol $\rightarrow$ is often used to indicate an assignment, as in

C $\rightarrow$ C + 1

associative store (**content-addressable store**) A storage device in which a LOCATION is identified by what is in it rather than by its position. The location contains a particular item of data – a *search word* – for which a search can be conducted by the storage device. This can be achieved in various ways. The desired data in the location is in close association or proximity to the search word. Associative stores are used, for example, as part of a VIRTUAL STORAGE facility.

asymmetric digital subscriber line /ay-să-**met**-rik/ *See* ADSL.

asynchronous /ay-**sink**-rŏ-nŭs/ Involving or requiring a form of timing control in which a specific operation is begun on receipt of a signal to indicate that the preceding operation has been completed. Operations do not therefore occur at regular or predictable times. *Compare* synchronous.

asynchronous transmission A form of transmission in which data is sent as it becomes available. The time at which the start of transmission occurs is arbitrary, although the rate at which the bits comprising the data are subsequently transmitted is fixed.

ATA (**Advanced Technology Attachment**) A standard INTERFACE used to connect disk drives and other storage devices to a computer. Although in use from the mid-1980s the first standard, ATA-1, was only approved in 1994; the latest version is ATA-7. Early versions of ATA were more widely known as IDE or EIDE. ATA is also known as *Parallel ATA (PATA)* to distinguish it from SERIAL ATA, which has superseded it in most new machines.

Atlas *See* second generation computers.

ATM *See* Adobe Type Manager.

attachment *See* e-mail.

audit trail A record showing the occurrence of specified events in a computer system. For example, an entry might be made in an audit trail whenever a user logs on or accesses a file. Examination of the audit trail may detect attempts at violating the security of the system by, say, unauthorized reading or writing of data in a file.

availability The actual AVAILABLE TIME expressed as a percentage of planned available time. Any shortfall will be the result of faults or breakdowns. In a modern computer system an availability that is not in the high nineties (say less than 98%) is cause for concern.

available time The amount of time in a given period that a computer system can be used by its normal users. During this period the system must be functioning correctly, have power supplied to it, and not be undergoing repair or maintenance.

AVI (**audio video interleave**) A Windows standard format for audio and video. Files of this type have the avi file extension.

back end 1. A program that is used to convert output from a general-purpose program into a specific form. For example, a graphics program might produce picture information in a quite general form, which is then processed by a back end specific to the sort of output device required. There would be a different back end for each device – plotter, VDU, graphics printer, etc. 2. *See* engine.

background printing The printing of a document at the same time as other processes are being carried out.

background processing The processing of computer programs with a low priority in a way that does not allow interaction with the user through the keyboard. Programs run in this way are called *background programs* and are processed at times when programs with a higher priority are not using the computer. *See also* foreground processing.

backing store (**backing storage; secondary store**) Storage devices in which programs and data are kept when not required by the PROCESSOR (or processors) of a computer. Programs can only be executed, however, when they are in the computer's MAIN STORE. A program plus associated data is therefore copied from backing store into main store when required. The storage CAPACITY of main store is much smaller than the capacity available as backing store. Only the programs that are currently being executed are (or need to be) in main store. The use of backing store means that a computer has access to a large repertoire of programs and to large amounts of data, and can call on these as and when it needs them.

The time taken to retrieve a particular item from backing store and transfer it directly into main store must be brief. RANDOM ACCESS to these items is thus necessary rather than SERIAL ACCESS. MAGNETIC DISKS, CDS, and DVDS provide random access and are extensively used for on-line backing storage. MAGNETIC TAPES are serial-access devices and are used as off-line backing store. Programs and data can be transferred from tape to disk for processing, or can be copied from disks to tape.

backslash The \ character on a keyboard. In MS-DOS and Windows systems the backslash is used to represent the root directory and also to separate directory names and filenames in a pathname.

backup A file, device, or system that can be used as a substitute in the event of a loss of data, development of a fault, etc. A backup file, for example, is a copy of a file that is taken in case the original is unintentionally altered or destroyed and the data lost. It is not stored ON-LINE. When a copy is made it is said to *back up* the original version. Such a backup is a type of DUMP.

Backus-Naur form /**bak**-ŭs nor/ *See* BNF.

backward compatibility (**downward compatibility**) Compatibility with earlier versions or models of the same product. A program is backward compatible if the files and data created by previous versions of the program can be used; an operating system is backward compatible if existing application programs continue to operate correctly. A computer is backward compatible if it can run the same applications and operating systems as the previous model. Sometimes, as new technology or techniques are developed, backward compatibility cannot be totally maintained, software and hardware both become obsolete, and the user has to upgrade the system. *See also* upward compatibility.

bad sector A SECTOR on a MAGNETIC DISK that has become unusable for holding data. This can be for a variety of reasons, usually some form of hardware fault. The management of such sectors is one of the functions of a computer's operating system, which can find and mark bad sectors on a disk so that they can be ignored.

bak A file extension commonly used to indicate that the file is a backup. In many applications a backup file is often produced automatically when files are saved; it is usually the previous version of the file.

band printer *See* line printer.

bandwidth /**band**-witth/ The amount of data that can be transmitted in a fixed time period. In digital transmissions it is measured in bits (or bytes) per second and in analog transmissions as cycles per second, or hertz.

banner A small advertising region on a Web page that contains a link to the advertiser's own Web site. It generally spans the width of the page but is not usually more than an inch tall.

bar code A pattern of parallel lines of variable width and spacing that provides coded information about the item on which it appears. The codes are read by relatively simple equipment, known as *bar-code readers*, using optical or magnetic sensing techniques. Bar-code readers commonly use a laser beam that scans backward and forward, with a photoelectric cell to detect the signal reflected from the code. Bar codes are found, for example, on goods sold in supermarkets, where they are used to identify the product and its cost at the checkout point and to update stock and sales records. The first digit indicates the country of origin (0 = USA, 5 = UK), and

other digits give the manufacturer code and identify the product.

barrel printer (**drum printer**) *See* line printer.

base (**radix**) The number of distinct digits (and possibly letters) used in a particular NUMBER SYSTEM. DECIMAL NOTATION has base 10, BINARY NOTATION has base 2, OCTAL NOTATION has base 8; the digits used in these three systems are 0–9, 0 and 1, and 0–7 respectively. HEXADECIMAL NOTATION has base 16 and uses the digits 0–9 and the six letters A–F. The base of a number can be indicated by means of a subscript as in

101_{10}, i.e. 101 in decimal notation
101_2, i.e. 101 in binary notation, i.e. 5_{10}

base address, base address register *See* relative address.

Basic /**bay**-sik/ A group of similar high-level PROGRAMMING LANGUAGES whose name is derived from the words beginners all-purpose symbolic instruction code. The original Basic was developed in the mid-1960s for INTERACTIVE computing, which was a new concept at the time. Many computer manufacturers and software houses subsequently developed their own dialects, and the emergence of the microcomputer increased the diversification to the point where Basic was often extremely difficult to move from one computer to another. Both ANSI and ISO standards have been defined for Basic.

Basic is a simple language, is easy to learn, and is of particular importance to beginners. It allows easy modification of programs since the program text is kept in main store during execution. Basic is normally interpreted although it may also be compiled (*see* interpreter; compiler).

All dialects of Basic should include the following components:
REM introduces a COMMENT line;
INPUT performs input from the keyboard or a file;
READ performs input from a DATA statement elsewhere in the program;
DATA introduces a line of data values;

PRINT performs output to the screen or a file;

IF introduces an IF THEN ELSE STATEMENT;

FOR introduces a LOOP;

NEXT terminates a LOOP;

LET introduces an ASSIGNMENT STATEMENT – may usually be omitted;

DIM declares ARRAYS;

GOSUB calls a SUBROUTINE;

RETURN returns from a SUBROUTINE to the statement after the GOSUB;

STOP interrupts execution of the program;

END terminates the program;

SQR square root function;

INT function truncating real numbers to integers – the fraction is discarded:

ABS function returning the absolute value of a number, i.e. without a sign;

SGN function returning –1 for negative numbers and 1 for positive numbers.

Apart from these components there are many statements and commands that occur in some dialects and not in others, and some that mean entirely different things in different dialects. *See also* Q Basic; Visual Basic.

bat The file extension of an MS-DOS BATCH FILE.

batch file A text file used to contain a list of MS-DOS commands, which may also need parameters. If this file has the bat extension then, when the filename and values for the parameters (if necessary) are typed at the command prompt, the instructions are processed in order. AUTOEXEC.BAT on a Windows system is an example of a batch file. Windows NT, 2000, and XP also support an enhanced type of batch file that uses the extension cmd.

batch processing A method of organizing work for a computer in which items of work are queued up, and the OPERATING SYSTEM takes one job at a time from the queue and processes it. Each job must be entirely self-contained – it must not require any intervention from the person who submitted it. This is because the order in which the jobs are run is at the discretion of the operating system, which will take into ac-count the estimated time to run the job and its demands on resources such as main store or tape drives. If a job is particularly large it may not be run until the middle of the night. Batch processing was formerly the usual method of working, when computer time was a very expensive commodity and had to be used efficiently. Such considerations no longer apply and modern computer systems, in particular microcomputers, do not rely on the use of batch processing. The term has come to be used more informally, to describe the BACKGROUND PROCESSING of one or more jobs under the control of a job control language or batch file, possibly at a set time rather than immediately the job is submitted.

baud /bawd/ A unit for measuring the speed at which SIGNALS travel in a computer or in a communication system such as a telephone link. One baud is normally assumed to be equal to one BIT per second, the signal being a sequence of the binary digits 0 and 1. One baud may however be equal to one symbol per second or one digit per second. In general the baud is equal to the number of times per second that the signaling system changes state. Signaling speeds generally vary from a few tens of thousands of baud (for, say, data transmission by MODEM) up to tens or hundreds of millions of baud (for, say, data transmission by satellite). The unit is named for the French engineer J. M. E. Baudot (1845–1903), who first developed the teleprinter.

bboard /**bee**-bord, -bohrd/ *See* bulletin board.

BBS Bulletin-board system. *See* bulletin board.

bcc (**blind carbon copy; blind courtesy copy**) A directive in an e-mail program specifying where to send additional copies of the e-mail without the knowledge of the main recipient. Copies are usually sent to bcc addressees for information purposes only and the recipients are not necessarily expected to take any further action. The mail header does not contain bcc addressees (*compare* cc).

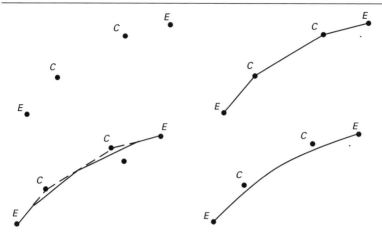

Formation of a Bezier curve

BCD *See* binary coded decimal.

benchmark A way in which the performance of computer systems – hardware and software – can be compared. It is normally in the form of a specially designed test or problem. For example, a system may be subjected to a known workload and the time taken to complete it is measured; this time can then be compared with the times achieved by other systems. The systems are said to have been *benchmarked*. A multi-user system may be benchmarked by another computer simulating the action of tens or hundreds of individual users all going through a predetermined sequence of actions.

When using a benchmark to evaluate two or more computer systems, it is important that a prospective purchaser uses a benchmark that reflects the sort of use to which the system will be put.

beta /**bay**-tă, **bee**-/ Describing new software or hardware that is in its final stages of development, containing most if not all of its intended functionality. It may not be totally reliable but is ready for testing in controlled situations by a few customers or individuals to iron out any unidentified problems before general release.

Bezier curve /**bez**-ee-ay, **bay**-zee-/ A type of curve used in computer graphics. A simple Bezier curve is defined by four points. Two of these are the end points of the curve. The other two are control points and lie off the curve. A way of producing the curve is to first join the four points by three straight lines. If the midpoints of these three lines are taken, together with the original two end points, a better approximation is to draw four straight lines joining these points. The midpoints of these lines can then be used to get an even better approximation, and so on. The Bezier curve is the limit of this recursive process. Bezier curves can be defined mathematically by cubic polynomials. A similar type of curve obtained by using control points is known as a *B-spline*.

bidirectional printer /bÿ-dă-**rek**-shŏ-năl/ *See* character printer.

binary /**bÿ**-nă-ree, -nair-ee/ Composed of or relating to two components. In computing the term refers to the base 2 numbering system, in which values are expressed as combinations of the digits 0 and 1. Thus, 2 in decimal is written as 10 in binary. *See* binary code.

decimal digit	0	1	2	3	4	5	6	7	8	9
BCD code	0000	0001	0010	0011	0100	0101	0110	0111	1000	1001

Binary coded decimal

binary code A rule for transforming data, program instructions, or other information into a symbolic form in which only the two binary digits 0 and 1 are used. Once encoded there is no way of distinguishing an instruction from a piece of data. The process of transforming letters, digits, and other characters into sequences of binary digits is called *binary encoding*, and the representation used or produced is called a *binary encoding*. The BINARY NOTATION used to represent numbers in a computer is a binary encoding. The representation of characters using, for example, the ASCII scheme is another example. *See also* code.

binary coded decimal (BCD) A CODE by which each decimal digit, 0, 1, 2, ..., 9 is transformed into a particular group of 4 binary digits (or bits). In the standard form of BCD, each decimal digit is represented by the group of 4 bits whose value is equivalent to the decimal digit, as shown in the table. The decimal number 58.271 would thus be coded
      0101 1000.0010 0111 0001
The form of a decimal number in BCD is usually different from its representation in BINARY NOTATION.

binary data Any data that uses BINARY NOTATION. All computer data is stored in this way. The term is also commonly used to denote data that is stored using all the bits available: for example, a series of the numbers 0 to 255 decimal in a system with 8-bit words. It is distinguished from data that uses only the numbers 32 to 126 decimal, which is generally represented by appropriate ASCII characters.

binary digit Either of the two digits 0 and 1. *See* bit.

binary encoding *See* binary code.

binary file A file made up from a sequence of 8-bit data or executable code. Such files differ from text files, which can be read by eye, in that they are usually in a form readable only by a program.

binary fractions Fractions expressed in binary notation. Binary fractions are usually handled in FLOATING-POINT NOTATION. They can however be represented in FIXED-POINT NOTATION, e.g. .0101 or 11.101, where the positional value decreases from left to right by powers of 2: the leftmost bit after the point has the value

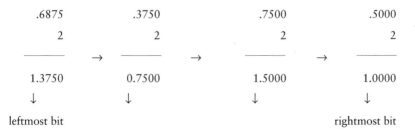

Conversion of decimal fraction .6875 to binary equivalent, .1011

2^{-1} (i.e. ½ or 0.5), the next bit to the right having the value 2^{-2} (i.e. ¼ or 0.25), and so on. The fraction .011 therefore has the value

$$(0 \times 2^{-1}) + (1 \times 2^{-2}) + (1 \times 2^{-3})$$

The decimal conversion is thus

.25 + .125 = .375

Conversion of a decimal fraction to binary is shown in the diagram. It is done by repeated multiplication by 2, the bit resulting before the point being removed to form the bits of the binary fraction. The process is continued until a zero fraction is obtained or the required accuracy is obtained. *See also* binary notation.

23 ÷ 2 = 11 remainder 1 (rightmost bit)
11 ÷ 2 = 5 remainder 1
5 ÷ 2 = 2 remainder 1
2 ÷ 2 = 1 remainder 0
1 ÷ 2 = 0 remainder 1 (leftmost bit)

Conversion of decimal 23 to binary equivalent, 10111

binary logic   1. The electronic components used in a computer system to carry out LOGIC OPERATIONS on binary variables (i.e. quantities that can take either of two values). The components consist of LOGIC GATES and LOGIC CIRCUITS.
2. The methods and principles underlying this implementation of logic operations.

binary notation (binary system)   The number system employed in computing to represent numbers internally. It uses the two digits 0 and 1 and thus has a BASE 2. These digits are known as *binary digits* (or BITS), and a *binary number* is composed of a sequence of bits, e.g. 1001. Like decimal notation the binary system is a positional notation: the value of a binary number depends not only on the bits it contains but also on their position in the number. The positional value increases from right to left by powers of 2. For example, the binary number 10100 has the value

$$1 \times 2^4 + 0 \times 2^3 + 1 \times 2^2 + 0 \times 2^1 + 0 \times 2^0$$

Conversion from binary to decimal can be achieved by multiplying each bit in the number by the relevant power of 2 (the least significant (rightmost) bit having a positional value of 20, i.e. 1), and adding them together. The above binary number, 10100, is thus equivalent to the decimal number 20.

Conversion from decimal to binary is achieved by repeated division of the decimal number by 2. The sequence of remainders forms the bits of the binary number, the first remainder being the least significant bit, and so on. Division stops when the quotient becomes 0. The steps in such a conversion are shown in the diagram. *See also* binary fractions; floating-point notation.

binary number   *See* binary notation.

binary operator (dyadic operator)   *See* operator.

binary representation (bit representation)   The representation used within a computer for numbers, CHARACTERS, and MACHINE INSTRUCTIONS, and having the form of distinctive sequences of the two BITS 0 and 1. *See also* binary code; binary notation; complement; floating-point notation; character set.

binary search   *See* searching.

binary signal   *See* digital signal.

binary system   Any system involving just two possible values, two alternatives, or two items. One example is the binary number system, i.e. BINARY NOTATION, in which numbers are represented by means of the two digits 0 and 1. These two digits are used in computing not only in binary notation but to represent the two possible values, alternatives, etc., of any binary system or situation, including the two possible directions of magnetization of a spot on a magnetic disk or tape or a low or high signal fed to a logic gate. *See also* bit.

binary tree   *See* tree.

bind   1. To assign values to the PARAME-TERS used in a procedure, subroutine, or function at the time that the procedure, subroutine, or function is called; this binding remains in force throughout the CALL. 2. To associate the VARIABLES used in a program in a HIGH-LEVEL LANGUAGE, or the symbolic addresses or labels used in a program in ASSEMBLY LANGUAGE, with particular ADDRESSES in computer memory.

BinHex /**bin**-heks/   1. A method for the encoding of binary data as ASCII text.
2. An Apple Macintosh program to perform such encoding and also the reverse decoding on data files.

biometrics /bÿ-ŏ-**met**-riks/   The study, measurement, and analysis of human biological characteristics. In computer security, biometric principles are used in high-level techniques to identify people by recognizing one or more specific attributes, such as fingerprint, voice, or iris patterns. The ideal biometric will unambiguously prove the identity of a person and ensure no one can impersonate him or her.

BIOS /**bÿ**-os/ (**basic input–output system**) A set of low-level instructions stored in read-only memory (ROM) on the main motherboard of a computer, providing basic control over the keyboard, monitor, mouse, disk drives, and other parts of the computer. When the computer is switched on these routines perform basic tests on the hardware and load and execute the operating system. They are transparent to the user but nonetheless are critical to performance. Many computer systems use Flash EPROM chips, which can be reprogrammed with an updated BIOS to fix bugs or to support new hardware.

bipolar technology /bÿ-**poh**-ler/   *See* integrated circuit.

B-ISDN   *See* ISDN.

bistable /bÿ-**stay**-băl/   *See* flip-flop.

bit   1. Either of the two digits 0 and 1 used in computing for the internal repre-sentation of numbers, CHARACTERS, and MACHINE INSTRUCTIONS. The bit is the smallest unit of information and of storage in any BINARY SYSTEM within a computer. The word is a contraction of 'binary digit'. *See also* binary notation.
2. A measurement of the capacity of a number, register, memory location, etc. in terms of the number of binary digits it contains. For example, an 8-bit number has eight binary digits and so can represent 256 (28) different numbers (e.g. 0 to 255, –128 to 127), whereas a 16-bit number can represent 65 536 (216) different numbers (e.g. 0 to 65 535, –32 768 to 32 767). *See also* word.

bit density   *See* density.

bitmap /**bit**-map/   A data structure in memory that represents information as a collection of individual bits. Simple graphics files are held in this format. The picture is made up of a set of dots and spaces, which are represented by bits in the file. *See also* computer graphics.

bitmapped font   A font in which each character is composed of a pattern of dots and spaces. A set of patterns must be held for each size of the font.

bit rate   The number of BITS transmitted or transferred per unit of time. The unit of time is usually the second; the bit rate is then the number of BITS PER SECOND (bps).

bit representation   *See* binary representation.

bits per second (**bps**)   The number of bits of digital data transferred per second. Modern communication devices, such as modems, are now so fast that rates are often described in multiples of bps, for example Kbps or Mbps.

bit string   A string of bits, e.g. 0111.

black box   A self-contained unit in, for example, a computer system or a communications system, whose function can be understood without any knowledge or reference to its electronic components or circuitry. The notion of a black box is useful to someone with little or no electronics training who is trying to find out, say, what a particular peripheral device does. The term is also used of a software module that can be used by other programs without knowledge of its internal workings.

blind carbon copy   *See* bcc.

blind courtesy copy   *See* bcc.

BlackBerry   *Trademark* A PDA manufactured by Research In Motion Ltd. First released in 1999, it became very popular because of its wireless E-MAIL facilities.

blink   To flash on and off. Various objects, such as the cursor and warning messages, are made to blink on the computer screen to catch the eye.

block   A group of RECORDS that a computer can treat as a single unit during transfers of data to or from BACKING STORE. This unit of transfer is sometimes called a *physical record*. The data records in the block are then called *logical records*. The number of data records, i.e. logical records, in a block is called the *blocking factor*. Blocks have either a fixed length or a variable length.

A stream of data recorded on magnetic tape is divided into blocks, and is subsequently read block by block (*see* tape unit). Successive blocks are separated by *interblock gaps* (IBGs), in which nothing is recorded. Recording in blocks is used for convenience in handling the tape and for error management: any errors occurring in the recording and reading of data are dealt with block by block. The equivalent recording subdivision on a magnetic disk is a SECTOR.

block diagram   A diagram that represents graphically the interconnection between elements of a computer system. The elements may range from electric circuits to major units of hardware. They are shown as labeled rectangles or other geometric figures, and are connected by lines. The whole diagram can represent any level of description, from an electric circuit to an over-all computer system.

blocking factor   *See* block.

blog   (weblog)   An online journal maintained on and publicly accessible from the WORLD WIDE WEB. Blogs began to appear in the mid-1990s and by the early 2000s had become so numerous as to form a distinct community on the web. A blog primarily reflects the interests and knowledge of its author; however, some have become respected sources of information or influential sources of opinion. Special software helps to create and maintain blogs, for example allowing readers to comment on their content, and blogs often use web feeds to notify users of new content. The activity of maintaining a blog is called *blogging*, and the totality of all blogs on the Web is called the *blogosphere*.

Bluetooth   *Trademark* A method for electronic devices, including computers and associated hardware, to exchange data over short distances by radio waves. When two Bluetooth-enabled devices come within range (typically 10 meters) they exchange information to determine whether they are compatible and can exchange data; if so they automatically form an appropriate link, known as a *piconet*. A laptop and a desktop computer, for example, will form a small network, and a cordless-phone handset will begin using a base; however, the laptop will not link to the handset because the two devices will determine that they are incompatible. All this happens automatically, with no user action required. The Bluetooth specification is controlled by the Bluetooth Special Interest Group, an association of IT, cellular telephone, and other electronics companies established in 1998.

bmp   The file extension of a type of RASTER GRAPHICS data file widely used on Windows systems.

BNF   (Backus-Naur form)   A symbolic

notation in which the SYNTAX of a programming language can be expressed.

board *See* circuit board; printed circuit board.

boldface A typeface used for emphasis, in which the characters print darker and heavier. In this dictionary the headwords are printed in boldface type.

bomb 1. A program residing on a computer unknown to the user; typically, the program will have been planted in order to damage or destroy the system. *See* virus; worm.
2. (**bomb out**) Of a computer or program, to fail abruptly without warning to the user and without giving the user a chance to recover or partially recover such items as data or files.

bookmark 1. A marker placed at a specific place in a document that enables the user to return quickly to that place at a later date.
2. A link set up by the user to a Web page on the Internet. The URL is stored in a local file and the bookmark can be clicked on to retrieve the Web page automatically.

Boolean algebra /boo-lee-ăn/ An extension of the principles of algebra into the field of logic. It was first put forward in 1847 by the British mathematician George Boole and is now of particular importance in computing. Boole applied the methods of algebra to problems in logic involving statements and conclusions that can be given a truth value – either *true* or *false* – and thus have a binary nature. These true or false expressions can be combined by the logical operations *and* and *or* and negated by the operation *not* to give a conclusion whose truth or falsity can be determined from the rules of Boolean algebra.

Boolean algebra is basic to many aspects of computing. For example it is used in high-level languages in conditional statements:
IF NOT(a<b) AND (x=0) THEN …
or logical assignments and expressions:
answer := (char = 'Y') OR (char = 'y')

It is also used in the design of the electronic circuits – LOGIC GATES and logical circuits – that control the flow of signals. *See also* truth table; Boolean expression.

Boolean expression (**logical expression**)
An expression that is formed according to the laws of BOOLEAN ALGEBRA. It contains variables that take the values *true* or *false* and that are linked by LOGICAL OPERATORS. An example is
a AND (b OR NOT c)
a, b, and c are variables or expressions (such as x > 1) that return the values *true* or *false*; AND, OR, and NOT are operators. A Boolean expression can be used to form a FUNCTION whose value is either *true* or *false* depending on the combination of values assigned to the variables. This function is known as a *Boolean function* or *logical function*. A Boolean function can be represented in a TRUTH TABLE. It can also be transformed into a LOGIC DIAGRAM of logic gates.

Boolean function *See* Boolean expression.

Boolean search A search that uses Boolean operators. Boolean searches are commonly used in database searching and in Internet SEARCH ENGINES.

boot (**boot up**) *See* bootstrap.

boot disk A floppy disk or CD-ROM containing all the necessary files for an operating system to start on a computer. The boot disk is inserted into the appropriate drive and the computer switched on. The files are copied into memory and the boot up sequence started. A boot disk can be used to start up the operating system in the event of failure of the hard drive.

bootstrap In general, a means or technique enabling a system to bring itself into some desired state. The word is used in several ways in computing. Most commonly, a bootstrap is a short program whose function is to load another longer program into a computer. For example, when a computer is first switched on its MAIN STORE will be empty except for those parts fabricated in ROM (read-only memory). A bootstrap stored in ROM is capable of reading from backing store (which is NONVOLATILE) the OPERATING SYSTEM, or some part of it, without which the computer cannot operate. The computer is then said to be *booted up* or *booted*, and a program can then be loaded into the machine.

bot A program that performs a repetitive, monitoring, or time-consuming action. It is an abbreviation of 'robot'.

bounce The return of a piece of e-mail because of an error in its delivery.

bounds *See* array.

bpi Bits per inch. *See* density.

bps *See* bits per second; bit rate.

branch 1. *See* jump.
2. A set of instructions that are executed between two BRANCH INSTRUCTIONS.

branch instruction A MACHINE INSTRUCTION that controls the selection of one set of instructions from a number of alternative sets during the execution of a program. A branch instruction is usually regarded as being the same thing as a jump instruction. *See* jump.

breadboard A CIRCUIT BOARD on which experimental arrangements of electronic components can be built and tried out. The arrangements can be easily modified.

break To halt program execution temporarily at a given spot (usually for debugging purposes), or to interrupt a program by pressing the *break key* (or its equivalent).

breakpoint /**brayk**-point/ A point in a program at which execution is halted temporarily. An examination can then be made of the values of program VARIABLES at that point. Breakpoints may sometimes be conditional – execution only pauses if certain conditions are true. There may be a count involved so that the breakpoint only operates after it has been passed a certain number of times. Breakpoints are normally used only in DEBUGGING.

bridge A device that connects two networks into one large logical network so that information can flow between them. The networks may be using different protocols, and they are joined together at the data link layer (OSI layer 2).

broadband /**brawd**-band/ 1. (wideband) Describing communications systems supporting a wide range of frequencies, on which multiple messages can be carried by the transmitting medium at the same time. Such systems are ideal for digital transmissions.
2. Denoting a method of INTERNET ACCESS by telephone line that uses these techniques to achieve much greater data transfer rates than are possible with traditional MODEMS. The most popular form of broadband is ADSL.

broadcasting *See* multicasting.

browse To scan files, documents, or the Internet for items of particular or general interest. In browsing, information tends to be checked and gathered but not altered.

browser A program that interprets and displays hypertext (HTML) in a readable form on the screen. Such programs are especially useful on the WWW and examples are Internet Explorer and Mozilla Firefox. Web browsers can use HYPERLINKS to jump from document to document, and can often download and transfer files, provide access to newsgroups, and play audio and video clips. *See* Internet Explorer; Mozilla; Firefox.

B-spline /**bee**-splÿn/ *See* Bezier curve.

bubble-jet printer A type of nonimpact printer. Ink is surface boiled in fine tubes to form bubbles that are then shot from nozzles to form characters on paper. Bubble-jet is a trademark of Canon Inc. *See also* laser printer; ink-jet printer.

bubble memory An obsolete type of memory in which data in binary form is represented by the presence of absence of minute magnetized regions (*bubbles*) within an oppositely magnetized magnetic material. Magnetic fields are used to form the bubbles and to move them along a path through the surface of the stationary magnetic medium. Bubble memory is a form of SERIAL-ACCESS memory but the data path is constructed so that the ACCESS TIME to a particular data item is very short. It is usually NONVOLATILE and has high storage capacity. Bubble memory has had only limited application and has been superseded by other faster devices, such as FLASH MEMORY.

bubble sort *See* sorting.

buffer **1.** A temporary store for data in transfer, normally used to compensate for the difference in the rates at which two devices can handle data during a transfer. It allows the two devices to operate independently, without the faster device being delayed by the slower device. Buffers generally form part of the MAIN STORE of a computer, holding data that is awaiting processing or is awaiting transfer to a magnetic disk or an output device such as a printer or a VDU. A buffer may also be built into a peripheral device: many printers have a buffer to compensate for the relatively slow speed of printing compared with the speed with which the information to be printed is received.
2. Any device, circuit, etc., that is inserted between two other devices to compensate for differences in operating speeds, in timing, in voltage levels, etc.

bug An ERROR in a program or a system. It is usually a localized error occurring when a working version of the program or system is being produced, rather than an error introduced during design. *See also* debugging.

bulk storage A medium on which large amounts of data can be stored, usually with relatively slow access times. Examples are hard disk, CD, DVD, and magnetic tape.

bullet A character, such as a large dot or small square, used to introduce the start of a paragraph to add emphasis to the text.

bulletin board (bboard) A general facility on a computer NETWORK allowing any user of the network to leave messages that can be read by all the other users. This can be contrasted with E-MAIL messages, which are read only by the addressees.

bundled *See* unbundling.

bundling *See* unbundling.

bureau An independent agency providing computer or related services to the public. It offers both software and hardware facilities. The user usually simply supplies the input data for a specified job, such as files to be printed or typeset, and pays for the results.

burn To write information to a CD or DVD.

burster A device that separates into sheets CONTINUOUS STATIONERY produced as output from a PRINTER. The paper is split at the perforations across its width. With multipart stationery a burster frequently also acts as a DECOLLATOR, separating the copies and possibly sorting them into stacks.

bus A set of conducting wires – a pathway – connecting several components of a computer and allowing the components to send SIGNALS.

bus network A form of local area network (LAN) in which all signals are carried over a main bus and are available to all nodes (computers) on the system. In a bus network each node monitors line activity and only accepts messages addressed to itself.

button 1. An area (e.g. a small box) in a GUI FORM that, when clicked on with the mouse or entered (if it has the focus) by pressing the enter key, will implement a command. 2. A movable piece on a mouse that, when pressed, will activate a particular function.

byte A unit of data or memory consisting of 8 bits.

C A PROGRAMMING LANGUAGE developed in the early 1970s by Dennis Ritchie at Bell Laboratories in the USA for systems development, and in particular for writing the operating system UNIX in order to make it PORTABLE. C has the CONTROL STRUCTURES usually found in HIGH-LEVEL LANGUAGES but has features that make it suitable for writing SYSTEMS SOFTWARE. Its popularity increased with the spread of the UNIX operating system, but it is by no means confined to UNIX machines. *See also* C++.

C++ /see-dub-'l-**pluss**/ A PROGRAMMING LANGUAGE developed from C in the 1980s. It retains the syntax of C while adding comprehensive OBJECT-ORIENTED PROGRAMMING features as well as other enhancements. It has become an important language for software development.

C# /see-**hash**/ A programming language based on C++ and JAVA. It was introduced by Microsoft as part of the .NET software-development platform and became an ISO standard in 2003.

cable television *See* CATV.

cache memory /cash/ (**cache**) Extremely fast SEMICONDUCTOR MEMORY that is used in computer systems in association with MAIN STORE. Its ACCESS TIME is much shorter than that of main store and it is therefore used to increase the accessibility to data required by the processor. It is a temporary store, i.e. a BUFFER, that is continually updated so that it contains the most recently accessed contents of main store.

When the processor requires an instruction not currently in the cache, then a whole block of memory (the size depends on the particular system) is copied from main store into the cache; the next instructions for execution are then likely to be in the cache as well as the one currently required. This is a very successful strategy and significantly increases execution speeds. The contents of a cache may be modified by the program, so it will be necessary to copy them back into main store before refilling the cache with another block. *See also* L1 cache; L2 cache.

The same principle is used to speed other data transfers. For example, a *disk cache* holds copies of recently accessed disk sectors in RAM and can save many time-consuming disk reads.

CAD /kad/ (**computer-aided design**) The application of computer technology to the design of a product. CAD is used especially in architecture and electronic, electrical, mechanical, and aeronautical engineering. Designs can be created by computer using information fed in by experts and also acquired from other sources, for example specifications of component sizes or building regulations. During the design process the design itself is displayed on a screen and can be tested and modified by the technical designer. The computer can analyze various characteristics of the design, the results being fed back to the designer. The final design is normally drawn by a flatbed PLOTTER, with design specifications, etc., listed on the printout.

The CAD output, the design of, say, a PRINTED CIRCUIT BOARD, may then be passed to other systems for computer-aided manufacture (CAM) and computer-aided testing (CAT). The combined process of computer-aided design and manufacture is known as *CADCAM*. The whole procedure – computer-aided design, manufacture, and testing – is often referred to as *CADMAT*.

CADCAM /**kad**-kam/ *See* CAD.

CADMAT /**kad**-mat/ *See* CAD.

CAE (**computer-aided engineering**) The use of computers in engineering design, testing, and research.

CAI (**computer-aided (or assisted) instruction**) *See* CAL.

CAL (**computer-assisted learning**) Any use of computers to aid or support education and training. CAL can test attainment at any point, provide faster or slower routes through the material for people of different aptitudes, and can maintain a progress record for the instructor.

This application is also known as *CBL (computer-based learning)*, *CAI (computer-aided (or assisted) instruction)*, and *CMI (computer-managed instruction)*.

calculator An electronic device that can perform simple arithmetic, and often other operations, on numbers entered from a keyboard. Final solutions and intermediate numbers are generally presented on LCD or LED DISPLAYS. Present-day calculators range from very cheap devices that can add, subtract, multiply, and divide numbers to those that can perform complex mathematical and statistical operations and may be programmed. Add-on memory modules containing specialist programs – for navigation, say – can be used with the more expensive calculators, as can small printers.

call An action whereby a program or a section of a program – a PROCEDURE, SUBROUTINE, or FUNCTION – is brought into effect. Control is transferred to the ENTRY POINT of this program or program section, causing the program or program section to be immediately executed by the computer. The program or program section is said to have been *called*. The program that issued the call is the *calling program*.

A piece of code, known as the *calling sequence*, is required to perform a call. The calling sequence includes some means of passing the necessary data to the program or program section (*see* parameter), and of returning control to the calling program following execution of the program or program section. When a whole program is to be called this is done by the OPERATING SYSTEM.

callback A network security feature used to verify users. The user's name and password are checked and then the network disconnects and returns the call to a predefined authorized number.

calling program *See* call.

calling sequence *See* call.

capacity (**storage capacity**) The amount of information that can be held in a storage device. It is usually measured in BYTES or in BITS. For example, the capacity of a CD is 700 megabytes, while the capacity of a hard drive is several tens of gigabytes.

caps lock A keyboard toggle key that when on makes all alphabetic characters typed appear as capitals (upper case). When off all alphabetic characters are lower case. Other characters are unaffected by the caps-lock key.

capture To transfer received data into a file.

carbon copy *See* cc.

card A PRINTED CIRCUIT BOARD, often of a fairly small size.

card reader A device that reads data encoded on cards and converts it into binary code for processing by a computer. The data is represented by magnetic patterns in, say, a MAGNETIC STRIPE on a plastic card.

caret /**ka**-rĕt/ The symbol ^ on the keyboard obtained by pressing the shift+6 keys. This symbol is used to represent exponentiation in some programming languages, as in 3^4 meaning 3^4, or to represent the control key in some computer documentation, as in ^S meaning ctrl-S.

carriage return (CR) The ASCII control character, with the decimal value 13, that is used to affect the format of printed or displayed output by aligning the print head or pointer at the beginning of the current line.

cartridge A removable module containing a magnetic disk, magnetic tape, integrated circuitry, printer ink, or some other computer-related device. It is usually designed so that the contents remain permanently inside the cartridge (or attached to it) and are not handled by the operator. The modular form thus protects and facilitates the use of the contents. *See also* tape cartridge; disk cartridge; ROM cartridge.

cascading style sheets (CSS) A specification used with HTML and XML documents. HTML and XML documents can have style sheets that define, amongst other things, information on font style, weight, and size. The style sheet may be a separate file or may be embedded in the document.

case The status of an alphabetic character, denoting whether it is capitalized (upper case) or not (lower case).

case-insensitive *See* case-sensitive.

case-sensitive Depending on the case of a character. A case-sensitive action will differentiate between upper- and lower-case letters, so that the word 'Mat' is totally different from 'mat' or 'MAT' and so on. A *case-insensitive* action will treat such upper- and lower-case forms as equivalent. Passwords, for example, may be case-sensitive in some systems and case-insensitive in others.

case statement A conditional CONTROL STRUCTURE that appears in many programming languages and allows a selection to be made between several choices; the choice is dependent on the value of some expression. For example, Pascal has

case expression of
  selector 1: choice 1
  selector 2: choice 2
  selector 3: choice 3
  ...end

If the expression evaluates to one of the selector values, then that choice is executed. Standard Pascal does not define what happens if the expression matches no selector, but nonstandard extensions and other languages offer a special selector to handle this eventuality.

The case statement is a more general structure than the IF THEN ELSE STATEMENT, which allows a choice between only two alternatives. It can in fact be written in the form of a NESTED if:
  if ... then ...
  else if ... then ...
       else if ... then ... (and so on)
A case statement could be used in selecting, say, a month in the year or one age group out of six age groups. It is often employed in programming MENUS. A similar control structure appears in some other languages as a SWITCH STATEMENT.

cassette /ka-**set**, kǎ-/ A container holding MAGNETIC TAPE and from which the tape is not normally removed. The casing protects the tape and makes it easier to handle. The tape is wound between two small reels. *See also* tape cartridge.

CAT /see-ay-**tee**, kat/ (**computer-aided testing**) A method used by engineers to assess and test designs, especially those produced by CAD systems. CAT is also used to automate software regression testing.

cathode-ray tube (CRT) *See* display.

CATV (**community access television**) More commonly known as *cable television*, the use of a broadband signal sent over fiber-optic or coaxial cables to distribute multichannel television program broadcasting. It is also an increasingly popular way to access the WWW and other forms of multimedia information and entertainment services.

CBL (**computer-based learning**) *See* CAL.

cc (carbon copy; courtesy copy) The directive in an e-mail program specifying where to send additional copies of the text. Copies are usually sent to cc addressees for information purposes only and the recipients are not expected to take any further action. The mail header contains a list of all cc addressees.

CCD *See* charge-coupled device.

CD (compact disk) A metal (aluminum) disk 120 mm in diameter on which digital data is recorded by mechanical means (as opposed to magnetic recording). The standard type of CD is the well-known audio disk used for sound recording. The disk has a spiral track containing a pattern of minute depressions. It is read by a low-power laser. The depressions (known as 'pits') have a lower reflectivity than the flat areas (known as 'lands'). The pattern of pits and lands correspond to 0 and 1 values of bits. CDs of this type are produced on a blank disk using a high-powered laser operating in short bursts to 'burn' the pits. Copies can be produced by taking impressions from a master. Such CDs are also used as a method of distributing data for use on computers (e.g. software). In addition the disks can be used as extra storage available to the processor. This is known as *CD-ROM* (compact disk read-only memory). CD readers on personal computers are generally designated as 36-speed, 52-speed, etc., referring to the access time in comparison to that of a standard audio compact disk. CD-ROM is widely used for

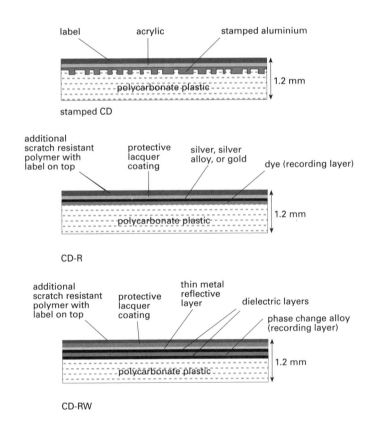

CD types

computer games and multimedia publications. It is also one of the safest methods of storing electronic information. Because the data is encoded as depressions in a metal surface it is less susceptible to external influences than data stored magnetically and is less likely to change with time. The data on a standard CD is regarded as permanent and cannot be changed or overwritten. A CD can store about 700 megabytes of data.

Producing an original standard CD requires specialized equipment using a high-power laser and is usually done by commercial companies. Different forms of compact disk have been introduced that can be produced using low-power lasers and relatively cheap equipment. *CD-R* (compact disk recordable) is a form of CD in which the metal disk has a thin gold layer covered with a photosensitive dye coating and a protective lacquer layer. The CD writer (which usually doubles as a CD reader) contains a laser that acts in short bursts to change the dye layer, reducing the reflectivity of the underlying gold. These regions correspond to the pits in a standard CD; the unchanged regions correspond to the lands. Once the data has been recorded it cannot be changed; i.e. CD-R is an example of a WORM device (write once, read many times). It is, however, possible to add extra data to a CD-R later. This is known as *multisession recording*. CD-R is a cheap and reliable method of archiving data.

CD-RW (compact disk rewritable) is a further development in which the data on a CD can be erased and overwritten (i.e. the CD is reusable in a similar way to a magnetic disk). This type of CD has a reflective layer consisting of a silver-based alloy. The laser writing the data causes localized heating of small areas, which changes the structure from a crystalline to an amorphous form. These amorphous regions have lower reflectivity and correspond to the pits in a standard CD. The data can be erased by heating to a higher temperature to restore the crystalline form. This type of CD was formerly called *CD-E* (compact disk erasable).

CD-E (**compact disk erasable**)   *See* CD.

CD-I (**compact disk interactive**)   A standard for combining audio, video, and text on a compact disk using optical-disk technology. A CD-I player needs no external computer. It plugs directly into a TV and stereo system and comes with a remote control to allow the user to interact with software programs sold on disk, thus providing a games-playing capability. Many features are included such as music, special audio effects, graphics, text, animation, and video.

CD-R   *See* CD.

CD-ROM   *See* CD.

CD-RW   *See* CD.

Celeron /sel-ĕ-ron/   *See* Pentium.

cell   **1.** A LOCATION in memory or a REGISTER. It is usually capable of holding a single item of information in binary form such as an integer or instruction, but may hold only a single bit.
2. A box on the screen in a spreadsheet program or in a spreadsheet-like grid when used in any other type of program, which contains a number, a formula, or a string of text.

cell address   **1.** The reference number of a particular memory location that stores one byte of information.
2. (**cell reference**) The row and column identifiers used to specify a cell in a spreadsheet or in a spreadsheet-like grid.

center   To align text around a horizontal midpoint on a line or vertical midpoint on a page or other defined area in text formatting.

central processor (**central processing unit; CPU**)   The principal operating part of a computer. It consists of the ARITHMETIC AND LOGIC UNIT and the CONTROL UNIT, i.e. the units in which program instructions are interpreted and executed. Sometimes MAIN STORE is considered a component of the central processor.

Over the years, the functions in the larger or more complex computers have become distributed among various units, each able to handle one or more tasks quite independently. In these systems the term central processor (or central processing unit) is inappropriate, and the word PROCESSOR (or processing unit) is used.

Centronics interface /sen-**tron**-iks/ An old but still widely used interface standard designed for parallel data exchange between computers and peripherals. Originated by Centronics Corporation for dot matrix printers, it provides for the ability to send 8 bits of data (usually one character) simultaneously together with control and status information.

CGA (**color graphics adapter**) One of the early graphics adapters introduced by IBM in 1981 and capable of several character and graphics modes. The quality was quite poor with a resolution of, for example, 320 × 200 pixels with four colors or 640 × 200 pixels with two colors. *See also* EGA; VGA.

CGI 1. (**common gateway interface**) A set of rules defining how a Web server communicates with a Web client. CGI programs are usually small and perform tasks such as counting the number of hits for a Web page, accepting and filtering data from a form, or reading an image map. All the processing takes place on the Web server, not on the computer doing the browsing. CGI scripts are often written in Perl.
2. *See* computer graphics interface.

chad /chad/ The piece of paper, plastic, etc., that is removed when a hole is punched in a data medium. For example, the punching of sprocket holes along the edges of continuous stationery produces chads.

chaining 1. An arrangement whereby one item in a sequence contains the means for locating the next item. For example, a FILE may be organized such that each entry contains the ADDRESS of the next entry in a sequence. The entries can then be dispersed randomly within a storage device. One particular entry may belong to more than one sequence.
2. A technique whereby the final action of a program is to load its successor into main store. For example, a suite of demonstration programs could be chained together with the last one chaining to the first, thus providing a continuous demonstration.

chain printer *See* line printer.

channel 1. A route along which information can be sent. It may for instance be a telephone link between two computers or a cable connecting a terminal to a computer. It may also be the route followed by data between a user's program and a file on backing store. 2. *See* web feed.

character A symbol used in representing data and in organizing and possibly controlling data. A character may be one of the 26 letters of the Roman alphabet, a letter combined with a diacritical mark, one of the digits 0–9, a punctuation mark, bracket, plus sign, etc. Note that **A** and **a** are different characters, and that a computer distinguishes between the capital letter O and zero, 0, and between the lower case letter l and one, 1. Modern computers are increasingly able to handle the large number of symbols needed to represent all the world's writing systems. *See also* character set; byte.

character encoding 1. The process of transforming letters, digits, and other characters into an encoded form. *See* code.
2. The symbolic representation used for the characters in a given CHARACTER SET when encoding them. The symbols are usually the two binary digits 0 and 1. ASCII and UNICODE are two widely used character encoding schemes.

character generator A ROM-based circuit or a program that converts codes for display on a screen. A valid ASCII code is converted into a set of numbers that will cause a VDU to illuminate the correct pattern of dots to display the character. This is much faster than the process of drawing the character by a series of graphics commands, and formerly made text-based VDUs the preferred systems for most tasks. However, the number of characters that could be generated was very limited. Such speed considerations are not relevant to today's faster computers, whose GUI operating systems use graphics commands to generate a wide range of characters in a variety of typefaces, styles, sizes, etc. Nevertheless, all computers still have a character generator so they can communicate with a user when the GUI is not loaded or is not working properly.

character printer   A PRINTER that produces a single character at a time, printing them in the order in which they appear in a line. The printing mechanism – known as the *print head* – prints as it moves from left to right or right to left along each line. Such printers have now largely been superseded by LASER PRINTERS and INK-JET PRINTERS. *See also* daisywheel printer; thermal printer.

character recognition   *See* OCR; MICR. *See also* OMR; document reader.

character set   A collection of different CHARACTERS that can be used for some purpose. A very simple example is the letters of the Roman alphabet. The characters that are valid within a given programming language also form a set. Yet another example is the collection of characters that a computer can handle. This latter set would have to include the characters of the one or more programming languages used on the computer.

The character set handled by a particular computer nearly always includes the ALPHANUMERIC CHARACTERS (letters and digits), together with the SPACE CHARACTER, SPECIAL CHARACTERS, and CONTROL CHARACTERS. Typical special characters are shown in the table; they include punctuation marks, brackets, and the symbols used in ARITHMETIC and LOGIC OPERATIONS. Modern character sets also include many accented letters, symbols from non-Roman alphabets, Chinese and Japanese ideograms, etc. The alphanumeric, space, and

,	;	:	.	?	!
(	)	[	]	{	}
"	'	#	%	&	|	\
+	–	*	/	>	<	=

Typical special characters used in a character set

special characters are known as graphic or printable characters, i.e. they are symbols that can be produced by printing, writing, etc. In contrast the control characters each produce a particular effect, such as a new line or backspace.

Within a computer all these characters must be represented in binary form. There are two widely used encoding schemes for characters; these are ASCII and UNICODE. The 128 different bit patterns used in ASCII and the many thousand used in Unicode are themselves character sets. *See also* EBCDIC.

character string   *See* string.

charge-coupled device   (CCD)   A device made from a connected array of semiconductor components in which the charge at the output of one component provides input to the next and so on. CCDs are sensitive to light, with both monochrome and color arrays being available, and are used extensively in scanners and digital and video cameras.

chat   A real-time conversation with another person or persons through a computer network, often the Internet. Once participants are logged in, text can be typed and entered by one user to appear on the screens of other users enabling a text 'conversation' to take place. Chats can be private or open. Such exchanges take place in a *chat room*.

check   A process for validating the accuracy of an item of information, which may be a piece of data or part of a program. It is thus a means of detecting ERRORS. A check is performed, for example, following the transfer of the item to or from a magnetic disk, following its transmission across a computer network, or following its computation.

A single *check digit* may be used in performing the check. This digit is added to one end of the item of information when it is encoded, and is derived arithmetically from the digits making up the item. It can be used subsequently to determine whether the item contains an error (or errors). In some cases a group of characters may be used to perform a check rather than a single digit. *See also* parity check; modulo-*n* check.

check box   A small, usually square, selection box used for interactive control in graphical user interfaces. The check box will contain a check or cross if it has been selected and clicking on the box will reverse the selection. There may be several check boxes in a form for the selection of several different features at the same time.

check digit   *See* check; modulo-*n* check.

checkpoint   A point in some processing activity, or a place in a program, at which a DUMP – a copy – is taken of data associated with the active program. It is hence the point or place from which the program can subsequently be restarted. The use of checkpoints guards against system failure during very long program executions: if anything goes wrong, the execution can be continued from the last checkpoint once repairs have been made. The frequency of checkpoints depends on the reliability of the system.

checksum   /**chek**-sum/   *See*   modulo-*n* check.

chip   1. A small section of a single crystal of semiconductor, usually silicon, that forms the substrate (i.e. base) on which an INTEGRATED CIRCUIT is fabricated.
2. *Informal name for* integrated circuit.

chipset   /**chip**-set/   A group of MICROCHIPS performing one or more related functions and designed to work as a single unit. Often a chipset will fit on a single chip. The term commonly refers to a set containing a

particular CPU and its supporting circuitry.

chip socket   A device to allow easy replacement of chips (INTEGRATED CIRCUITS) on a PRINTED CIRCUIT BOARD. The chip socket is soldered to the circuit board and the chip is pushed into the socket, which has a small hole for each of the chip's legs. With larger chips care is needed to avoid bending the legs of the chip on insertion.

CIM   1. (**computer-integrated manufacturing**) The use of computers in manufacturing. Such topics as advanced control techniques including robotics, manufacturing systems design and control, systems modeling, and analysis and management accounting are all included.
2. (**common information model**) A framework for describing and sharing management information that is both system and application independent.

cipher, ciphertext   /sȳ-fer/   *See* cryptography.

CIR (**current instruction register**)   *See* instruction register.

circuit board   A rigid board of insulating material on which an electric circuit has been or can be built. The term is commonly used to refer to a PRINTED CIRCUIT BOARD. Circuit boards come in a variety of sizes, some of which are standardized.

circular shift   *See* shift.

CISC   /sisk/ (**complex instruction set computer**)   A CPU designed with a full set of assembly calls. Systems based on this philosophy can make software simpler and thus smaller, but are generally slower in the execution of each instruction.

class   A fundamental structure in OBJECT-ORIENTED PROGRAMMING. A class contains all the *methods* (i.e. FUNCTIONS and PROCEDURES) and data necessary to represent some object. A games program, for example, might contain a class called 'Auto' that models the characteristics of an automobile, with such methods as 'accelerate', 'brake', 'turn', etc. The accelerate method might use data items in the class, called PROPERTIES, that define such characteristics as maximum speed and maximum rate of acceleration to determine how long it takes the automobile to reach a given speed. Because multiple instances of the Auto class can be declared, each automobile can be given its own characteristics by assigning different values to this data. The rest of the program interacts with a class entirely through its public methods and properties; the internal workings of the class, which might include private methods and properties, are a BLACK BOX. The use of classes can considerably simplify complex programming tasks.

clear 1. To set the contents of a register, counter, or storage location to zero.
2. To erase all the characters from a VDU screen, leaving it blank.

CLI *See* command line interface.

click To press and release a mouse button with the CURSOR pointing at an ICON or other object on the screen. In general clicking selects the object, but in some cases, such as BUTTONS and hyperlinks on Web pages, an action is initiated. *Compare* double click.

clickstream In Web advertising, the sequence of 'clicks' or pages requested as a visitor explores a Web site.

click through (click-through; click-through) The action of directing a Web browser to a particular Web site by clicking the mouse on a link from another site or in an e-mail. When the link is an advertisement, it is usually paid for at a fixed rate per click through: i.e. the payment depends directly on how successful the advertisement is at attracting people to the site. In E-MAIL MARKETING the *click-through ratio* measures the success of a marketing e-mail by dividing the number of click throughs it produces by the total number of copies sent out.

click trail The record of a Web surfer's route through the Web sites visited. Web browsers generally maintain such a record as a convenience to the user. However, the possibility of this data being analyzed by third parties has raised privacy concerns, especially if it is obtained illicitly through MALWARE.

client 1. A desktop computer or workstation attached to a network and relying on one or more SERVERS for resources such as files, devices, and other services.
2. A computer system or process that requests a service of another computer system or process. In client–server computing, the 'front-end' client program that the user runs contacts and obtains data from a server software program on another computer, often at long distances. Each client program is designed to work with one or more specific kinds of server programs and each server requires a specific kind of client. For example, a workstation requesting the contents of a file from a file server is a client of the file server.

client-server (client/server) Describing a program whose operation is divided between two computers on a network. These are the CLIENT, which interacts with the user by receiving commands and presenting results; and the SERVER, which manages the application's data and does most of the computation or data processing in response to requests from the client. Often the server is a dedicated machine that serves multiple clients or even multiple client-server programs of a similar type. A single server running a DATABASE MANAGEMENT SYSTEM, for example, can handle multiple clients running a range of programs that utilize its databases. The principal advantages of client-server applications are: reduced cost, because the computational workload is concentrated in the server and so the clients need not be powerful and expensive machines; increased security and robustness, because the server manages all access to shared data; and reduced network traffic, because the clients send requests to the server rather than manipulating the data directly.

clip art Simple artwork in digital form designed for use in publications or Web pages. The use of clip art can save time and make art available for nonartists. Clip art includes both subject-related illustrations and visual elements such as lines, bullets, and text separators. It can be purchased and is also available free for personal use from many Web sites.

clipboard A temporary storage area inside a computer. It is used to copy or move data from one area of a document to another, or from one application to another. *See* cut; paste.

clock An electronic device (normally a quartz oscillator) that provides a series of pulses at extremely regular intervals of time. The interval between successive pulses, i.e. their rate of repetition, is known as the CLOCK SPEED. In computers the clock speed is measured in megahertz or, if appropriate, gigahertz: i.e. there are at least one million, or one billion, pulses per second.

Because of its constant rate, the signal from a clock is used to synchronize related pieces of computer equipment. Their operations can therefore be controlled so that events take place in sequence at fixed times. For example, clock pulses can be fed to all the FLIP-FLOPS in a computer, causing them to change state at the beginning (or the end) of each clock pulse. The primary clock rate controls the fastest parts of the computer, while slower components are timed by numerous submultiples of this rate.

clock speed In a computer, the number of pulses per second generated by an oscillator setting the tempo for the processor. Clock speed is usually measured in MHz (megahertz: millions of pulses a second), or GHz (gigahertz: billions of pulses a second). A clock speed of up to 3 GHz can now be obtained and is determined by a quartz-crystal circuit, similar to those used in radio communications equipment.

clone A computer based on a well known and corresponding PC, but made by a different company, usually with cheaper components.

closed source *See* open source.

cluster 1. A group of terminals, magnetic disks, processors, whole computers, etc., that are configured to work together and which perform as a single unit in interactions with the rest of the computer or network. Clustering improves both system speed and fault tolerance. 2. The base unit of disk-space allocation in a FILE SYSTEM. A cluster consists of one or more SECTORS.

cmd The file extension of a type of BATCH FILE used in Windows NT, 2000, and XP.

CMI (**computer-managed instruction**) *See* CAL.

CMOS /see-mos/ (**complementary metal oxide semiconductors**) Any of a family of LOGIC CIRCUITS that are all fabricated with a similar structure by the same INTEGRATED-CIRCUIT techniques. They are a type of MOS technology and are characterized by very low power requirements. Compared with TTL and ECL they have high PACKING DENSITIES but low SWITCHING SPEEDS.

CMYK system A scheme for specifying colors in terms of four primary colors. C stands for cyan (aqua), M stands for magenta (pink), Y stands for yellow, and K stands for black. A given color is specified by a combination of fractions of these components. The CMYK color model is usually used in commercial color printing of books, magazines, etc. *See also* RGB system.

coaxial cable /koh-**aks**-ee-ăl/ (**coax**) An electric cable having two or more conducting paths, one conducting path being surrounded by but insulated from another conducting path, which may in turn be surrounded by and insulated from a third conductor (and so on). The outermost cable is often earthed. Coaxial cable provides a continuous path along which electrical signals (usually of high frequency) can be conveyed from one point to another in a system.

Cobol /**koh**-bol/ A high-level PROGRAMMING LANGUAGE whose name is derived from the words common business oriented language. It was developed specifically for administrative and financial use in the late 1950s and early 1960s. It was adopted by the US Department of Defense, and is still the most widely used commercial data-processing language. There have been several international standards as the language has developed.

code 1. A rule for transforming data or other information from one symbolic form into another. In the codes employed within a computer, the symbols used are generally restricted to the two BITS 0 and 1. The process of transformation is called *encoding*, and the symbolic representation used or achieved is known as an *encoding*. The process of reconverting the coded information into its original form is called *decoding*. Like encoding the decoding process follows a strict rule. Both processes are thus algorithmic in nature. They are carried out by an *encoder* or a *decoder*. These may be in the form of a piece of hardware or a piece of software.

Different kinds of codes are used in computing for different purposes. For example, when characters – letters, numbers, etc. – are fed into a computer they are transformed into a CHARACTER ENCODING, such as UNICODE, so that they can be manipulated by the computer equipment. Again, ERROR-DETECTING and ERROR-CORRECTING CODES are used to improve the reliability of data when it is transferred to or from disk or when it is transmitted.
2. Any piece of program text written in a programming language.
3. The particular language or type of language in which a piece of program text is written, e.g. machine code, C++ code, source code, etc.
4. The encrypted form of a message. *See* cryptography.

5. To represent data or a program in a symbolic form.

codec /**koh**-dek/ 1. (coder–decoder) An integrated circuit or chip that performs analog-to-digital conversion and vice versa. The most common use for such a device is in a modem.
2. (compression–decompression) An algorithm or specialized computer program or hardware that reduces the size of large files.
3. A combination of these functions, particularly used for creating and compressing digital data for video or audio signals from analog data being received in real time and subsequently expanding it for playback. MPEG is a popular codec for computer video.

cold start *See* restart.

collating sequence An ordering of characters in a character set used within a computer. It is employed, for example, in sorting into alphabetic or alphanumeric order.

collision In an Ethernet network, the result of two devices on the same network attempting to transmit data at exactly the same time. The network should detect the collision and discard both pieces of data.

color graphics adapter *See* CGA.

color monitor A monitor that displays colors. The highest resolution can be obtained using a monitor in which dots of three different phosphors, one each for red, blue, and green, coating the inside of the screen are caused to glow by electronic signals representing different intensities of the three colors. By this method the three basic colors can be blended into millions of different colors. A monitor using this technique is known as an *RGB monitor*.

com /kom/ 1. In MS-DOS, the file extension of a command program file. This is a type of executable binary file limited to one 64-kilobyte segment and is now seldom used except to provide backward compati-

bility. COMMAND.COM is the main program of commands for the MS-DOS operating system.
2. A top-level Internet domain name, generally describing addresses operating in a commercial environment.

COM /kom/ **1.** (**communication port**) A port on the central unit for a cable connection to a serial device such as a modem.
2. (**component object model**) A Microsoft standard designed to allow implementation of shared software components across diverse programs running on Microsoft Windows platforms by specifying the system their INTERFACES must use. ActiveX controls and DirectX are examples of COM objects.
3. (**computer output on microfilm**) Computer output recorded in miniaturized form on microfilm, either on a reel of film or on card-sized sheets of film known as *microfiche* (or *fiche*). The term COM also applies to the techniques used to produce this form of output. The information is considerably reduced in size before the recording process, enabling a large amount of information to be stored. Special optical viewers must be used to enlarge the information on the microfilm so that people can read it. New film or fiche must be produced if additions or corrections are required.

COM has been extensively used since the 1960s, for example, in libraries to catalog books by author and subject. Microfilm and microfiche also have applications in areas that do not require a computer, e.g. in the recording of books, newspapers, and other documents.

combinational circuit *See* combinational logic.

combinational logic A simplified form of BOOLEAN ALGEBRA that is used in the design of LOGIC CIRCUITS. The circuits in combinational logic are known as *combinational circuits*. They contain only LOGIC GATES such as AND, OR, and NOT gates. The behaviors of these components are described by TRUTH TABLES. It is then possible to describe the whole circuit by drawing a truth table (as is done in the diagram at logic circuit). The output of the circuit at a particular time can thus be determined by the combination of inputs.

combo box A type of CONTROL on a GUI FORM. It combines the functions of a LIST BOX and a TEXT BOX: the user can either select its content from a list of options or type it in if something not on the list is required.

command *See* command language.

command language A kind of programming language by means of which a user can communicate with the OPERATING SYSTEM of a computer. Statements in such a language are called *commands*. They are requests from someone using a terminal or a microcomputer for the performance of some operation or the execution of some program, examples being 'list', 'sort', 'delete', 'load', 'run'. JOB CONTROL LANGUAGES are a type of command language; in turn, command languages are a type of SCRIPTING LANGUAGE.

command line interface (CLI) A user interface to an operating system or application program in which the user responds to a visual prompt by typing in a command on a line, receives a response back from the system, and then enters another command, and so on. The Command Prompt in a Windows operating system is an example of a command line interface. Today most users operate with a GRAPHICAL USER INTERFACE (GUI) such as those offered by Windows, Apple Macintosh, and others.

comment Part of the text of a program that is included for the benefit of the human reader and is ignored by the COMPILER. The way in which a comment is constructed depends on the programming language. It might be contained in brackets of some sort, for example
    { ... }   in Pascal
    /* ... */   in C and C++
Some languages prefer end-of-line comments, which are introduced by a specified character or characters (such as // ; — or !)

and are automatically terminated at the end of a line. Other languages, including Fortran and some varieties of Basic, restrict comments to be whole lines starting with a special LABEL, such as C (Fortran) or REM (Basic).

common gateway interface *See* CGI.

common information model *See* CIM.

common LISP A standardized version of LISP, a programming language that was designed for easy manipulation of data strings, developed in 1959 by John McCarthy. It is commonly used for artificial intelligence (AI) programming and is one of the oldest high-level languages still in relatively wide use.

communication card (**communication board**) *See* expansion card.

communication channel (**transmission channel**) A CHANNEL, i.e. an information route, for data transfer. *See also* communication system.

communication line (**transmission line**) Any physical medium used to carry information between different locations. It may, for example, be a telephone line, an electric cable, an optical fiber, a radio beam, or a laser beam. *Compare* data link.

communication network *See* communication system; network.

communications *See* communication system.

communications program A software program that enables a computer to connect with another computer and to exchange information.

communications protocol 1. The collection of settings required for serial communications with a remote system. There are a multitude of standards applicable to different areas of communication, such as file transfer (e.g. XMODEM and ZMO-

DEM), handshaking (e.g. XON/XOFF), and network transmissions (e.g. CSMA/CD).
2. A set of standards designed to enable computers to communicate and exchange data with one another. The common standard is the 7-layer set of hardware and software guidelines known as the OSI (Open Systems Interconnection) model.

communication system Any system by which information can be conveyed from one point – the *source* – to another point – the *destination* – with due regard to efficiency and reliability. There may be more than one source and/or more than one destination, in which case the system is called a *communication network*. The information is sent from source to destination via a COMMUNICATION CHANNEL. In general the channel will distort the information (due to NOISE) and will produce ERRORS in it. In order to reduce the effect of noise, the information is converted into an encoded form before transmission through the channel and is subsequently decoded at its destination. *See also* network.

community access television *See* CATV.

compact disk *See* CD.

compatibility 1. The ability of a computer to execute program code originally produced for another computer. This generally occurs for successive computers in a family of machines. Since later computers are almost always more capable (have a larger instruction set and/or more memory), a computer that is able to run a program of a less capable earlier machine is said to be *backward compatible*.
2. The ability of a piece of hardware (e.g. a storage device or a terminal) to be used in place of the equipment originally specified or selected. If the equipment substituted fits into the computer without any modifications and works straight away, it is said to be *plug compatible*.

compilation *See* compiler.

compilation error An error, generally a

SYNTAX ERROR, detected when a program is compiled. *See* error.

compilation time The time taken to perform the compilation of a program. *See* compiler; compile time.

compiler A program that takes as input a program written in a HIGH-LEVEL LANGUAGE, such as C++ or Fortran, and translates it into MACHINE CODE, or occasionally another form of object code. Each STATEMENT in the high-level language is converted into many MACHINE INSTRUCTIONS, sometimes hundreds. The input to the compiler is called the *source program* and the output is called an OBJECT MODULE. The object module is usually stored on magnetic disk separately from the source program. The translation process is known as *compilation*, and the program that has been translated is said to have been *compiled*. The entire program must be compiled before it can be executed. Complex programs are usually written in several modules, each of which is compiled separately. The object modules are then combined by a LINK EDITOR to produce the OBJECT PROGRAM.

A compiler not only translates program statements but includes LINKS for PROCEDURES and FUNCTIONS from the system library and allocates areas in main store. The object program can subsequently be loaded into and run by a computer. The compiler can also produce a LISTING of the source program, if required, and reports the errors (mainly SYNTAX ERRORS) found in the source program during compilation; it normally indicates the position and nature of these errors.

Each high-level language that can be used on a particular computer requires its own compiler or INTERPRETER. Program run times are much faster when a compiler is used rather than an interpreter. Once compiled, the same program can be run any number of times. In contrast, an interpreter must be used each time the program is run. *See also* assembler.

compile time The period of time during which a program in a high-level language is translated into MACHINE CODE, so that the program can subsequently be executed by a computer. It is at compile time that information about the program, such as a compiler listing or a SYMBOL TABLE, may be produced to aid later DEBUGGING. *See also* compiler; run time.

complement The outcome of a NOT OPERATION on a digit, integer, truth value, proposition, or formula. The complement of the binary digit 0 is 1 while the complement of the binary digit 1 is 0. The complement of the truth value *true* is *false* and vice versa. The complement of the proposition 'this object is spherical' is 'this object is not spherical'.

In general, the complement of a digit, d, can be found from the formula
$$(B - 1 - d)$$
where B is the BASE of the number system (equal to 10 in decimal notation, 2 in binary notation). The complement of the decimal digit 6 is thus 3.

The complement of a whole number, i.e. an integer, is obtained by replacing each digit in the integer by its complement. In binary notation this gives the *one's complement*. (In decimal notation it gives the *nine's complement*.) The one's complement of 0110 is thus 1001. In binary notation the *two's complement* is formed by the addition of 1 to the one's complement. The two's complement of 0110 is thus 1001 + 1, i.e. 1010. To determine the original integer from its one's or two's complement, the same process is repeated: the complement is taken (for the one's complement) or the complement is taken and 1 is added (for the two's complement).

One's complement and two's complement are both used in computing. A one's complement is produced by a NOT operation on the binary number concerned, converting 1s to 0s and 0s to 1s; a two's complement is produced by a further addition of 1. Two's complement arithmetic is simpler than one's complement arithmetic and is used both in the representation of negative numbers and in subtraction. Each positive number is represented in its usual form in binary notation except that it always has at least one leading 0. Each negative number is represented by its two's complement and always has at least one leading 1. For example, the decimals +34 and –34 are represented in 8-bit two's complement notation as

$$00100010$$
$$11011110$$

respectively. There is no need for a separate plus or minus sign. The leftmost (most significant) bit indicates the sign (0 for +, 1 for –) and is called the *sign bit*. The largest positive number in an *n*-bit word has the form

$$01111\ldots \text{ and is equal to } +(2^{n-1} - 1)$$

The largest negative number has the form

$$10000\ldots \text{ and is equal to } -(2^{n-1})$$

In the case of subtraction, one number can be subtracted from another by adding its two's complement. For example, the subtraction 27 – 19 (decimal) is equivalent in binary to

$$11011 - 10011$$

and in two's complement arithmetic is achieved, using an 8-bit word, by the sum

$$00011011 + (11101100 + 1) =$$
$$00001000$$

The 'carry' drops off the end and is ignored. The leading zero indicates that the result is a positive number.

component object model *See* COM.

compound document A document that contains one or more embedded files, such as a graphics image or a spreadsheet. The embedded file is physically saved with the document and increases its size. *See also* OLE.

compression The process of reducing the space taken up by data. File compression is achieved by using a process that reduces file size by eliminating redundancies in the file, such as repeated characters. A popular method of compression is *zipping*, which is implemented by such programs as PKZIP. The data in compressed files cannot be used without first decompressing the file. However, some operating systems allow files, directories, or whole disks to be compressed in a fashion that is invisible to the user.

CompuServe /**kom**-pyoo-serv/ An Internet Service Provider, now part of AOL.

computed GOTO *See* GOTO statement.

computer /k**ŏm**-**pyoo**-ter/ A device by which data, represented in an appropriate form, can be manipulated in such a way as to produce a solution to some problem. It is able to perform a substantial amount of computation with little or no human assistance. In general when the word computer is used by itself it refers to a *digital computer*. The other basic form of computer is the ANALOG COMPUTER, which is far less versatile and thus finds fewer but more specialist applications.

A digital computer accepts and performs operations on *discrete data*, i.e. data represented in the form of combinations of CHARACTERS. Before being fed into the computer the characters consist of digits, letters, punctuation marks, etc. Inside the computer all these characters are in binary form, encoded as combinations of binary digits, or BITS (*see* binary representation). ARITHMETIC and LOGIC OPERATIONS are performed on these strings of bits following a set of instructions. The instructions form what is known as a PROGRAM, which is stored along with the data in the MEMORY of the computer.

The main components of a (digital) computer are:
(a) devices for the INPUT and OUTPUT of data and programs;
(b) memory (MAIN STORE and BACKING STORE) in which to store the data and programs;
(c) a CONTROL UNIT and an ARITHMETIC AND LOGIC UNIT (ALU) for processing the data following the sequence of program instructions; the control unit and ALU are generally combined into a CENTRAL PROCESSOR.

The instructions of the stored program are executed one after another; a device in the control unit (the INSTRUCTION ADDRESS REGISTER) indicates the LOCATION in memory from which the next instruction is to be taken. The instructions indicate how items of data, specified by their locations in memory, are to be used, or what information is to be read from the input devices or written to the output devices.

Computers range widely in performance, size, and cost. They are often classified as MICROCOMPUTERS, MINICOMPUTERS, or MAINFRAMES. All three groups may be used as general-purpose machines, capable of solving a wide variety of problems. Alternatively they may be designed for a special purpose or for a limited range of problems.

computer-aided design *See* CAD.

computer-aided engineering *See* CAE.

computer-aided testing *See* CAT.

computer architecture *See* architecture.

computer-assisted learning (computer-assisted instruction) *See* CAL.

computer-based learning *See* CAL.

computer game Any game recorded on CD or DVD or accessed from such online sources as the Internet that can be used on a computer. Computer games are played by manipulating a mouse, joystick, or games pad, or by using the keyboard or other device in response to the graphics on the screen.

computer graphics The input, processing, and output of pictorial information using a computer. The information may be a reproduction of a picture or photograph or a diagram ranging from a simple graph to a highly complicated engineering design, molecular structure, etc. It may be in one or more colors.

The output of computer graphics may be displayed on a VDU screen or may be in the form of a permanent record produced by a PRINTER or PLOTTER. Information is input to the computer by various means, such as a SCANNER, DIGITIZING PAD, MOUSE, or LIGHT PEN. In the case of generated graphs, etc., the computer can draw on data from other programs such as spreadsheets. The computer can be made to ma-

nipulate the information, for example to straighten lines, move or erase specified areas, expand or contract details. Images may be two-dimensional, i.e. flat, or may appear three-dimensional. A series of images can be combined in ANIMATION.

There are many different formats for storing images but they fall into two main classes. In *raster graphics* the picture is stored as a series of dots (or pixels). The information in the computer file is a stream of data indicating the color of a dot. Images of this type are sometimes known as *bitmaps*. This format is used for high-quality artwork and for photographs. Diagrams are more conveniently stored using *vector graphics*, in which the information is stored as mathematical instructions. For example, it is possible to specify a circle by its center, its radius, and the thickness of the line forming the circumference. More complicated curves are usually drawn using BEZIER CURVES. Vector images are easier to change and take up less storage space than raster images.

computer graphics interface (CGI) A standard used for software applying to computer graphics devices such as printers and plotters.

computer instruction *See* machine instruction.

computer-integrated manufacturing *See* CIM.

computer literacy The understanding of the principles of using a computer and associated software. In-depth knowledge is not required, but an ability to understand how computers relate to everyday life is important.

computer modeling The development using a computer of a description or mathematical representation (i.e. a *model*) of a complicated process or system. This model can then be used to study the behavior or control of the process or system by varying the conditions in it, again with the aid of a computer.

computer network *See* network.

computer-orientated language *See* low-level language.

computer science The study of computers and their application. This is a broad discipline covering the theory and practice of hardware and software design.

computer system (**computing system**) A self-contained set of computing equipment consisting of a COMPUTER, or possibly several computers, together with associated SOFTWARE. Computer systems are designed to fulfil particular requirements.

computer word *See* word.

concatenation /kŏn-kat-ĕ-**nay**-shŏn/ An operation by which two STRINGS are joined together in a specified order. The result is itself a string whose length equals the sum of the lengths of the two strings. If one string, A, is given by 'com' and another string, B, by 'puter', then the concatenation of A and B yields 'computer'. The concatenation of A and B is denoted by, for example,
A + B   A & B   A ‖ B
depending on the programming language used.

conditional Taken in some but not all circumstances. *See* control structure; jump.

conditional jump (**conditional branch**) *See* jump.

conferencing /**kon**-fĕ-rĕn-sing/ *See* teleconferencing.

configuration /kŏn-fig-ŭ-**ray**-shŏn, -fig-yŭ-/ The particular pieces of hardware making up a computer system, and the way in which they are interconnected. *See also* reconfiguration; configure.

configure To select various optional settings of a hardware device or a software program so that it will run to best advantage on a particular computer. A new disk, for example, might need a switch set to indicate whether it is the only disk in the computer, or, if not, whether it is the main or a subsidiary disk. A program may need to know the display's resolution, how many different colors it supports, what sort of printer is available, how many disks are available, and so on. Once the program has been configured, the initial settings will be stored and used the next time the program is run. All kinds of programs, from word processing programs on microcomputers to mainframe operating systems, need to be configured. *See also* reconfiguration.

connector 1. A device for connecting electrical components. Most connectors are available as male (plug) and female (socket) versions and can be used to connect cables together or cables to devices. 2. In software, a logical construction, such as AND or OR, used for example to connect conditions in a database query. 3. In flowcharting, a circular symbol used to indicate a break, as to another section or page.

connect time The time a user is connected to a remote system. This time is often used to determine user charges.

console /**kon**-sohl/ A part of a computer used by an operator to communicate with the computer and monitor its operation. Nowadays the console is usually, and for microcomputers always, a VDU and a KEYBOARD considered as a single input/output device, or possibly a TERMINAL. In a system with many such devices attached, the console is the one that interacts with the computer when it is first switched on, before the operating system has been loaded and normal multiuser access enabled. For convenience and security it is normally situated near the computer's hardware.

constant A value or an item of data that is fixed or that cannot change. It may, for example, be a numerical constant, like π, whose value must be specified, or it may be a LITERAL.

content-addressable store /kon-tent/ *See* associative store.

content delivery The service of copying Web site pages to servers at major Internet access points in different parts of the world. This speeds up the delivery of pages as page content can be obtained from the closest server to the user. ISPs often hire the services of a company that provides content delivery.

context-sensitive help A help system in which different help messages are displayed according to the particular feature that is being used in an application. *See also* help.

contiguous /kŏn-**tig**-yu^-ŭs/ Touching or immediately adjacent to, with no intervening gaps. For example, contiguous sectors on a disk are sectors that come one after the other.

continuous stationery Paper that is perforated at regular intervals across its width enabling it to be fan-folded into a stack of 'pages'. It has a row of regularly spaced holes – known as *sprocket holes* or *tractor holes* – down each side so that it can be fed automatically through a PRINTER by means of TRACTOR FEED. Before printing, the paper may be blank or may be preprinted forms. It may be *multipart* stationery, consisting of two or more lots of paper treated to produce multiple copies and folded into a single stack.

control An area of a GUI FORM used to receive data or commands from and present data to the user. From the programmer's point of view, a control is an OBJECT that adds functionality to the form. The basic types of control include TEXT BOXES, LIST BOXES, COMBO BOXES, BUTTONS, and RADIO BUTTONS. More specialized types include calendars, clocks, and invisible controls that perform such functions as connecting the form to a database. Each type of control has many options that can be set by the programmer to govern the size, appearance, behavior, etc., of each instance.

control bus *See* bus.

control character A member of a CHARACTER SET whose occurrence in a particular context produces a particular effect. It may start, stop, or modify the recording, processing, transmission, or interpretation of data. For example, there are control characters for 'delete', 'carriage return', 'horizontal tab', 'start of text', and 'end of text'. *Compare* graphic character.

control key *See* keyboard.

control store *See* microprogram.

control structure A structure in a high-level programming language used to express the flow of control in a program, i.e. the order in which instructions are executed. There are three basic ways in which control can flow. Instructions can be executed in strict order, giving a *sequential* flow; this is the normal flow of control and requires no special control structure. Use of a *conditional* control structure permits the execution of one of a specific number of possible instruction sequences; this is commonly achieved by an IF THEN ELSE STATEMENT or by the more general CASE or SWITCH STATEMENT. Use of an *iterative* control structure allows a sequence of instructions to be executed repeatedly; this is generally achieved by some form of LOOP. For each of these three types of flow control there is one entry point and one exit point.

control total A number produced by adding together corresponding fields in all the RECORDS in a file. It is used for checking purposes (*see* check). A control total may or may not have a sensible meaning in the outside world. A *hash total* is a meaningless control total. It is used solely for verifying the records in the file, and hence for checking the reliability of the associated program and computer.

control unit The part of a computer that supervises the execution of a PROGRAM. It is a component of the processing unit and is wholly electronic. Before a program can be executed it must be translated into a sequence of MACHINE INSTRUCTIONS. The control unit receives the machine instructions in the order in which they are to be executed. It interprets each instruction and causes it to be executed by sending command signals to the ARITHMETIC AND LOGIC UNIT (ALU) and other appropriate parts of the computer. The data required for the execution is moved between MAIN STORE, ALU, and other portions of the machine in accordance with the sequence of instructions and under the direction of the control unit.

The control unit contains REGISTERS, COUNTERS, and other elements enabling it to perform its functions. In its simplest form it has an INSTRUCTION ADDRESS REGISTER, an INSTRUCTION REGISTER, and a register to decode the operation part of the instruction. It can then operate in a two-step *fetch-execute cycle*. In the *fetch step* the instruction is obtained from main store and loaded in the instruction register; the decoder can then determine the nature of the instruction. In the *execute step* the indicated operation, or operations, are carried out, including the necessary references to main store to obtain or store data. In some cases no reference to memory is required, as when a JUMP INSTRUCTION occurs. In other cases an additional step is required, as in INDIRECT ADDRESSING when two (or more) memory references are needed.

Control units nowadays can be very much more complex than this, containing additional registers and other functional units that can, for instance, prepare the next instruction for execution before the current one is complete (*see* pipeline processing). At present the functions of most control units are accomplished by means of MICROPROGRAMS.

conversational mode *See* interactive.

cookie A small file of information, usually containing personalized details, sent by a Web server to a Web browser and sent back to the server whenever the browser makes further requests. In simple terms information is put on a user's hard disk by a Web site so that it can remember something about the user at a later date. According to the browser being used cookies can be refused or can be stored for a short or long time. In general, cookies help Web sites to serve users better, but their potential and actual misuse is a current cause for concern.

cooperative multitasking *See* multitasking.

coprocessor /koh-**pros**-ess-er/ A microprocessing element designed to extend the capabilities of the main MICROPROCESSOR in a microcomputer. For example, it may have better mathematical processing capabilities, including high-speed floating-point arithmetic and calculation of trigonometric functions. Another design could provide superior graphics display capabilities (*see* graphics accelerator). A coprocessor extends the set of instructions available to the programmer. When the main processor receives an instruction that it does not support, it can transfer control to a coprocessor that does. More than one coprocessor can be used in a system if the main processor has been suitably designed.

copy To READ data from a source, leaving the source data unchanged, and to WRITE the same data elsewhere. The physical form of the data in this destination may differ from that in the source, as occurs when data is copied from magnetic disk to main store.

copy protection Methods used by software companies to prevent unauthorized duplication of their programs and so prevent software piracy.

CORBA /kor-bă/ (**common object request broker architecture**) A system that allows for the sharing of objects, such as files, applications, and pieces of programs, over the Internet, regardless of what programming language they are written in or what operating system they are running on. CORBA was developed by an industry consortium known as the Object Management Group (OMG). *See also* DCOM; interface.

core store   A form of magnetic storage widely used for MAIN STORE in computers built between the mid-1950s and mid-1970s. It has been displaced in present-day machines by SEMICONDUCTOR MEMORY – RAM and ROM. Core store consisted essentially of tiny rings of ferrite (the brown material used to coat magnetic tapes), which could be magnetized in either of two ways and were thus each capable of storing one BIT of information.

corrective maintenance   MAINTENANCE that is performed after a fault has been found – in hardware or software – in order to correct that fault. *Compare* preventative maintenance.

corrupt   To alter data stored in a computer system, usually accidentally, and hence introduce errors. The process is known as *data corruption*. It results in a loss of data INTEGRITY.

counter   An electronic device whose 'state' represents a number, and that on receipt of an appropriate signal causes the number represented to be increased (or decreased) by unity. Only a fixed group of numbers can be represented. Once all these numbers have occurred, one after the other, the sequence starts again. The counter is usually able to bring the number represented to a specified value, such as zero. In a computer system a counter can be used as a *timer* if the signal received is a clock pulse (*see* clock), the counter normally counting down from a selected value and raising an INTERRUPT when it has finished.

courtesy copy   *See* cc.

CP/M /see-pee-**em**/   *Trademark* An OPERATING SYSTEM produced by Digital Research in the late 1970s. It was intended for use on microcomputers with floppy disks and the 8-bit Intel 8080 microprocessor, or either of the later Intel 8085 or Zilog Z80 8-bit chips. A version of CP/M for use with

the 16-bit microprocessor chips in the Intel 8086 family was called CP/M-86, but it was never widely used due to the popularity of MS-DOS. It is now obsolete.

cps /see-pee-**ess**/ **(characters per second)**   A measure of the rate of output of, say, a character printer, or sometimes a measure of the rate at which data is processed or transferred.

CPU (central processing unit)   *See* central processor.

CR   *See* carriage return.

cracker   *See* hacker.

crash   1. **(system crash)** A system failure that requires at least operator intervention and often some maintenance before system running can resume. The system is said to have *crashed*.
2. **(program crash)** A failure in an individual program that causes it to exit prematurely, to be aborted by the operating system, or to HANG so that the user must abort it. In all cases any changes to data files that have not been saved will be lost.
3. *See* head crash.

crawler (spider)   A program that visits Web sites, particularly ones that are new or updated, and collects information from pages to enter in the index of a search engine.

CRC (cyclic redundancy check)   An algorithm used to derive a check number from data being transmitted. The CRC number is transmitted with the data. By recalculating it at the receiving end and comparing it to the original, errors in transmission can be detected. CRC methods are also used in other circumstances, for example in the production of CDs so that they are virtually error free.

crop   1. To insert marks on a document to be printed showing where the paper can be trimmed to produce the correct page size.

2. To reduce the size of an image by cutting off one or more margins. Most graphics applications allow images to be cropped.

cross assembler An ASSEMBLER that runs on one computer, producing an object module to run on a different computer. It is usually used to generate software for EMBEDDED COMPUTER SYSTEMS.

cross compiler A COMPILER that runs on one computer, producing an object module to run on a different computer. It is usually used to generate software for EMBEDDED COMPUTER SYSTEMS.

cross-post To copy a message from one communications channel, such as a newsgroup or e-mail system, to another.

CRT (cathode-ray tube) *See* display.

cryptography The protection of a message so as to make it unintelligible to anybody not authorized to receive it. The sender of a message renders it into an unintelligible form by processing it. This processing is known as *encryption*. Many techniques are known for the conversion of the original message, known as *plain text*, into its encrypted form, known as *cipher* (or *cypher*), *ciphertext*, or *code*. Computers usually accomplish this by applying an encryption algorithm to the plain text to produce a pseudorandom cipher, the encoding being governed by a user-specified KEY. The original message is recovered by processing the encrypted message. When this is done by an authorized recipient holding the secret key to the encryption, the processing is known as *decryption*. Recovery of the message is expected to be impossible without prior knowledge of the key.

CSLIP /**see**-slip/ (**compressed serial line Internet protocol**) A version of SLIP in which the Internet address information is compressed making transmission quicker.

CSS *See* cascading style sheets.

current instruction register (CIR) *See* instruction register.

cursor A symbol on a VDU screen that indicates the 'active' position, for example the position at which the next character to be entered will appear. The cursor's shape usually varies with the content: for example, the active position in a word-processing document is often indicated by a thin flashing vertical line, whereas a WINDOW displaying the contents of a directory as ICONS might use an arrow.

cursor keys Keys on a KEYBOARD that can be used to move the CURSOR to a new position on a display screen. They include the ARROW KEYS.

cut To delete selected text from a document displayed on a screen. The text is retained on the CLIPBOARD until another piece of text is cut or the computer is closed down. It can then be pasted – i.e. inserted into the document at a different place or inserted into another document.

cut sheet feed A technique for feeding paper into a printer, the mechanism used being called a *cut sheet feeder*. The feeder picks up a single sheet from a pile of sheets in a hopper and feeds it into the printer. In more sophisticated feeders there may be a number of hoppers, for example one for headed top sheets, one for plain second sheets, and one for envelopes. A multipage letter complete with envelope may thus be printed. *See also* tractor feed; friction feed.

cybercafe /**sÿ**-ber-ka-fay/ **1.** A place, usually a coffee shop or restaurant, where Internet access is made available to the general public for a charge.
2. A virtual cafe on the Internet where chats can take place and messages can be posted using BBSs.

cybercash /**sÿ**-ber-kash/ *See* e-money.

cybernetics /**sÿ**-ber-**net**-iks/ The study of thought processes and control systems, which draws an analogy between brains and electronic circuits. It is based on theories of communication and control and the extent to which they can be comparatively applied to living and nonliving systems.

cyberspace /sÿ-ber-spayss/ A term originated by the author William Gibson in his novel *Neuromancer* and now used to describe the whole gamut of information resources obtainable through the Internet. Cyberspace is a shared virtual universe whose inhabitants, contents, and spaces comprise data that is seen, heard, and felt and through which a virtual-reality user can move. *See also* virtual reality.

cycle 1. A sequence of events that is repeated regularly and in the same order. A piece of hardware may operate in such a way, examples being the STORAGE CYCLE of a storage device and the fetch-execute cycle of a CONTROL UNIT.
2. (**cycle time**) The minimum period of time required to complete such a sequence of events, i.e. the time between the start of one sequence and the start of the next. The term is often used specifically to mean STORAGE CYCLE, i.e. the minimum time required between successive accesses to a storage device.

cycle time *See* cycle.

cyclic redundancy check *See* CRC.

cylinder *See* disk pack.

cypher /sÿ-fer/ *See* cryptography.

DAC /dak/ *See* D/A converter.

D/A converter (**digital to analog converter; DAC**) A device for converting a DIGITAL SIGNAL into an equivalent ANALOG SIGNAL, i.e. for converting the output from a computer (in the form of a series of binary values) into a continuous representation. *See also* A/D converter.

daemon /**dee**-mŏn, **day**-/ A dormant program or process that is part of a larger program or process and that becomes active when certain conditions occur.

daisywheel printer /**day**-zee-hweel, -weel/ An obsolete type of impact character printer, i.e. a PRINTER that prints a character at a time by mechanical impact. The printing mechanism involved a rimless 'wheel' consisting of spokes that extended radially from a central hub; solid characters were embossed on the ends of the spokes. This *print wheel* was rotated until the required character was opposite the printing position, and a hammer then struck it against the inked ribbon and paper. The print wheel was interchangeable, enabling different character sets to be

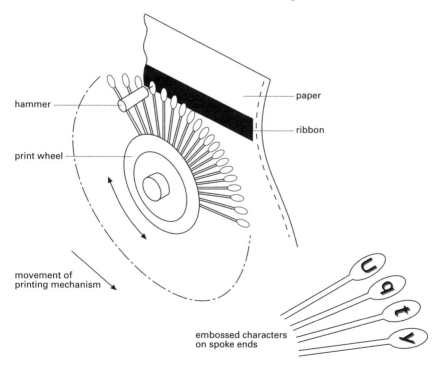

Daisywheel printer

used. The quality of the print was high, and daisywheel printers were used for printing letters, documents, etc.

DASD (direct access storage device) A storage device from which data can be accessed in any order rather than sequentially. Magnetic disk drives, CDs, and DVDs are examples of DASDs, whereas tape units are not.

DAT /dat, dee-ay-**tee**/ *See* digital audio tape.

data /**day**-tă, **dat**-ă, **dah**-tă/ In computing, the basic facts – numbers, digits, words, characters – that are fed into a computer system (in the required form) to be stored and processed for some purpose. In this case data is regarded as the INPUT to a computer system as opposed to the OUTPUT of the system, i.e. the results obtained from processing. It is possible, however, for the results of one process to serve as the data – the input – to another process.

The word 'data' is also used as distinct from program instructions. In this case data refers to all the OPERANDS that a program handles, i.e. the numbers or quantities upon which ARITHMETIC and LOGIC OPERATIONS are performed.

DATA *See* Basic.

databank /**day**-tă-bank, **dat**-ă-, **dah**-tă-/ A system that offers facilities to a community of users for the deposit and withdrawal of data on a particular topic, such as trade statistics or share prices. The user community is usually widespread and the databank itself may be a public facility. Access to a databank may, for instance, be by some form of NETWORK, such as the Internet. The data to be accessed may be organized as a DATABASE or as one or more FILES.

database /**day**-tă-bayss, **dat**-ă-, **dah**-tă-/ A set of DATA FILES – i.e. an organized collection of related data – that is defined, accessed, and managed by a set of programs known as a DATABASE MANAGEMENT SYSTEM (DBMS). The RECORDS in a database are addressable and can be accessed in any order. The organization of the records in a database, and the methods used to access the data, are more sophisticated than those of a data file with no associated DBMS.

Databases are usually more complex than data files, and are often collections of data previously held in many separate files. They are not necessarily large and complex however: DBMS software is available on small as well as large computers. Storing data as conventional files can limit the data's use to one particular application. A database provides data that is available to all users of the system, and may be shared by a number of different applications. Often, a dedicated computer called a *database server* provides database services to all users on a network.

A database together with its DBMS and hardware is often referred to as a *database system*. A data item in a database is typically associated with many other items, some of which may be in the same record. Data is usually retrieved by giving values of specified items in order that the database system should respond with the values of specified associated items. For example, a system might retrieve the registration numbers of all cars of a given color, make, and year of registration. *See also* database language.

database language Any of a group of languages used for setting up and communicating with DATABASES. The instructions given in a database language are executed by the DATABASE MANAGEMENT SYSTEM. Each language consists of at least one *data description language (DDL)* and at least one *data manipulation language (DML)*. The DDL is used to define the structure of the RECORDS within the database files and the relationships between them. It can be used to specify what type of information is to be stored – numbers, characters, dates, etc. – as well as which values are to be used for indexing. The DML is then used to copy the data into the database files and manipulate and retrieve it in various ways. That part of the DML used for retrieving the data may be called the *query language*. An important database language is SQL.

database management system (DBMS; **database system**) A collection of programs that handle and control all accesses to a DATABASE and that maintain the integrity of the database, i.e. the correctness of the stored data. It achieves this by carrying out instructions couched in a DATABASE LANGUAGE while checking that their effects do not violate the integrity rules; any that do are aborted. A DBMS thus has features in common with OPERATING SYSTEMS and COMPILERS. The DBMS allows simultaneous access to the database by a number of users. A good DBMS is characterized by the ease and speed with which complex searches and retrievals are carried out and the flexibility and power of the way databases are specified. A single DBMS is often capable of managing more than one database.

database system 1. *See* database management system.
2. A DATABASE MANAGEMENT SYSTEM together with its associated hardware (storage devices, dedicated computers, etc.) and the database or databases that it manages.

data bus *See* bus.

data capture A process by which data can be extracted during some operation or activity and can then be fed into a computer. The equipment involved may be connected directly to a computer, allowing the incoming data to be monitored automatically. For instance, the use of point-of-sale terminals in a supermarket involves data capture: although the prime objective of the operation is the sale to the customer, the type of product sold can be recorded and hence stock control can be improved.

data cleaning (**data vetting**) The process of checking RAW DATA for completeness, consistency, and validity. Any bad characters, out-of-range values, and inconsistencies are either removed or brought to the attention of someone for a decision.

data coding The use of standard abbreviations or simplified representations in the recording of data on documents, the input

of data via a keyboard, etc. This reduces the work involved and the chance of error. For example, the letters Y and N can be used for 'yes' and 'no'.

data collection The process by which data from several locations is collected together before it is fed into a computer.

data corruption *See* corrupt.

data description language (DDL) *See* database language.

data dictionary A set of descriptions of the data components of some computer-based system. It is normally held in the form of a FILE or DATABASE. It gives information about the nature of the data – meaning, relationships with other data, format, etc. – and its use. A database, for example, will have a data dictionary. A data dictionary is an important tool in the effective planning of a computer-based system, and in the overall control, storage, and use of data in the operational system.

data element (**data item**) An individual unit of data.

data encryption standard (DES) An algorithm used for the encryption of data. It was developed in the 1970s, but the increased processing power of modern computers has made it insecure and it is no longer an official standard. However, it is still in widespread use.

data entry The process by which an operator feeds data into a computer by means of an input device. *See also* direct data entry.

data file A FILE containing data, i.e. the numbers, text, etc., upon which operations are performed by a computer. A data file is often organized as a set of RECORDS. *Compare* program file; database.

data flowchart (**system flowchart**) *See* flowchart.

data integrity *See* integrity.

data link The physical medium – telephone line, optical fiber cable, etc. – by which two locations are connected for the purpose of transmitting and receiving data, together with the agreed procedures (i.e. the PROTOCOL) by which the data is to be exchanged and any associated devices or programs.

data logging 1. The recording of all data passing through a particular point in a computer system. The point chosen can, for example, be in a pathway to or from a VDU. The record – or *log* – that is produced can be used to reconstruct a situation when, say, a fault occurs in the system or an unexpected result is obtained.
2. The regular sampling of a number of quantities, such as temperature, flow rate, and pressure, by a device known as a *data logger*. The information is stored within the logger and periodically transmitted to a computer for analysis.

data manipulation language (DML) *See* database language.

data mining The analysis of large quantities of data to discover previously unknown patterns. The data may be held in a database or a DATA WAREHOUSE, and many database management systems provide such facilities as OLAP that assist data mining. The use of store loyalty cards, for example, means that a record of each customer's purchases can be built up over a period of time; data mining can then identify any patterns, such as favored brands, amount usually spent on each visit to the store, etc., which can be used to generate marketing literature and even special offers tailored to each customer.

data preparation The conversion of data into a coded form that can be read by a machine and hence fed into a computer. The necessity for data preparation has been largely removed with the advent of DIRECT DATA ENTRY. The operation of a keypunch to encode data on punched cards was an example.

data processing (DP) The operations conducted mainly in business, industrial, and government organizations whereby data is collected, stored, and processed on a routine basis in order to produce information, regularly or on request. The operations almost always (but not necessarily) involve computers. A DP system usually handles large quantities of data organized in a complex way. Typical applications include production of payslips, accounting, market research and sales forecasting, stock control, and the handling of orders.

data protection The protection of data handled in a computer. The term is usually applied to confidential *personal data*, i.e. data concerning a living person who can be identified from that information, and possibly including some opinion expressed about that person. Credit ratings produced by banks, donations received by charities, transactions by mail-order firms, and personal records kept by government agencies and the police all involve data concerning the individual; for ease of access, ease of updating, saving of space, and other reasons, this kind of information is now usually stored in computers.

Legislation exists in many countries to protect personal data when it is stored in computers. The aim is to control the potential for misuse of such information. Personal data could, for example, be extracted from one stored record and correlated with data concerning the same person from another file. The combination of information that could result is considered an infringement of privacy.

data retrieval The process by which data is selected and extracted from a FILE, a group of files, a DATABASE, or some other area of memory. *See also* information retrieval.

data structure Any of several forms in which a collection of related data items can be organized and held in a computer. Examples include ARRAYS, RECORDS, FILES, STRINGS, and TREES. Various operations can be performed on the data structure as a whole and on its individual elements, the choice depending on the type of data structure and the programming language.

data tablet   *See* digitizing pad.

data transfer   The movement of data from one point to another. These points may, for instance, be storage locations within a computer system or at either end of a long-distance TRANSMISSION LINE.

data transfer rate   The rate at which data can be moved between two points, for example from a magnetic disk to main store or from one location on a NETWORK to another.

data transmission   *See* transmission.

data type   Any one of the kinds of data (e.g. integers, real numbers, strings) that can be identified by a VARIABLE, an ARRAY, or some other more complex data object in a programming language. The choice of data type defines the set of values that the variable, etc., is allowed to take and the operations that can be performed on it.

data vetting   *See* data cleaning.

data warehouse   A sophisticated database system holding a large amount of data that can, in fact, be made up of several databases spread over several computers at different sites. Fast data searching and filtering techniques are used to supply information to the user, who is unaware of the complexity of the system. In general data warehouses hold a historical record of an organization's data, which is augmented periodically with copies of the current state of the 'live' data. They are optimized for OLAP rather than OLTP operations.

daughterboard   /**daw**-ter-bord, -bohrd/ A circuit board that can be plugged into another board, thus increasing its functionality.

dB   *See* decibel.

dBase   /**dee**-bayss/   *Trademark*   A relational database management system devel-oped by Ashton Tate. The first version was dBase II, which in the early 1980s became the first widely used DBMS on microcomputers. It lost its market dominance in the 1990s and is no longer widely used.

DBMS   *See* database management system.

DCE   (**data communication equipment**) An intermediary device that enables data terminal equipment (DTE) to access a communications line, as defined in RS232 and X25 specifications. One example is a modem, which represents the provider side of a data communication interface.

DCOM   (**distributed component object model**)   A version of Microsoft's component object model (COM) specification. It deals with the communication of components over Windows-based networks when different parts of an application are split and run across different machines on the network.

DDE   1. (**dynamic data exchange**) A technique used by Windows and some other operating systems that enables two or more applications running simultaneously to link data and commands together. If data is changed in one application, then the same data will automatically be updated in the other application. DDE has been superseded on Windows systems by OLE and COM.
2. (**direct data entry**) The input of data and updating of files directly in an on-line system, usually by typing the data at a keyboard or sometimes by using a data-capture device.

DDR SDRAM   *See* SDRAM.

DDS   (**digital data storage**)   *See* digital audio tape.

debugger   *See* debug tool.

debugging   The identification and removal of BUGS – i.e. localized errors – from a program or system. The bugs in a program may, for example, be SYNTAX ERRORS uncovered by the compiler, RUN-TIME ER-

RORS detected during execution of the program, or logical ERRORS, which are more difficult to identify. There are various diagnostic aids that can be used in debugging a program, including TRACE programs or a DEBUG TOOL. Again, a DIAGNOSTIC ROUTINE may be entered as a result of some error condition having been detected – in either software or hardware. *See also* error-detecting code; error-correcting code.

debug tool (debugger)   A special program – a SOFTWARE TOOL – that assists in the DEBUGGING of programs by allowing the internal behavior of the programs to be investigated. Typically a debug tool allows the user to follow a program's execution in the source code, either by stepping through the source code line by line or by inserting BREAKPOINTS to stop execution at selected points. While execution is halted the values of program VARIABLES can be examined and perhaps modified.

decibel /**dess**-ă-bel/ (dB)   A unit of comparison against a known reference for currents, voltages, or power levels and also the strength of sound, equal to one tenth of a bel. A bel is equal to the logarithm to the base ten of the ratio of the measured quantity to the known reference.

decimal notation /**dess**-ă-m⌄l/ (decimal system; denary notation; denary system) The familiar number system, using the 10 digits 0–9. It thus has a BASE 10. The value of a number is determined not only by the digits it contains but also by their position in the number. The positional value increases from right to left by powers of 10. For example, in the decimal number 473, 3 is in the units place, 7 is in the tens place, and 4 is in the hundreds place.

decision table   A table that indicates all the conditions that could arise in the description of a problem, together with the actions to be taken for each set of conditions. It thus shows precisely what action is to be taken under a particular set of circumstances. Decision tables can be used in specifying what a program is to do (but not how it is to achieve this).

	N	N	Y	–	–	–
rain	N	N	Y	–	–	–
snow	N	N	N	Y	–	–
fog	N	N	N	N	Y	Y
temp. (°C)	>8	<8	–	–	>0	<0
take bike	×					
take car		×	×			
take train				×	×	
stay home						×

Decision table for traveling to work

A decision table usually has four parts (see diagram). One part (top left) lists the possible conditions, while another part (bottom left) lists the possible actions. The remaining parts show the conditions under which each action is selected. The top right section specifies the conditions by means of a Y (yes), N (no), – (don't care), or a BOOLEAN EXPRESSION that resolves to Y or N; the bottom right indicates, by means of a cross, the particular action to be taken for each set of circumstances.

decision tree   A binary tree in which every nonterminal node represents a decision. Control will be passed either to the left or right subtree (yes or no) of the node according to the decision taken. A *leaf node* represents the final outcome of all the decisions taken.

declaration   A statement in a computer program that introduces an entity for part of the program, gives it a NAME, and establishes its properties. The entities that can be named include VARIABLES, ARRAYS, PROCEDURES, and FILES. The declaration is fundamental to most programming languages, the form it takes depending on the language.

decoder, decoding   *See* code.

decollator /dee-kŏ-**lay**-ter/   A device that separates the copies of multipart CONTINUOUS STATIONERY produced as output from a PRINTER. *See also* burster.

decompression /dee-kŏm-**presh**-ŏn/   The expansion to its original form of data that has been stored in a compressed form.

decryption *See* cryptography.

dedicated Committed entirely to a single purpose or application. For example, a computer can be dedicated to controlling a machine tool, running a DBMS and storing its associated databases, or synthesizing musical sounds.

default option A predetermined action to be performed, or a value to be used, if no specific action or value has been indicated. For instance, a printer might assume a page length of 11 inches or a C++ compiler might assume that the user did not want a compiler listing. Both these defaults could be overridden by specific instructions to the contrary.

defragmentation /dee-frag-men-**tay**-shŏn/ The rearrangement of data on a disk drive so that each file's data occupies contiguous SECTORS. Following defragmentation, an entire file can be read with the minimum number of physical movements of the disk's read–write heads. *See also* fragmentation.

degradation /deg-ră-**day**-shŏn/ **1.** The spreading out of the shape of a pulse of an electrical signal. It usually occurs when data is transmitted but can also take place if too many peripherals are attached to an output terminal. The consequence is that data interpretation may be difficult because pulses become hard to count.
2. A decrease in performance or service of a computer system.

deletion /di-**lee**-shŏn/ Removal or obliteration of an item of data or of a collection of data. Data that has been fed into a system by KEYBOARD and displayed on a screen can be deleted by operating a function key, such as the backspace or delete keys. With magnetic disk, deletion of data is achieved by overwriting with new data or null characters. However, in some cases, especially the deletion of whole files, the data is noted as no longer needed but not actually overwritten until that part of the disk is needed for another file. This increases the computer's performance but

can be a security risk. Special programs exist to ensure that data is overwritten when its file is deleted.

delimiter /di-**lim**-i-ter/ A character or group of characters used to mark the beginning and end of a program STATEMENT, CONTROL STRUCTURE, or some other item of data. Different characters may be used for the beginning delimiter and the end delimiter. Many languages use *begin* and *end* to delimit complex statements. Most languages use single or double quotation marks to delimit character STRINGS. *See also* separator; terminator.

demodulation, demodulator /dee-moj-ŭ-**lay**-shŏn, dee-**moj**-ŭ-lay-ter/ *See* modulation.

demonstration software **1.** A scaled-down version of a program used for marketing purposes.
2. A prototype that shows some of the expected functionality, the on-screen look, and the data-handling capabilities of software under development.

demountable disk /dee-**mown**-tă-băl/ *See* exchangeable disk store.

denary notation /dee-nă-ree/ (**denary system**) *See* decimal notation.

density **1.** A measure of the amount of data that can be stored per unit length or per unit area of a storage medium. It may be quoted in terms of *bit density*, i.e. the number of BITS per unit length or per unit area. The density of a MAGNETIC DISK can be quoted as the maximum number of bits per SECTOR, the number of sectors per track, and the number of tracks per disk or per inch.
2. *See* packing density.

deposit To place a value in a LOCATION in memory or in a REGISTER in a processor, i.e. to store a value.

DES *See* data encryption standard.

descending order The arrangement of a

set of data in order starting with the largest character value, as in Z to A or 50 to 1.

desktop The general screen background on which windows appear and icons can be placed by the user in Apple Macintosh and Windows systems.

desktop computer A compact microcomputer in which the main components, such as the CPU, keyboard, and monitor, can be placed on top of a desk but are too large to be portable.

desktop publishing (**DTP**) The use of a computer system to produce professional-standard page layout, including the choice of fonts and type size and the inclusion of pictures. Illustrations can be created in graphics software programs and artist's images, photographs, and other pictures can be scanned, converted to a suitable format, then displayed, scaled, and cropped. DTP software usually produces its output in a suitable form for printing, such as PostScript files.

desktop video The use of a personal computer to display video images from a video tape, a DVD, a file stored on the computer or downloaded from the Internet, or directly from a video camera.

destination The drive, folder, or directory to which a file is being sent, or the memory address or location in a document or program to which data is being sent.

developer A company or individual that produces software. This involves the analysis of requirements, defining the application structure, programming, testing, modification, documentation, implementation, and issue.

device A peripheral or other active part of the electronics of a computer system.

device driver A short program that provides the interface between a computer and a device such as a printer, disk drive, or mouse and controls the detailed operation of the device. *See also* plug-and-play.

DFS *See* distributed file system.

DHCP (**Dynamic Host Configuration Protocol**) A protocol that allows a computer on a network to obtain various configuration parameters from a server. This server (a *DHCP server*) can ensure that these parameters allow the computer to operate properly in the particular network environment and that it does not interfere with other machines on the network. The most important use of DHCP is to allocate IP ADDRESSES dynamically from an organization's available pool. The alternative is to configure each machine manually to use a specific IP address, which is not only less efficient (more IP addresses are required) but is also prone to such errors as giving two machines the same address.

DHTML *See* dynamic HTML.

diagnostic routine Part of a program, or a sequence of instructions called by a program, that is entered as a result of the detection of some condition causing an ERROR. A diagnostic routine may analyze the cause of the error, or provide information that can be used for this purpose. It might attempt to isolate the cause of the error to a particular piece of hardware or software.

dialog box A small box that is displayed on the screen and is an essential part of GUI programming. It contains a message that needs a reply from the user. This may be as simple as clicking with the mouse on a Yes/No selection, or entering a few characters of data as required.

dial-up A temporary connection between machines, established over a standard telephone line.

dictionary *See* data dictionary.

Difference Engine A mechanical computing device designed by Charles Babbage in 1823 to produce mathematical tables but never completed. Some of the ideas used in the Difference Engine were incorporated by Babbage in the design of his ANALYTICAL ENGINE.

digit /**dij**-it/ Any of the numerals (or possibly letters) used in a particular NUMBER SYSTEM. Decimal notation uses 10 numerals:

0, 1, 2, 3, 4, 5, 6, 7, 8, 9

Binary notation uses the numerals 0 and 1. Octal notation uses the numerals 0 to 7. Hexadecimal notation uses 16 digits: 10 numerals (0–9) and 6 letters (A–F).

digital audio tape (DAT) A magnetic audio-tape storage medium for recording digitally coded audio information. The series of *DDS* (digital data storage) formats (DDS, DDS2, DDS3, DDS4, and DDS5) builds on DAT and adapts it for computer storage. Tapes using the DDS standards are used for backing up computer systems. They are 4 mm wide and varying lengths from 90 m upwards offer different capacities, with each standard providing a greater recording density than its predecessor.

digital computer A computer that accepts and performs operations on *discrete data*, i.e. data represented in the form of combinations of digits, letters, or other CHARACTERS. Compared with an ANALOG COMPUTER, a digital computer is much more versatile and hence much more widely used. It is thus generally referred to simply as a computer. *See* computer.

digital data transmission A method of transmitting digital data in which the data is represented by discrete discontinuous signals. These signals can be sent using either DC or AC transmission systems. DC transmissions are fast and have a low error rate. Different voltages (or currents) are used to represent the values (usually 0 or 1) and signals from many sources can be multiplexed using digital techniques. AC transmissions use analog signals and a modulator is used to convert from the digital signal to an analog signal.

digital logic 1. The electronic components used in a computer system to carry out LOGIC OPERATIONS on discrete variables. Usually only binary variables are involved (i.e. quantities that can take either of two values), and the term *binary logic* can then be used. The components consist of LOGIC GATES and LOGIC CIRCUITS.
2. The methods and principles underlying this implementation of logic operations.

digital recording The storage of information such as text, graphics, video, or sound in binary coded form (a string of 1s and 0s) that can be physically represented on a particular medium. The term is used especially for such recordings of video and sound, to contrast them with recordings that use analog technologies.

digital signal An electrical SIGNAL whose voltage at any particular time will be at any one of a group of discrete (distinct) levels. The voltage therefore does not vary continuously (unlike an ANALOG SIGNAL). In general there are just two discrete levels: the voltage jumps back and forth between the two levels over a period of time; such a signal is often called a *binary signal*. (A digital signal is therefore usually but not necessarily a binary signal.) The LOGIC GATES and LOGIC CIRCUITS used in computers handle two levels of voltage, and thus their inputs and outputs are binary signals; the high level of voltage is usually used to represent binary 1, with the low level representing binary 0.

digital signature 1. A security measure used especially on the Internet. Two keys, one private and one public, are utilized respectively to encrypt and decrypt messages. The consequence is that, if recipients of a message are able to decrypt it using the appropriate public key, they can be sure that the message did indeed originate from the claimed sender.
2. A binary number used as an identification code for devices such as ROMs, printers, and graphics cards.

digital to analog converter *See* D/A converter.

digital versatile disk *See* DVD.

digital video disk   *See* DVD.

digital video interface (DVI)   A hardware-based technique for compressing and decompressing full-motion video, audio, and computer data on a CD. The technique is also referred to as *digital video interactive*. DVI is a hardware-only CODEC.

digitize /**dij**-i-tÿz/   To transcribe any continuously varying (analog) source of data, such as sound, into a digital (binary coded) form so that it can be directly processed by a computer. The term is also applied to the scanning of images or OCR of text.

digitized signal   The representation of a continuously varying signal in a digital form. The digitized signal has values that are identical to those of the continuously varying (i.e. analog) signal but only at discrete (separate) instants of time. As a result, the voltage or current changes in steps between discrete (distinct) values. For example, if the voltage of an analog signal is measured at discrete intervals, then the sequence of measured values is a digitized signal; the analog signal is said to have been *sampled* so as to produce the digitized signal.

digitizer   A device to change information into a digital form.

digitizing pad (**digitizing tablet; data tablet; graphics tablet**)   A flat surface that can be placed on a desk and is used together with a penlike device for the input of data to a computer graphics system. (Larger surfaces are also available and are known as *digitizing tables* or *boards*.) The position of the pen on the digitizing pad can be accurately and rapidly located by any of a variety of methods; it is measured in terms of the *x* and *y* coordinates of the point of contact, i.e. the horizontal and vertical distances of the pen from one corner of the digitizing pad. The position is thus in a digital form that can be fed into the computer. When the pen is moved by hand over the surface, closely spaced positions can be measured. A DIGITIZED SIGNAL representing the path of the pen is thus generated. The digitizing process is activated by one or more switches or buttons on the pen. The thickness, color, and/or opacity of lines is governed by the pressure applied.

DIL switch /dil/ (**dual in-line switch**)   A device similar in form to a DIP, but instead of an integrated circuit the package contains a row of small switches making or breaking the circuit between opposite pairs of legs. DIL switches are used for setting the default state of some printers, terminals, etc.

DIM /dim/   *See* Basic.

dimension   *See* array.

DIMM /dim/ (**dual in-line memory module**)   A memory board on which memory chips are mounted and separate connector pins are on both sides of the circuit board. This increases the amount of memory that can be plugged into a single connector and also increases the size of the data path for faster data transfers. *See also* SIMM.

dingbat /**ding**-bat/   **1.** A small graphical element used for decorative purposes.
2. A font that is composed of a set of symbols rather than alphanumeric characters.

DIP /dip/ (**dual in-line package**)   An INTEGRATED CIRCUIT contained in a rectangular plastic or ceramic package with a row of metal legs down each of the long sides (see diagram). The legs can either be soldered into holes in a PRINTED CIRCUIT BOARD or inserted into a CHIP SOCKET. A *SIP* has only one row of pins. See illustration overleaf.

direct access   RANDOM ACCESS to a storage device, such as a magnetic disk or CD.

direct-access file   *See* file.

direct addressing (**absolute addressing**)   An ADDRESSING MODE in which the address specified in a MACHINE INSTRUCTION is the actual address to be used, i.e. the MACHINE ADDRESS of the location to be accessed is given explicitly. The address specified is

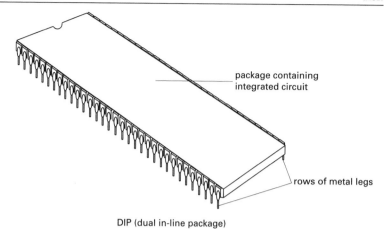

package containing integrated circuit

rows of metal legs

DIP (dual in-line package)

then called a *direct address. See also* address.

direct data entry   *See* DDE.

direct memory access (DMA)   A method of transferring data between main store and other storage devices without involving the CENTRAL PROCESSOR. Without DMA, every byte or word of data has to be individually transferred from main store to the central processor, and then written to the destination device (or vice versa). With DMA, the central processor initiates the transfer by informing an *I/O processor* of the source and destination of the data and the total number of bytes or words to be transferred. The I/O processor then performs the transfer, and informs the central processor (probably by means of an INTERRUPT) when the transfer is complete. The central processor can continue to operate during the DMA.

directory   A FILE in a computer system containing a list of filenames, their locations on BACKING STORE, and their size, as well as other information such as creation date, author, date of last access, and FILE PROTECTION code. A computer system may have many directories, usually organized in a HIERARCHICAL FILE SYSTEM where directories can contain other directories (*subdirectories*) as well as data files. On a MULTIACCESS SYSTEM each user normally has a directory, and possibly subdirecto-ries, and there are one or more shared directories. Units of removable media (CDs, DVDs, floppy disks, etc.) each have their own directory. Directories are used by the operating system to locate files when given their names, and by computer users to keep track of what files are available. *See also* folder.

DirectX /dă-**rekt**-eks/   A Windows application program interface (API) for creating and manipulating graphic images and multimedia effects. It is used in games and in active Web pages. Games specifications will often require a certain level of DirectX to be implemented before the game can be played.

disable   To switch off a device or prevent the operation of a particular function of a device or a particular feature of a program. *Compare* enable.

disassembler   A program that attempts to translate machine code back into ASSEMBLY LANGUAGE, i.e. it performs the reverse function of an ASSEMBLER. It is used as an aid to DEBUGGING. It is only the one-to-one relationship between machine instructions and assembly language instructions that makes this process possible. It does not work with high-level languages.

discrete data   *See* computer.

disk   *See* magnetic disk; optical disk.

disk array A hardware device consisting of several HARD DISKS controlled by a *disk array controller*, which handles all interaction with the rest of the computer system. The array is thus a BLACK BOX and the logical structure it presents to the computer need not match its physical structure; for example it can appear to be one large disk. An important use of disk arrays is RAID, a service that is usually implemented by the disk array controller.

disk cache *See* cache.

disk cartridge An obsolete type of HARD DISK consisting of a single disk permanently housed inside a protective plastic cover. A disk cartridge was used in a specially designed DISK DRIVE, from which it could be removed and replaced by another cartridge. It was thus an EXCHANGEABLE DISK STORE.

disk controller A printed circuit board containing the electronics to control the detailed operation of magnetic disks and carry out requests from the OPERATING SYSTEM to read or write data. A single disk controller can control one or more disks.

disk crash *See* head crash.

disk drive (**disk unit**) A peripheral device that has a mechanism for rotating one or more MAGNETIC DISKS at constant high speed, and devices known as *read/write heads* (plus associated electronics) for writing and reading data on the spinning disk(s). Data is recorded on both sides of a disk, along concentric tracks in the magnetic coating. Items of data are stored and retrieved by the process of RANDOM ACCESS.

Two basic types of disk drive are available. One type uses HARD DISKS and is designed for high performance and large storage CAPACITIES. The other type is designed for low purchase price and uses inexpensive storage media in the form of FLOPPY DISKS.

In disk drives using hard disks, the disks are *fixed* in position and, with the read/write heads and supporting machinery, are hermetically sealed inside the device. In such a controlled environment precision engineering is possible that gives fast access times and very high recording densities. A single disk 3.5 inches in diameter can now hold many gigabytes of data. However, hard disks are fragile and must be protected from physical damage.

In floppy disk drives the disks are removable and do not require such careful protection. Disk plus protective envelope is fed into the disk drive by hand through a slot, and is automatically mounted on the rotation mechanism. Rotation speeds are considerably lower than for hard disk drives. The read/write head or heads operate through slots in the envelope. Floppy disks were formerly used extensively on microcomputers but have now largely been superseded by recordable CDs and DVDs.

In a disk drive there is usually one read/write head per recording surface, either touching it (in the case of a floppy disk) or very close to it (in the case of a hard disk). The head is normally mounted on an arm that moves radially (i.e. towards and away from the disk center). To read or write data, the head is accurately positioned by the disk drive over the required track, and then waits until the right sector rotates into place underneath it. When the head is to write data, it receives an electrical signal coded with the data and converts it into magnetized patterns along the specified sector of a specified track. (The head is an electromagnetic device.) The data is encoded in one of the appropriate disk FORMATS. For reading data, the head senses the magnetized patterns in the specified sector and produces a corresponding electrical signal.

A large disk will probably be used to store many FILES. The allocation of storage space on the disk for each file is handled automatically by the operating system.

diskette /diss-**ket**/ *See* floppy disk.

disk format *See* format; magnetic disk.

disk pack An obsolete form of disk storage used in a specially designed DISK DRIVE, from which it could be removed and replaced by another pack of the same type. It

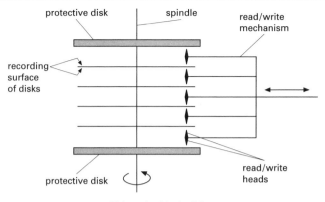

Disk pack with six disks

was thus an EXCHANGEABLE DISK STORE. It was an assembly of identical MAGNETIC DISKS permanently mounted on a single spindle. The disks were usually 14 inch (350 mm) hard disks. Their number varied from 5 to 12 per pack. A similar-sized disk used for protective rather than recording purposes was fitted to the top and bottom of the assembly; in a 6-disk pack there were thus 10 recording surfaces. When not mounted on a disk drive the assembly was kept in plastic covers to protect it from damage and dust. The bottom cover was removed just before mounting the pack; the top cover could only be removed when the pack was mounted. Once the pack was mounted, the read/write mechanism moved automatically into position and the pack was set into rotation. A read/write head was provided for each recording surface. The heads were carried on a single comblike mechanism that was moved radially between the disks. The heads thus all moved together, and could be positioned over the same track number on each of the recording surfaces. This set of tracks could therefore be written to and read without moving the heads; the tracks formed and were referred to as a *cylinder*. The ACCESS TIME to related records could thus be minimized by writing them in a cylinder.

disk unit A disk drive. The term disk unit is sometimes used to mean the rotational mechanism plus read/write heads and associated electronics as a whole,

while the term disk drive is restricted to the rotational mechanism.

display 1. To make information visible on a screen.
2. The device that enables information – textual or pictorial – to be seen but not permanently recorded. The most widely used technology involves a *cathode-ray tube (CRT)*: in most cases the image is formed and changed as a beam of electrons continually traces a pattern of horizontal lines (a *raster*) on the screen, as occurs in a domestic TV; in some CRT devices the image is formed by 'drawing' individual lines on the screen, of any length and at any angle.

The CRT display is rapidly being displaced by the *flat panel display*, a thin panel with a FLAT SCREEN that is both more compact and more rugged. Many of these devices use ACTIVE MATRIX DISPLAYS; another popular technology is the *plasma panel* (or *gas panel*) display, where electrical discharges through a gas form the image on the panel. *See also* LCD; LED display.

distributed file system A file management system in which files may be kept on several computers connected by a LAN or WAN and a number of users may have shared access to these files.

distributed processing The processing of data in a system in which a number of independent but interconnected computers can cooperate. The system itself is known as a *distributed system*, an example of which is a LOCAL AREA NETWORK. The processors involved may be situated at different places and are connected by communication lines. They normally have their own peripherals – console, disks, printers – so that 'local' data can be processed and 'local' decisions made. Data with a wider application can be exchanged over the communication lines.

distributed system   *See* distributed processing.

dithering   A technique for mixing the pixels of available colors to simulate missing colors in a picture. For example, it is used to simulate shades of gray in monochrome (black and white) pictures. There are several ways of dithering. The technique is also used in anti-aliasing in order to make jagged lines appear smoother on the screen.

DLL (**dynamic linked library**)   A Microsoft Windows standard for sharing code between programs. A *DLL file* (usually given the extension dll) consists of a collection of PROCEDURES and FUNCTIONS. Rather than copies of these being incorporated in every program that uses them, a single copy is shared between all programs and only loaded into MAIN STORE when required during execution.

DLT (**Digital Linear Tape**)   *Trademark* A type of TAPE CARTRIDGE that was the tape backup medium of choice for microcomputer networks from the mid-1990s. A single DLT cartridge can hold up to 40 Gb of uncompressed data; usually data compression allows up to 80 Gb to be stored. Its successor, *SDLT* (*Super DLT*), can store up to 300 Gb (600 Gb compressed) with future capacity increases planned. The DLT format was developed in the 1980s by Digital Equipment Corporation and is now owned by Quantum Corporation. *See also* LTO.

DMA   *See* direct memory access.

DNS   *See* domain name system.

doc /dok/   The file extension that identifies a word processing document. It is used as the default by Microsoft Word for document files.

docking station   A set of equipment consisting of a monitor, a keyboard, and a power supply into which a laptop can be slid, effectively combining to form a desktop computer.

document   Any piece of data created by a program and saved as a file. The term is used especially of text formatted in some way. Sometimes document information such as the author, date of last change, number of words, etc., is included.

documentation   The written information describing the operation, structure, and use of hardware or software. It may be in the form of printed manuals. Alternatively it may be held on BACKING STORE and be accessible from a computer CONSOLE or TERMINAL; this is called *on-line documentation*. Software documentation normally includes a tutorial guide to new users, a reference section, an explanation of error messages, an installation guide, and perhaps a small detachable summary card. Hardware documentation is similar but includes a technical specification and a trouble-shooting and maintenance-engineers' manual. Good documentation is crucial to the success of any computer system.

document object model   *See* DOM.

document reader   An INPUT DEVICE that reads data directly from a paper document and feeds it, in coded form, into a computer. The document reader has to recognize characters on the document, or to sense marks on it. It operates by OCR (optical character recognition), MICR (magnetic ink character recognition), or OMR (optical mark reading). The documents can generally be read not only by the input device but also by people, and may involve more than one reading process, for instance OCR and OMR.

do loop   *See* loop.

DOM /dee-oh-**em**, dom/ (**document object model**)   The specification of the logical structure of a Web page or other HTML or XML document in terms that programs can use to extract information from it or to change it. Web pages, etc., are held to consist of various kinds of objects, for example text, images, and links. The DOM defines the attributes of each kind of object, how they and the objects themselves can be manipulated, and how to describe the objects' relationships to each other. Dynamic HTML relies on DOM to change the appearance of Web pages dynamically after they have been downloaded by a user's BROWSER.

domain name   The human-oriented version of the address of a network connection on a TCP/IP network, particularly the Internet. Domain names are hierarchical with parts separated by dots. For example, www.nasa.gov is the domain name of the US space agency. The highest subdivision of the domain name is called the *domain*. This may identify the type of organization (e.g. .gov for government, .com for commercial companies, .edu for educational establishments, or .org for noncommercial organizations). It may also identify the country in which the organization is located (e.g. .uk for United Kingdom or .fr for France). Each domain name has an associated unique IP ADDRESS, to which it is translated by DNS.

domain name system (**DNS**)   A cooperative Internet service that provides a database of host and domain names. DNS is used to match domain names to their associated IP ADDRESSES. When a request is made to a particular Web site it is sent to the nearest DNS server, which, if it cannot translate the name, will ask another DNS server, and so on. On finding a match the IP address is returned to the browser, which then tries to connect to that IP number.

dongle /**dong**-găl/   An anti-piracy device for commercial software consisting of an EPROM in a D-25 connector shell. Programs that require it will run only while it is plugged into the printer port of the computer.

do-nothing operation   *See* no-op instruction.

DOS /doss/ (**disk operating system**)   **1.** The official name for MS-DOS as licensed to IBM in the 1980s for use on its PCs. DOS is now often used as a synonym for MS-DOS.
2. Any disk operating system, i.e. an operating system that supports using disks to store data.

DOS box *See* MS-DOS.

dot matrix printer   A PRINTER that creates each character as an array or matrix of dots. The dots are usually formed by transferring ink by mechanical impact. The device is almost always a CHARACTER PRINTER, i.e. it prints a character at a time. It typically has a print head containing 9, 18, or 24 thin rods, known as *needles*. The needles are selectively operated as the head moves along the line, thus building up each character as required. In many printers the number of characters per inch (cpi) can be varied. Generally between 100–1000 characters per second can be printed, with slower speeds giving better quality output.

Dot matrix printers can print a large selection of shapes and styles of letters and digits, and may also print Arabic characters or the idiograms of oriental languages. In addition to characters it is possible to produce diagrams, graphs, etc.

dot pitch   The measure in millimeters (the smaller the number the better) of the sharpness of a monitor display. Most CRT monitors will have a dot pitch of between .25 and .28. The difference between a dot (as in dot pitch) and a pixel is that a pixel is mapped to dots on the screen. When a monitor is set to lower resolutions, one pixel encompasses many dots.

double click   To press and release a mouse button twice in quick succession with the CURSOR pointing at an ICON or other object on the screen. In general double-clicking initiates the default action for that object: programs are executed, data files are opened in their associated application, etc. *Compare* click.

double-length word   *See* precision.

double precision   *See* precision.

downlink   The transmission link from a communications satellite to a ground station. *Compare* uplink.

download /**down**-lohd, *n.* down-**lohd**, *vb.*/ To send programs or data from a central or controlling computer to a remote terminal, client computer, or other device, such as a printer. A common example is to download files from the Internet onto a user's PC. *Compare* upload.

downloadable font   A SOFT FONT that is stored on a disk and can be downloaded (sent) to a printer's memory. Downloadable fonts are often used with laser printers.

downtime /**down**-tÿm/   The time, or the percentage of time, during which a computer is not available for use. If the computer has been taken out of service for regular PREVENTATIVE MAINTENANCE, then this is *scheduled downtime*. If the computer breaks down and has to be taken out of service for repairs, then this is *unscheduled downtime*.

downward compatibility   *See* backward compatibility.

DP   *See* data processing.

dpi (**dots per inch**)   A measurement of the resolution of a RASTER GRAPHICS image or the maximum number of dots per inch a printer can print, a monitor display, or a scanner resolve. A 600 dpi printer can print 600×600 or 360 000 dots on one square inch of paper. Higher dpi ratings give better quality but lead to larger files and longer printing times.

draft quality   A printing mode available on most printers where quality of output is sacrificed for increased printing speed.

drag   To hold the mouse button and move the mouse pointer to a new position. This can be used to highlight text in a document.

drag and drop   To move or copy an object using the mouse by pointing at the object, holding down the mouse button, moving the mouse pointer (and the object) to the new location, and then letting go of the mouse button.

DRAM /**dee**-ram/   *See* dynamic RAM.

driver   A software program that acts as a translator between an application and some piece of hardware. It contains information needed to recognize and control the device that the computer and the programs using it do not already have. In Windows-based computers a driver is often packaged as a dynamic link library (DLL) file. In Macintosh computers most hardware devices do not need drivers.

drum plotter   *See* plotter.

dry run   Execution of a program for purposes of checking that the program is behaving correctly rather than for producing useful results. The results of execution are compared with the expected results. Any discrepancies indicate an error of some sort, which must be removed before the program is put into productive use.

DSL (**digital subscriber line**)   A method for moving data over private telephone lines. A DSL uses copper wires similar to those used by regular telephone lines but is much faster than a standard telephone connection. A DSL circuit must be configured to connect two specific locations in a similar way to a leased line. *See also* ADSL.

DSP   *See* digital signal processor.

DTD (**document type definition**)   In SGML and XML, a part of a document that defines its structure, or a separate file that defines the structure of a set of similar documents. It uses a special syntax to detail such matters as which ELEMENTS are allowed in the document, in what order they must appear, and how many of each type can be used together. HTML, for example, is defined in SGML by a DTD detailing the structure of Web pages. Documents that are correctly structured are said to *conform* to their DTD, and DTDs can be used to process or display any conforming document using appropriate software. In XML, DTDs are being superseded by SCHEMAS. The abbreviation DTD is sometimes used for 'document type declaration' – a statement at the start of the document assigning the document type definition.

DTP   *See* desktop publishing.

DTV   *See* desktop video.

dual-core processor   A CHIP that contains two MICROPROCESSORS. A computer equipped with a dual-core processor can perform MULTIPROCESSING. Functionally equivalent to two single-core chips, a dual-core processor is faster, uses less power, and takes up less space. Dual-core processors began to appear in mainstream PCs in 2005 and are expected to be the forerunners of multi-core processors.

dual in-line memory module   *See* DIMM.

dual in-line package   *See* DIP.

dumb terminal   *See* terminal.

dump   To copy or make a copy of some or all of the contents of a storage device at a particular instant in order to safeguard against loss of data, to check for errors in a program, or for some other purpose. The data copied is called a *dump*. For example, in a system handling large numbers of users' files stored on magnetic disks, the contents of the disks may be dumped periodically on magnetic tape. This provides a reference copy of the data in the event of, say, accidental overwriting or damage of the disks. Such a dump, which can be used to restore the data automatically, acts as a BACKUP. Again, when a system CRASH occurs, a human-readable version of the contents of main store and possibly backing store might be produced. This can be studied to try and determine the immediate cause of the crash. Dumping can also be performed at some selected point in the execution of a computer program, usually during TESTING or DEBUGGING of the program, or may occur when a program crashes or is aborted.

DSL TYPES						
Type	Description	Maximum send speed	Maximum receive speed	Maximum distance	Lines needed	Phone support at same time
ADSL	Asymmetric DSL	800 Kbps	8 Mbps	5500 m	1	yes
SDSL	Symmetric DSL	2.3 Mbps	2.3 Mbps	6700 m	1	no
HDSL	High bit-rate	1.54 Mbps	1.54 Mbps	3650 m	2	no
ISDL	Integrated Services Digital Network DSL	144 Kbps	144 Kbps	10 700 m	1	no
MSDSL	Multirate Symmetric DSL	2 Mbps	2 Mbps	8800 m	1	no
RADSL	Rate Adaptive DSL	1 Mbps	7 Mbps	5500 m	1	yes

DVD STORAGE COMPARISONS			
Number of sides	Number of layers	Capacity	Movie time (approximate)
1	1	4.7 Gb	2 hours
1	2	8.5 Gb	4 hours
2	1	9.4 Gb	4.5 hours
2	2	17.0 Gb	more than 8 hours

duplex transmission /**dew**-pleks/ (**full-duplex transmission, FDX**) Transmission of data between two endpoints, the data being able to travel in both directions simultaneously. Duplex transmission is used over telephone networks and terminal communication lines, and any other situation in which it is desirable that both ends be allowed to transmit at once. *Compare* half-duplex transmission; simplex transmission.

DVD The successor of the CD, with a far greater storage capacity. The disks can have 4 layers, 2 on each side, with capacities ranging from 4.7 Gb for a single-layer single-sided disk to 17 Gb for a double-layer double-sided disk. Audio, video, and computer data can be recorded. Disks of this type have enough capacity to hold full-length feature films. DVDs can be read-only, recordable, and rewritable in the same fashion as CDs. Originally 'DVD' was an abbreviation for *digital video disk*, but later also for *digital versatile disk*. Because no consensus was reached DVD is now the official name of the format and not an abbreviation for anything.

DVI *See* digital video interface.

dyadic operator (**binary operator**) *See* operator.

dynamic 1. Changing or capable of being changed over a period of time, usually while a system or device is in operation or a program is running. For example, a *dynamic allocation* of a resource can be made while the system is running, and *dynamic RAM* is a type of semiconductor memory that requires its contents to be refreshed periodically since they degrade with time.
2. Taking place during the execution of a program, as happens with a *dynamic dump*. *Compare* static.

dynamic data exchange *See* DDE.

dynamic HTML (**dynamic hypertext mark-up language; DHTML**) An extended version of HTML designed to add additional animation features and more user interactivity to Web pages. It involves scripts in JavaScript or VBScript run on the user's computer to change elements on the screen and respond to user input.

dynamic link library *See* DLL.

dynamic RAM (**DRAM**) For years the most common type of relatively inexpensive main RAM. Dynamic refers to the memory's method of storage, basically storing the charge on the capacitor, which leaks the charge over time and must be refreshed about every thousandth of a second. It is slower than SRAM.

dynamic Web page *See* server-side scripting.

Easter egg An unexpected message, picture, extra function, etc., hidden in a program and designed to appear when a combination of circumstances occur, for example pressing a particular set of keys. Easter eggs are sometimes put in application programs for amusement by programmers.

EBCDIC /eb-see-dik/ (**extended binary coded decimal interchange code**) A CODE developed by IBM and used for the interchange of data between equipment manufactured by or associated with IBM. EBCDIC encoding produces coded characters of 8 bits, and hence provides 2^8, i.e. 256, different bit patterns. These 256 characters form a CHARACTER SET, consisting of letters, digits, special characters such as punctuation marks, and control characters.

echo To reflect transmitted data back to its point of origin. (As in ordinary speech the word is also used as a noun.) For example, characters typed on a keyboard will not appear on the display unless they are echoed. *See also* echo check.

echo check A way of establishing the accuracy achieved during the transfer of data between, say, devices in a computer system or computers in a NETWORK. When the data is received it is stored and is also sent back – echoed – to its point of origin. The returned data is compared with the original data and errors can then be detected. *See also* check.

ECL (**emitter-coupled logic**) A family of LOGIC CIRCUITS that are all fabricated with a similar structure by the same INTEGRATED-CIRCUIT techniques. The circuits are bipolar in nature and are all characterized by very high SWITCHING SPEEDS, but have high power requirements and a low PACKING DENSITY compared with CMOS. *See also* TTL.

e-commerce (**e-business**) Business activities such as advertising, buying and selling, and the servicing of customers conducted through the Internet.

edge connector Part of a PRINTED CIRCUIT BOARD (PCB) where a number of the metallic conducting tracks meet the edge of the board, at right angles, to form the male half of a plug and socket. The tracks are broadened, thickened, and usually gold-plated to provide a good electrical contact; the PCB itself provides the necessary strength and rigidity. A single edge connector may have a hundred or more individual connections, half on each side of the board. The female half of the connector, the socket, consists of a number of sprung metal contacts, embedded in plastic, which make contact with the corresponding tracks on the PCB when the two are pressed together.

EDI (**electronic data exchange**) A standard format for exchanging business data using computer systems; for example, sales information, orders, and invoices.

edit To create or change text, graphics, etc., in a file by deleting, moving, copying, adding, formatting, etc. *See also* word processing.

editor *See* text editor; link editor.

EDO RAM (**extended data out dynamic**

random access memory) A type of fast DYNAMIC RAM used in the 1990s.

EDP Electronic data processing. *See* data processing.

EDSAC /ed-sak/ Electronic Delay Storage Automatic Calculator. *See* first generation computers.

EDVAC /ed-vak/ Electronic Discrete Variable Automatic Computer. *See* first generation computers.

EEPROM /ee-ee-prom/ (**electrically erasable programmable read-only memory**) A type of rewritable read-only memory in which the chips can be erased with electrical signals, rather than ultraviolet light as in the case of EPROMs. EEPROMs do not have to be removed from the computer to be erased but do have to be changed in their entirety. They have a limited life. A FLASH MEMORY is a special form of EEPROM. *See also* EPROM.

EFTPOS /eft-pos/ (**electronic funds transfer at point of sale**) A system that debits buyers' and credits sellers' bank accounts instantly. Point-of-sale terminals are connected to the host computers of credit-card companies or banks. When a transaction is needed calls are initiated automatically to the host computer for verification and authorization and responses are sent back to the terminal.

EGA (**enhanced graphics adapter**) A video standard for IBM PC compatible computers introduced in 1984. It retained compatibility with the old CGA but with more graphics modes, having a resolution of up to 640 × 350 pixels with 16 colors. It was superseded by the VGA standard.

EIDE (**extended IDE**) The popular name for *ATA* (Advanced Technology Attachment), a series of hard-disk interfacing standards for PCs. Features include logical block addressing, native support for CD-ROM, tape, and floppy drives, fast transfer rates, support for hard drive sizes, and plug-and-play. Because of its lower cost,

EIDE is preferred to SCSI in many areas. Today virtually all PCs have two EIDE controllers built into the motherboard and so can support up to four EIDE devices.

EISA /ee-să/ (**extended industry standard architecture**) A 32-bit data path computer expansion bus. This maintained ISA standards but gave additional features. It is now obsolete.

electric cable *See* coaxial cable; twisted pair; ribbon cable.

electronic commerce *See* e-commerce.

electronic data interchange *See* EDI.

electronic filing A computer-based system for the storage, cataloging, and retrieval of documents. It plays a major role in an ELECTRONIC OFFICE. The 'objects' in an electronic filing system are usually stored on magnetic disk and can be organized using a variety of methods. They may be letters, complex reports, charts, graphs, pictures, etc., which are created, manipulated, or deleted as required.

electronic funds transfer The use of computers to bring about payments between individuals and/or organizations such as banks or companies.

electronic mail *See* e-mail.

electronic money *See* e-money.

Electronic Numbering *See* ENUM.

electronic office A computer-based system designed for office tasks. This may involve the use of ELECTRONIC FILING, WORD PROCESSING, DATABASES, COMPUTER GRAPHICS, E-MAIL, and TELECONFERENCING.

electronic publishing The publication of information by means of electronic media such as CD-ROM, DVD, or communications networks, such as the Internet.

electronic shopping Buying and paying

for goods and services through the Internet. *See also* e-commerce.

element In an SGML or XML document, a 'tagged' section that indicates a specific type of content. In this book, for example, the headword of each entry is contained in an 'hw' element, which for this entry is keyed <hw>element</hw>. Elements can contain text, other elements, or both.

elevator The small movable box in a scroll bar that, when dragged with the mouse, alters the part of the text or image that is displayed on the screen.

e-mail (electronic mail) Messages sent between users of computer systems, the computer systems being used to transport and hold the messages. Sender and recipient(s) need not be on the same computer to communicate, or be ON-LINE at the same time. The mail-handling program will have facilities for typing in messages, selecting one or more recipients, and checking for incoming mail. It will also be able to reply to messages, forward copies of incoming messages to other people, and store copies of sent and received messages for future reference or action. A single message may possibly combine text, graphics, and other forms of information and can contain one or more data files as *attachments*. It is also possible to send e-mail using certain types of cellular telephone. See illustration overleaf.

e-mail marketing The use of E-MAILS as a direct-marketing tool. Its main advantage is that is it very cheap since a single e-mail can be sent to thousands of recipients. Each e-mail therefore does not have to attract much extra business to justify its cost. It is easy to analyze an e-mail's effectiveness by measuring such responses as CLICK THROUGHS. However, many recipients find such e-mails highly irritating, especially when they are unsolicited (*see* spam), and software has been developed to filter them out. The use of e-mail marketing techniques for PHISHING and other types of fraud has given it a bad reputation.

embedded Describing a command, code, file, etc., that is inserted into another piece of data while remaining distinct from it in some way. For example, when a spreadsheet is embedded in a word-processing document, it is printed as if it were part of the word-processing document but is still edited using a spreadsheet rather than a word-processing application.

embedded computer system Any system that uses a computer as a component dedicated to a particular task. Examples are programmable washing machines, arcade video games, and satellite navigation systems. It is not possible for the operator of a washing machine to reprogram it to play games, whereas a general-purpose computer system can perform a series of unrelated tasks under the control of the operator.

e-money (electronic money; cybercash) Money exchanged through the Internet.

emoticon /i-**moh**-ti-kon/ A sequence of text-based characters often used in e-mails and online chat. When viewed sideways they are supposed to represent facial expressions to give the reader a sense of how a writer is feeling. For example :-) (known as a *smiley*) is used to show the writer is happy or has just written something intended to be humorous, while :-(shows that the writer is sad.

empty string (null string) A STRING with no characters. Not all programming languages cater for empty strings, and the concept, although useful, should be treated with care.

emulation The imitation of all or part of one computer system by another computer system such that the imitating system executes the same programs, accepting identical data and producing identical results (but not necessarily in the same way), as the system imitated. A device or program used in producing an emulation is called an *emulator*. A particular emulation could be used as a replacement for all or part of the system being emulated, and fur-

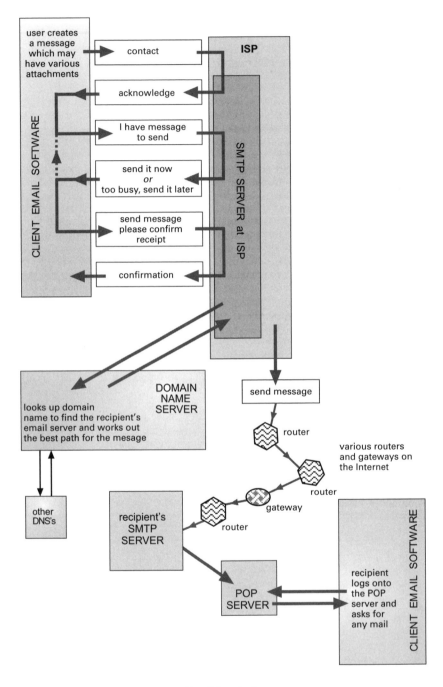

user creates a message which may have various attachments

CLIENT EMAIL SOFTWARE

contact

acknowledge

I have message to send

send it now
or
too busy, send it later

send message please confirm receipt

confirmation

ISP

SMTP SERVER at ISP

DOMAIN NAME SERVER

looks up domain name to find the recipient's email server and works out the best path for the mesage

other DNS's

send message

router

various routers and gateways on the Internet

router

gateway

recipient's SMTP SERVER

router

POP SERVER

recipient logs onto the POP server and asks for any mail

CLIENT EMAIL SOFTWARE

E-mail flowchart

thermore could be an improved version. For example, a new computer may emulate an obsolete one so that programs written for the old one will run without modification.

In contrast with a SIMULATION, an emulation is usually a realistic imitation; a simulation may be no more than an abstract model.

enable   To switch on a device or to select and activate a particular function of a device or a particular feature of a program. *Compare* disable.

encapsulated PostScript   *See* EPS.

encapsulation   **1.** In object oriented programming, the inclusion within a program object of all the resources needed for the object to function and interact with the rest of the program, which treats the object as a BLACK BOX.
2. In telecommunication, the inclusion of one data structure within another so that the first data structure is hidden for the time being. For example, in layered protocols each layer adds header information to the protocol data from the previous layer.

encoder   A means by which data is converted into a coded form. It may be a hardware device, such as a KEYBOARD, or a piece of software. *See also* code.

encoding   *See* code.

encryption /in-**krip**-shŏn/   *See* cryptography.

END /ee-en-**dee**/   *See* Basic.

end user   The person for whom a hardware or software product is designed as opposed to the developer, installer, and servicer of the product.

Energy Star   A symbol attached to systems and components with low power consumption.

engine (back end)   The part of an application that manages and controls data,

contrasted with the FRONT END, which interacts with the user. For example, a database engine is the part of a database application that manages the database itself; it might be used with a variety of front ends, each tailored to the needs of a specific application. (Often, the engine is a DBMS running on a server, while each client computer runs its own front end.) Similarly, Web search engines communicate with users directly via an HTML front end but can also be used to provide the search facility for other Web sites.

enhanced IDE   *See* EIDE.

ENIAC /en-ee-ak/   Electronic Numerical Integrator and Calculator. *See* first generation computers.

enter key   *See* return key.

entity   In an SGML or XML document, a means of representing of an object that is not simple text. Entities have the form &entity; and are often used to represent non-ASCII characters; for example, the entity α often represents a Greek alpha. Entities thus allow SGML and XML documents to use only ASCII characters, which makes them extremely portable. In the above example the correct character for a Greek alpha would only have to be substituted for the entity when the document was typeset or displayed on screen in a WYSIWYG form. Entities also have other uses: for example they can be simple shortcuts that expand into lengthy sequences of text (e.g. &sgml; could be a keyboarding shortcut for 'Standard Generalized Markup Language'), or they can represent other files that SGML- or XML-aware software will insert into the document in place of the entity.

entry point (entry)   The point to which control is passed when a program or a section of a program – a PROCEDURE, SUBROUTINE, or FUNCTION – is called. The entry point is the first instruction in the program or program section to be executed (or more specifically is the ADDRESS or LABEL of this instruction). There may be more than one entry point in a program or program section.

ENUM (Electronic Numbering) A protocol, defined in RFC 2916 (2000), that allows telephone numbers to be mapped to Internet URLS. It is used in VOIP.

environment 1. The combination of hardware and software in a computer.
2. A collection of system- and user-specified settings that can be used to define the context in which a program operates. Each setting is held in a separate named *environment variable*, whose value can be examined by the program. For example, the current USERNAME might be held in an environment variable called 'USERNAME', allowing the program to run differently depending on who is using it.

EPOS /ee-pos/ (electronic point of sale) A system that updates stock and cash records electronically direct from data acquired by the till operation.

EPROM /ee-prom/ (erasable programmable read-only memory) A type of SEMI-CONDUCTOR MEMORY that is fabricated in a similar way to ROM. The contents, however, are added after rather than during manufacture and then if necessary can be erased and rewritten, possibly several times. In the nonprogrammed state the contents are all usually set at binary 1. The desired contents are obtained electronically by a device known as a *PROM programmer*, which sets selected elements to binary 0. The contents are erased (i.e. reset to binary 1) normally by exposure to ultraviolet radiation, and can then be reprogrammed. It is only outside a computer that an EPROM can be programmed. Within a computer the contents cannot be changed; they can only be read. *See also* PROM; EEPROM.

EPS (encapsulated PostScript) An extension to the PostScript page-description language to handle embedded material, usually images. An encapsulated PostScript file is similar to a normal PostScript file except that the EPS file contains a specifica-tion of the size of the output (the bounding box). This extra information allows EPS files to be treated as graphics files by DTP software. Their content is incorporated in a normal PostScript file for printing, typesetting, etc.

EQ gate *See* equivalence gate.

equivalence A LOGIC OPERATION combining two statements or formulae, A and B, in such a way that the outcome is false when either A or B is true but not both; otherwise the outcome is true. The TRUTH TABLE is shown in the diagram, where the two truth values are represented by 0 (false) and 1 (true). The operation can be written as $A \equiv B$.

A	B	output/outcome
0	0	1
0	1	0
1	0	0
1	1	1

Truth table for equivalence gate and equivalence

equivalence gate (EQ gate) A LOGIC GATE whose output is low only when one of its (two or more) inputs is high and the rest are low; otherwise the output is high. It thus performs the operation of EQUIVALENCE on its inputs and has the same TRUTH TABLE. The truth table for a gate with two inputs (A, B) is shown, where 0 represents a low signal level and 1 a high level. *See also* nonequivalence gate.

erasable PROM *See* EPROM.

erase To delete data permanently from a storage medium in such a way that it cannot be recovered.

ergonomics /er-gŏ-**nom**-iks/ The science of how natural and easy things are to use. An ergonomically designed mouse or keyboard will cause less strain and stress to the user.

error A discrepancy between the value or condition that is calculated, observed, or measured and the value or condition that is known to be accurate or correct, or that has been specified as being accurate or correct, or that was simply expected. In computing, errors can be caused by some failure in the hardware of the system or by some condition occurring in the software.

Errors may arise, for example, as data is being transferred to or from magnetic disk, i.e. as data is being written or read. They may also arise during the transfer of data between two computers over a network. Errors may be caused by a flaw in the storage medium, by a fault in a device, or by NOISE in a communication channel. A binary 1 may then be incorrectly written, read, or transmitted as a binary 0, or a binary 0 as a binary 1, producing an error in the data. The detection, handling, and possible correction of such errors is an important aspect of computing. *See also* error-detecting code; error-correcting code.

In programming, errors may arise because the program fails to obey the SYNTAX of the programming language, i.e. the program breaks one of the rules defining how characters and sequences of characters can be combined. These are known as SYNTAX ERRORS and are normally detected when the program is compiled. Other programming errors may not become apparent until the program is executed. These are called RUN-TIME (or execution) errors, and may arise from, say, attempted division by zero or from OVERFLOW. *Logical errors* are another type, occurring because the logic or design of the program is incorrect; a JUMP to the wrong instruction is an example. Logical errors can arise if the problem that the program is designed to solve is not fully understood, so that the wrong ALGORITHM is selected or written. Yet again, errors may arise from the ROUNDING or TRUNCATION of numbers, producing *rounding* or *truncation errors*.

error-correcting code Any of a variety of CODES used to correct ERRORS that occur, for example, during the transfer of data. The data is sent along a channel as a coded signal, with each element of the code constructed by an encoder using a specific rule. A decoder at the other end of the channel detects any departure from this construction, i.e. any errors, and automatically corrects them – with a high probability of success. It is possible for errors to be corrected by an ERROR-DETECTING CODE, by a request for the retransmission of the affected data. Error-correcting codes are more complex and hence more costly to implement than error-detecting codes but reduce the necessity for retransmission.

error-detecting code Any of a variety of CODES used to check data for certain kinds of ERRORS that may have occurred, say, during transfer of the data or due to a writing or reading process. The information is encoded in such a way that a decoder can subsequently detect – with a high probability of success – whether an error has arisen. A fixed number of bits, k, is normally taken into the encoder at a time, and then output as a codeword consisting of a greater number of bits, n. Each codeword in a particular code is constructed following a specific rule. The decoder will subsequently take in n bits at a time and output k bits. On each codeword the decoder performs a CHECK for errors. If an error is detected then generally the sending device is requested to retransmit the affected set of bits. *See also* acknowledgment; error-correcting code.

error diagnostics Information that is presented following the detection of some condition causing an ERROR, for example after an attempt to execute instructions in a program that are invalid or that operate on illegal data. The information is mainly to assist in identifying the cause of the error. Error diagnostics can be produced, for instance, by the COMPILER, and the information would then generally concern SYNTAX ERRORS. Other kinds of errors are detected at run time. In this case the error diagnostics may be produced by a RUN-TIME SYSTEM. *See also* debugging.

error message A message that reports the occurrence of an ERROR. There may be some attempt to diagnose the cause of the error. The message appears on a VDU screen or is stored in an *error file*.

error rate   A measure of the proportion of ERRORS occurring in data transfers to or from a magnetic disk or magnetic tape or in data transmissions along a communication line. The error rate of, say, a magnetic tape system is usually given as the number of bits or bytes of data transferred per error, e.g. 1 error in 10^{14} bits (often expressed as an error rate of 10^{-14}). The error rate of a communication line may be expressed as the number of bits transmitted per error, e.g. 1 error per 10^5 bits. Since errors tend to come in bursts during transmission, the error rate is sometimes given as the percentage of error-free seconds.

escape key   A keyboard key that when depressed sends the escape character to the computer. This is often used in applications to negate the previously initiated action, to go back one level in the menu structure, or to exit the program.

e-text (**electronic text**)   A text-based work, such as a journal or book, that is available for reading on-line or that can be downloaded and read off-line.

Ethernet /**ee**-th'er-net/   A type of local area network (LAN) using a bus topology, developed by Xerox in 1976, then refined by Dec and Intel to form the standard IEEE 802.3. Data to be transmitted is split into packets. Standard Ethernet runs at 10 Mbps (10 million bits per second), fast Ethernet at 100 Mbps, and Gigabit Ethernet at 1000 Mbps. 10 Gigabit Ethernet, which runs at 10 000 Mbps, is now coming into use.

ETX/ACK /ee-tee-**eks** ak/   *See* flow control.

even parity   *See* parity check.

Excel (**Microsoft Excel**)   *Trademark* A SPREADSHEET application developed by Microsoft. First released for the Macintosh in 1985 and for Windows in 1987, it has been upgraded several times and the current version is Excel 2003 (Windows) and Excel 2004 (Macintosh). Excel's early success was important in establishing Microsoft as an applications developer and it is now the most widely used spreadsheet application. Excel forms part of Microsoft OFFICE.

exchangeable disk store (**demountable disk**)   An obsolete disk storage medium – either a DISK PACK or a DISK CARTRIDGE – that could be removed from a DISK DRIVE and replaced by another of the same type. The store could be mounted on other disk drives of appropriate design.

exclusive-OR gate (**XOR gate**)   *See* nonequivalence gate.

execute   To carry out a MACHINE INSTRUCTION or a computer PROGRAM. The process is described as *execution*. A program executes from the time that it is started until either it completes its task and returns control to the OPERATING SYSTEM, it encounters an error whose severity prevents it from continuing, or it is aborted by the user. *See also* fetch-execute cycle; control unit.

execution   *See* execute.

execution error   *See* run-time error.

executive   *See* supervisor.

exit   An instruction in a program or PROGRAM UNIT whose execution by a computer causes control to be transferred away from that program or program unit. It follows that to *exit* from a program or program unit is to transfer control away from that program or program unit. A *return instruction* is an exit that returns control to a specified instruction in the program that originally called the program or program unit.

expansion card (**expansion board; add-in card**)   A PRINTED CIRCUIT BOARD that can be plugged into an existing printed circuit board within a microcomputer in order to improve the performance and capability of the computer. An expansion card is inserted into a special *expansion slot*; several slots are usually provided. The cards are made either by the computer manufacturer or by specialist manufacturers, and may supply a particular need or offer a range of functions.

Memory cards provide additional memory (up to the maximum a computer can handle). *Graphics cards* give a faster display speed, greater resolution, and more colors than provided by default. *Interface cards* are used for communications involving MODEMS, BROADBAND, or LOCAL AREA NETWORKS. A *multifunction card* provides a range of functions on one circuit board.

expert system A computer program or system based on knowledge acquired from experts and used, for example, to assist in medical diagnosis, mineral prospecting, and fault finding. The programs are produced by means of techniques developed in the field of ARTIFICIAL INTELLIGENCE for problem solving, etc.

exponent *See* exponentiation; floating-point notation.

exponentiation /eks-pŏ-nen-shee-**ay**-shŏn/ The operation of raising a number or algebraic expression to a power, i.e. of multiplying that number or expression by itself a given number of times. The common arithmetic notation for exponentiation is A^n, where the superscript n indicates the number of times A is multiplied by itself. The symbol n is the *exponent* of A, and A is said to be raised to the *power n*. For example,
$$2^4 = 2 \times 2 \times 2 \times 2 = 16$$
The value of 2^1 is 2; the value of 2^0 is taken to be 1. Exponents are not restricted to integers but may be fractional or negative.

Exponentiation is supplied in some form in most programming languages. It is represented by means of an ARITHMETIC OPERATOR. The most common representations are ∧ (used for example in Basic) and

** (used for example in Fortran). 2^4 can thus be written as 2 ∧ 4 or 2**4. In ORDER OF PRECEDENCE, exponentiation is performed before multiplication, division, addition, and subtraction. Thus
$$3 ∧ 2 + 5 = 3^2 + 5 = 14$$
To indicate 3^{2+5}, it would be necessary to use brackets, i.e.
$$3 ∧ (2 + 5)$$

expression A programming construct where the values of one or more OPERANDS are altered or combined by OPERATORS or FUNCTIONS to yield (or *return*) a single value. This value is then generally used in some way, being output, assigned to a VARIABLE, or forming a PARAMETER of a function or PROCEDURE. The order in which operators are applied in an expression is determined by the way the expression is constructed and by the rules of the programming language being used.

extended ASCII *See* ASCII.

extended precision *See* precision.

extension The file format identification characters at the end of a filename. For example, the name file.txt has the extension 'txt' indicating a text file. A number of standard extensions are in use. MS-DOS limited extensions to 3 characters, but modern operating systems (including Windows) generally allow more.

external modem A MODEM housed in its own casing and connected to its computer by a cable, usually via an RS232C or USB interface. *Compare* internal modem.

extranet A private network that uses Internet protocols to share part of a business's information and operations with customers, suppliers, and so on. An extranet can be regarded as an extension of a company's INTRANET.

failure Termination of the normal operation of some portion of a computer system, or of the whole system. A failure is the result of a FAULT. The *mean time between failures (MTBF)* is often used as a measure of the RELIABILITY of a computer system.

FAQ /fak/ **(frequently asked question)** A list of questions about the basic points of some subject, together with answers. FAQs are often included in hardware and software documentation, and on the Internet they are provided in specialized newsgroups to help new subscribers or to give general information on Web pages.

FAT /eff-ay-**tee**/ *See* file allocation table.

father file *See* master file.

fault An accidental condition that prevents some portion of a computer system from performing its prescribed function. It usually causes a FAILURE, although it is possible for a system to be FAULT-TOLERANT.

fault-tolerant system A computer system that is capable of providing either a full or a reduced level of service in the event of a fault. For example, in a fault-tolerant system if a disk were to develop a faulty track, then one of a number of spare tracks would automatically be brought into use in its place with minimum impact on the overall performance of the system.

FDX *See* duplex transmission.

feasibility study A study carried out before development of a computerized system in order to establish that the proposed system is possible, practical, and can serve a useful purpose. It may be a purely paper exercise or may involve the construction of an experimental or prototype version. It may concentrate on specific areas or decisions where the feasibility is questionable or the potential risk is greatest.

feed *See* Web feed.

feedback The use of part of a system output as an input to the same system, sometimes intentional and at other times unwanted. This term usually applies to analog rather than digital computing. It is used as a form of correction opposing a change and is then called *negative feedback*. *Positive feedback* reinforces changes and can cause instability and other problems.

Ferranti Mark I /fĕ-**ran**-tee/ *See* first generation computers.

fetch To locate and load a MACHINE INSTRUCTION or item of data from main store. *See also* fetch-execute cycle; control unit.

fetch-execute cycle (**instruction cycle**) The two steps of obtaining a MACHINE INSTRUCTION from main store and carrying it out. The fetch-execute cycle is under the direction of the CONTROL UNIT and is repeated for every instruction.

fiber optics transmission TRANSMISSION of data using special glass (or plastic) fibers, known as *optical fibers*. An optical fiber is very thin, flexible, and is made of very pure material that absorbs only a tiny proportion of the light passing through it. It is constructed in such a way that light can travel down the length of the fiber (which can exceed several miles) with very little loss through the walls. This is

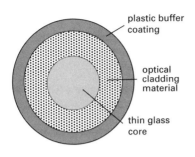

plastic buffer coating

optical cladding material

thin glass core

cross section of a single optical fiber

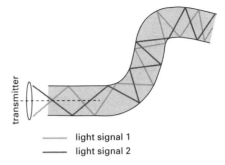

transmitter

 light signal 1
light signal 2

total internal reflection in an optical fiber

Fiber optics

achieved by making the refractive index of the optical fiber vary across its diameter: it is lowest on the outside of the fiber and increases either smoothly or in one or more steps towards the center. (The refractive index of a medium is a measure of the extent to which light bends on entering that medium.) The light is thus made to snake or to zigzag down the center of the fiber.

A simple fiber optics transmission system consists of a transmitter with a light source (such as a laser), a receiver with a light detector, and an intervening optical fiber. An electrical signal carrying the data to be transmitted is fed to the transmitter and causes the light emitted by the source to be modulated, i.e. encoded with the data in some way (*see* modulation); normally the light is emitted as a sequence of pulses. The light signal enters and travels along the optical fiber, and at the other end is detected and converted back into the original electrical signal.

In fiber optics transmission a sheaf of optical fibers is used so that many signals can be carried at the same time. Optical fibers have become cost-competitive with COAXIAL CABLE, with the advantages of having a much greater data-carrying capacity, much lower signal loss, immunity to electrical interference, and no signal interaction between the fibers.

fiche /feesh/ (**microfiche**)   *See* COM.

field   **1.** A portion of a RECORD. It contains a fixed or variable number of letters, digits, or other characters.
2. A portion of a MACHINE INSTRUCTION. It contains a coded representation of, say, the operation to be performed or the address of the operand or operands.

FIFO /fȳ-foh/ (**first in first out**)   *See* queue.

fifth generation computers   The types of computer currently under development. The features are still conjectural but point toward the 'intelligent' machine. *See also* first, second, third, fourth generation computers.

fifth generation language   A type of programming language that expresses a problem to be solved in terms of the logical relationships between and the constraints on the elements of the problem. Such languages remove entirely the need for the user to specify an ALGORITHM to solve the problem, and so carry further the abstraction already visible in FOURTH GENERATION LANGUAGES.

file   A named collection of information held temporarily or permanently on magnetic disk, CD, DVD, tape, etc. The filename is stored in a DIRECTORY and is used to identify the file to the OPERATING SYSTEM of the computer. The details of how a file is stored are handled by the FILE SYSTEM. Files may hold data, programs, text, or any other information, and can therefore be described as DATA FILES, PROGRAM FILES, etc. They can also be broadly classified as MASTER FILES, TRANSACTION FILES, or REFERENCE FILES. Some files are WORK FILES and have only a very brief existence.

A file is *sequentially accessed* if the bytes or RECORDS are read or written one after another starting at the beginning. A file organized to support sequential access is known as a *sequential file*. Sequential organization is used for any files for which there is no requirement to access a particular byte or record directly, but that are normally accessed by starting at the first byte or record and moving through the file one byte or record at a time. This is the case for many data files.

A file is *randomly accessed* if any byte or record can be accessed directly without having to pass over previous bytes or records in the file. A file organized to support random access is known as a *random-access file* (or *direct-access file*). In such cases a program needs a method of determining which byte or record to access to find the data it needs. This information must be already known to or computable by the program or held within the file itself. Three common random access methods are indexed (*see* indexed file), indexed sequential (*see* indexed sequential file), and HASHING.

file allocation table (FAT)   The area on a hard drive that uses the FAT FILE SYSTEM or its derivatives where details about the location of files are kept. These include which CLUSTERS hold each file's data, the order of clusters within the files, what free space is available, and where any flawed segments occur. The FAT file system was used by MS-DOS and is still supported by Windows.

file archive   A collection of magnetic disks, CDs, DVDs, or tapes on which are held rarely used FILES, which can be restored to the computer system if required. Files may be explicitly placed in the archive, or may be automatically moved there after they have remained unused for a given period.

file attribute   A property of a file, such as whether the file is hidden, read only, or archived.

file extension   *See* extension.

file format   A method of organizing data in a FILE. There are many file formats, some of which (e.g. TEXT FILES) are understood by many programs. However, sophisticated applications tend to use their own formats. A file may or may not contain an indication of its format.

file maintenance   The maintenance of both the integrity of FILES, i.e. the correctness of all the data values, and an efficient internal organization of files. It is performed by software. *See also* file management.

file management   The overall management of FILES, including their allocation to space on BACKING STORE, control over file ACCESS, the writing of BACKUP copies, the movement of files to the FILE ARCHIVE, and the maintenance of DIRECTORIES. It is performed by software, basic file management being done by the OPERATING SYSTEM of the computer. *See also* file maintenance.

filename   All files have names but different operating systems impose different rules on how names are constructed. Most operating systems ban the use of certain characters, for example a colon, and impose a limit on the length. Also many systems including Windows and UNIX allow filenames to have an EXTENSION that consists of one or more characters following the proper name.

file protection   The protection of FILES from mistaken or unauthorized ACCESS of information or, in the case of PROGRAM FILES, from mistaken or unauthorized execution. File protection may be concerned with the security of the contents of files, or with the protection of the media on which files are held. The former is implemented by software, the latter by operating procedures.

file server A computer on a network that acts as a storage device, on which any user can store files. File servers are often *dedicated*, i.e. they perform no other function.

file sharing The sharing of public or private computer data or space in a network with various levels of access privilege. File sharing can happen on a small LAN and has been a feature of large mainframe computers for many years. It allows users to read, write, modify, erase, copy, and print files according to their level of access privilege.

filestore /fȳl-stor, -stohr/ The portion of disk BACKING STORE used for storing permanent FILES.

file system 1. The method by which the storage of FILES on magnetic disk, CD, DVD, etc., is organized. Files must be stored in ways that make efficient use of the available space (e.g. files might not be physically contiguous), that provide quick access times (e.g. filenames might be indexed), and that take account of the characteristics of a particular storage device (e.g. CDs work differently to magnetic disks). There are several ways this can be done and each operating system supports one or more. For example, MS-DOS used the *FAT* (file allocation table) file system, whereas Windows XP uses *NTFS* (NT file system) or FAT for fixed disks, FAT only for floppy disks, and *CDFS* (CD file system) for CDs.
2. The part of an operating system responsible for maintaining a particular file system and hiding its organizational details from users, so that they can treat all files simply as named collections of information.

file transfer The movement of entire FILES from one computer system to another, typically across a NETWORK. In order to transfer a file successfully there must be cooperating programs running on each system: the sender has to be able to specify the filename, ultimate destination, and authorization for the transfer to the receiver, and

the receiver must agree to the transfer and inform the sender of its eventual success or failure. In MULTIPROCESSING SYSTEMS file transfer is normally achieved by BACKGROUND PROCESSING. *See also* file transfer protocol.

file transfer protocol (FTP) A protocol that allows a user on one host to access (read and possibly write to), and transfer files to and from, another host over the Internet. It is the simplest way to exchange files on the Internet and can be used with a simple command line interface from MS-DOS, with a Windows GUI program, or through most Web browsers. A password may be needed to gain access to some files. *See also* anonymous FTP; HTTP; SMTP.

file updating Inserting, deleting, or amending values in a file – usually a DATA FILE – without changing the organization of the contents. File updating can be achieved in two ways. In one, commonly used in DATA PROCESSING, the updating process is carried out automatically by program. In the other, a file is displayed on an INTERACTIVE device and the operator can then see to amend it, using an appropriate editing program.

filter A section of code or a program that takes input data and transforms it in some way. The output might be written to a file or passed to another program or section of code.

find *See* search.

Finder In Apple Macintosh computers, the desktop and file management system. It is also responsible for looking after the clipboard and scrapbook and all desktop icons and windows.

finger A piece of software that displays information about a particular user, or indeed all users, logged on to a local or remote system. It typically shows the full name, last name, last login time, idle time, terminal line, and terminal location (if appropriate). The most common use is to see if a person has an account at a particular

Internet site. Many sites do not allow incoming finger requests.

Firefox *Trademark* An OPEN SOURCE WEB BROWSER produced by the Mozilla Foundation. First released in 2004 and favored by technical users, it quickly became the second most widely used Web browser after INTERNET EXPLORER.

firewall Software or a combination of hardware and software that monitors and controls the data that passes across a network's connection to another network. It is often used to protect a private network or intranet that also has connections to the Internet from unauthorized access by users of other networks. The firewall is often installed in a dedicated computer separate from the rest of the network so that incoming users cannot directly access private network resources. Software-only firewalls can be installed on standalone PCs, and are now considered an essential safety precaution when using the Internet.

FireWire (**IEEE 1394**) *Trademark* A standard INTERFACE used to connect a computer to up to 63 external devices. Developed by Apple Computer, Inc., it became an IEEE standard in 1995. It supports PLUG AND PLAY and offers high data-transfer speeds, either 400 Mbps or 800 Mbps depending on the version. FireWire has long been the *de facto* standard method of transferring data from digital camcorders to computers. All Macintosh and an increasing number of Windows computers support FireWire.

firmware /**ferm**-wair/ Programs that do not have to be loaded into a computer but are permanently available in main store. They are held in ROM (read-only memory) and thus are fixed in content even when the power supply is removed.

first generation computers The earliest computers to be produced, in the 1940s and early 1950s, based on the technology of the time. Electronic valves were used in the circuitry and storage devices were usually delay lines, electrostatic devices, or rotating MAGNETIC DRUMS. Input/output was achieved mainly with PUNCHED CARDS, punched PAPER TAPE, and later on with MAGNETIC TAPE; simple PRINTERS were available. The use of valves meant that first generation computers were very large in size, generated considerable heat and thus required cooling, and did not operate very fast or very reliably. They were still able to perform impressive computations. Most of them embodied the concept of the STORED PROGRAM.

The following are important first generation computers.

ENIAC: design begun 1943 by J. W. Mauchly and J. P. Eckert at the University of Pennsylvania; operational 1946, but did not use a stored program and was regarded as a general-purpose electronic calculator.

EDVAC: design begun 1945 by J. von Neumann at the University of Pennsylvania, the proposal being the first written documentation of a computer with a stored program; not operational until 1952.

Manchester Mark I: design begun 1946 by T. Kilburn and F. C. Williams at the University of Manchester; regarded as the first operational stored program computer, with realistic problem-solving first achieved in April 1949.

EDSAC: design begun 1946 by M. V. Wilkes at Cambridge University; started operating in May 1949 as the first complete operational stored program computer.

Ferranti Mark I: commercial version of the Manchester Mark I, produced by Ferranti Ltd., Manchester, and delivered in 1951; it was the world's first commercially available computer. Ferranti's computer interests were sold, 1963, to ICT, which merged, 1968, with English Electric Leo Marconi to form ICL (International Computers Ltd.).

UNIVAC I: product of the US company Eckert-Mauchly Corp. and commercially available in 1951, shortly after the Ferranti Mark I. Eckert-Mauchly was formed in 1947 following development of ENIAC and later became Sperry-Univac.

LEO I: design begun 1947 by T. R. Thompson and J. Pinkerton for J. Lyons &

Co. (a large UK catering firm) and operational by 1953; enhanced version (LEO II) marketed by Leo Computers Ltd., formed 1954. Leo Computers merged with English Electric, 1963, and subsequently formed part of ICL.

See also second, third, fourth, fifth generation computers.

fixed disk, fixed disk drive *See* disk drive.

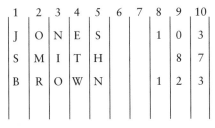

1	2	3	4	5	6	7	8	9	10
J	O	N	E	S			1	0	3
S	M	I	T	H				8	7
B	R	O	W	N			1	2	3

Fixed format

fixed format Data arranged so that particular items always occur in particular positions. In the example shown in the diagram, the name always starts in the first position, while the numbers are in positions 8, 9, and 10. Because the whereabouts of each item of information is strictly defined, there is no need for gaps between the items. The format of the example could be 'A name in columns 1 to 5, an age in positions 8 and 9, and the number in the family in position 10.' *Compare* free format.

fixed-point notation A representation of REAL NUMBERS that is in everyday use and in which a number is expressed in the form of a set of digits with the RADIX POINT fixed in position so that the digits have their correct positional value (*see* number system). Examples of fixed-point notation in decimal and binary are
4927.54, 101011.11, 0.0000001
Fixed-point notation is used in computing in situations, such as commercial data processing, where huge quantities of numbers are handled, making speed important, or where the rounding errors possible with

FLOATING-POINT NOTATION are unacceptable (e.g. in monetary calculations). However, the values involved must be neither very large nor very small. The range of values that a computer can handle efficiently is limited by WORD length: each item of information must fit into a word containing a fixed number of bits. A 32-bit word, for example, can have 2^{32} possible values, which can be used, say, to represent whole unsigned numbers from 0 to 4 294 967 295, or 0 to 429 496.7295 if an accuracy of one ten-thousandth is required. Although more than one word can be used to represent a number and hence give a wider range of values, this increases processing time. Where a wide range is needed, floating-point notation must be used.

flag A device, item of data, etc., that can be set to a particular value to indicate that some condition or state has been attained by a piece of equipment or by a program. For example, an *overflow flag* is a single bit that is set to 1 when OVERFLOW occurs in a register during an arithmetic operation; an *interrupt flag* is set when an INTERRUPT occurs.

The word *sentinel* is used as a synonym for flag but is usually used in the context of input and output. For example, an *end-of-data sentinel* indicates that all the data has been read. A flag or sentinel, once set, is subsequently used as a basis for a conditional JUMP and similar decision processes.

flame A strong opinion, criticism, or derogatory comment about something or someone in an e-mail, chat, or usergroup posting.

flash drive *See* flash memory.

flash memory (flash RAM) A nonvolatile memory that can be erased and reprogrammed in units called memory blocks. It is a variation of EEPROM, which is erased and rewritten at byte level and is thus slower than flash memory updating. Flash memory can be used to hold the BIOS in PCs but is not suitable for RAM. A popular use is in the *flash drive*, a tiny device that acts as a disk drive holding many megabytes of data. Flash drives usually connect to a computer through a USB port and can thus be easily unplugged. Flash memory is also used in digital cellular phones, digital cameras, cards for notebook computers, and many other devices.

flash RAM   *See* flash memory.

flatbed plotter   *See* plotter.

flatbed scanner   A type of scanner in which the page to be scanned is stationary behind a glass window while the head moves past the page, similar to a photocopier. In the flatbed scanner a series of mirrors keeps the image picked up by the moving head directed onto a lens that feeds the image to a bank of sensors. The advantage of flatbed scanners is that they are relatively inexpensive and thick documents, such as books, can be scanned.

flat panel display   *See* display.

flat screen   A type of SCREEN in the form of a thin flat panel rather than the protruding surface usually associated with the cathode-ray tube used in many display devices. Plasma panel DISPLAYS, ACTIVE MATRIX DISPLAYS, and LCDS have flat screens.

flexidisk /fleks-ă-disk/   *See* floppy disk.

flicker   The phenomenon whereby a display screen appears to change brightness. Flicker can be the result of several factors including a low refresh rate, ambient lighting, or low-persistence phosphors on the screen. It is somewhat subjective and most people perceive no screen flicker if the refresh rate is 72 Hz or higher.

flip-flop (bistable)   An electronic circuit that has two stable states, either of which can be maintained until the circuit is made to switch, or 'flip', to the other state. A change in output state requires a suitable combination of input signal(s) and existing output state; in effect, the receipt of a pulse will reverse the state. The two states are used in computer circuits to represent binary 1 and binary 0. Flip-flops can therefore store one BIT of information, and are widely used as storage elements, e.g. in ROM and RAM and in REGISTERS. Various types have been developed, differing slightly in function.

floating-point notation   A representation of REAL NUMBERS that allows both very large and very small numbers to be expressed in a convenient form. Examples are
$$1.3 \times 10^6, \; -2.5 \times 10^3, \; 0.65 \times 10^{-3}$$
Such numbers are encountered in science, math, and engineering and cannot be handled efficiently in a computer in FIXED-POINT NOTATION.

A floating-point number is represented by a pair of quantities called the *mantissa* and the *exponent*, and has the general form
$$m \times b^e$$
b is the base of the number system, equal to 10 in decimal notation; the mantissa, m, can be positive or negative; the exponent, e, is the power to which the base is raised, and is a positive or negative integer. In programming languages, whose character sets do not include superscripts, different conventions are used. A common one is to use an 'E' to stand for '10 to the power of', e.g.
$$2.7 \times 10^{-3}$$
is written
$$2.7 \; E \; -3$$
Inside a computer a floating-point number is stored and manipulated as a sequence of bits. Typically, the first bit is a *sign bit*, indicating whether the mantissa is positive or negative; this is followed by a fixed number of bits giving the sign and magnitude of the exponent; this in turn is followed by a fixed number of bits representing the magnitude of the mantissa. The radix point of the mantissa is usually taken to be before the leftmost bit, i.e. the mantissa is a fraction. There may be, say, 7 bits for the exponent and 40 for the mantissa. This gives a wide range of numbers but means that on small machines each floating-point number needs several WORDS of store, which degrades performance. Also, floating-point numbers cannot represent all possible real numbers within their range exactly; this leads to rounding errors.

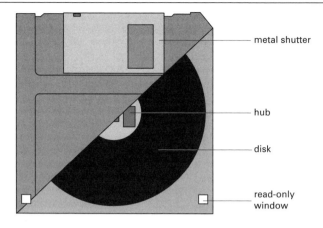

metal shutter

hub

disk

read-only
window

A 3.5 inch floppy disk with its envelope

floating-point number A number expressed in FLOATING-POINT NOTATION.

floppy disk (**diskette; flexidisk**) A relatively small flexible MAGNETIC DISK, enclosed within a stiff protective envelope. Floppy disks were formerly used in small computers as on-line BACKING STORE and off-line storage devices. They have now largely been superseded by writeable CDs, but are still used for storing or transferring small quantities of data. They have much smaller storage CAPACITIES and much longer ACCESS TIMES than HARD DISKS, but are considerably cheaper and need less careful handling.

The most common type of floppy disk is 3½ inches in diameter. The older 5¼ inch disks are now rarely seen. As with other magnetic disks, data is recorded in concentric tracks in the magnetic coating. The tracks themselves are divided into SECTORS. Most floppy disks have a capacity of about 1.4 megabytes.

The floppy disk remains inside a rigid plastic envelope when it is fed into a DISK DRIVE for reading and writing purposes. The disk rotates inside the envelope, becoming rigid as it spins. A sliding metal shutter covers a slot in the envelope, which allows the read/write head to make contact with the disk surface. The disk drive can be prevented from writing to the disk by means of the *write-*

protect hole in one corner of the envelope.

flops (**floating-point operations per second**) A commonly used measure of computer performance, made in terms of the number of arithmetic operations that can be performed on FLOATING-POINT NUMBERS in a period of one second. It is usually expressed in *megaflops* (a million flops), *gigaflops* (a billion flops), *teraflops* (a trillion flops), or *petaflops* (a thousand trillion flops).

floptical disk /**flop**-tă-kăl/ A type of magnetic disk the same size as a 3½″ floppy that used a laser beam to help position the read/write heads more accurately than was possible with an ordinary floppy. There was a series of small, precise concentric tracks stamped in its surface coating. These tracks were where the data was written and were thinner and more numerous than on a conventionally formatted floppy. The combination of more precision and more tracks allowed a floptical to hold up to 20 megabytes of data. Flopticals are now obsolete.

flowchart **1.** (**program flowchart**) A diagrammatic representation of the structure of a program, showing the actions performed by the program and the flow of control. It consists of a set of *boxes* of various standard shapes, interconnected by a set of *flow lines*. For clarity the lines may have arrows to indicate the flow of control. The different shapes of the boxes indicate either different kinds of activity or a decision to be made. The activity or decision is described within the boxes, typically in natural language (e.g. English). In addition there are circular connector symbols that connect flow lines and are used, for example, when a flowchart continues on another page.

2. (system flowchart; data flowchart) A diagrammatic representation of a complete

symbol	use
	for indicating start or finish of a flowchart
	for any kind of processing activity, excluding decisions
	for a processing activity dealt with elsewhere, e.g. subroutine, procedure
	where a decision is to be made, the two (or more) exits being labeled, e.g. 'yes' or 'no', in accordance with the possible choices
	for any input or output operation
	for indicating continuation on or from another part of the flowchart
description / medium	for any input or output of data, however achieved
content / medium	for data held in any on-line file, the medium usually being disk
program function	for processing of data
	for indicating continuation elsewhere

Symbols in program flowcharts (top) and system flowcharts (bottom)

computer system, showing both the operations involved and the flow of data in the system. It does not indicate how the processing operations are carried out – this information would appear in a program flowchart. The number of flow lines entering and exiting each symbol depends on the situation.

flow control   Procedures used to control the DATA TRANSFER rate, i.e. to limit the rate at which data is transferred to the rate at which it can be received – either at its final destination or at a series of points along its route.

There are three flow-control methods in common use. With *XON/XOFF* the receiver can temporarily halt the sender by transmitting an XOFF (transmitter off, ASCII code 19), and an XON (transmitter on, ASCII code 17) when once more ready to receive; this method requires full DUPLEX TRANSMISSION. With *ETX/ACK* the sender terminates a block of data with an ETX (end transmission, ASCII code 3), and the receiver responds with an ACK (acknowledge, ASCII code 6) only when it is ready for more data; this method can be used in HALF-DUPLEX communications but relies on the sender not sending too much at a time. Thirdly there is the use of *control signals*, where the receiver uses a separate channel to indicate to the sender whether it is ready for data.

focus   The state of the active item in a GUI FORM. Only one item per form can have the focus at any one time and it is this item that receives input from the keyboard when the form is active. The focus can be moved by clicking on a new item with the mouse or by pressing the tab key (move to next item) or shift+tab (move to previous item) on the keyboard.

folder   1. An alternative term for DIRECTORY on GUI systems.
2. A component of a storage facility found in some programs. For example, an e-mail client will often provide a system of user-configurable folders to store received and sent messages.

font   A set of printable or displayable characters. The design for a set of fonts is called the typeface and the variations of this design (bold, italic, etc.) form the typeface family. Many fonts have been designed and can be purchased for general use. Three main kinds of font are in use on computers. TRUETYPE fonts are the commonest on personal computers. The other kind are POSTSCRIPT fonts (also known as *Type 1 fonts*), which are preferred for output of PostScript printing files used in commercial printing. *OpenType* fonts are a development of TrueType fonts that can incorporate PostScript font data.

footprint   The area of floor or desk space occupied by a device, or the shape and size of panel opening into which it would fit.

FOR   *See* Basic.

foreground processing   In a MULTI-TASKING system, the processing of a computer program in a way that allows interaction with the user through the keyboard. Only one program per user can run in this way at any one time, and these are called *foreground programs*; all other programs are *background programs*. Foreground programs are usually granted a high priority and can thus preempt the resources – main store, processor, etc. – of the computer from background programs, which have a lower priority. Users can usually change the foreground program at will by selecting one of the background programs, which immediately becomes the foreground program.

for loop   *See* loop.

form   1. A page of printer paper. It may be a single sheet or may be sheets joined together in the form of CONTINUOUS STATIONERY. In either case it may be multipart so that copies can be produced. Individual sheets may be preprinted with headings or other information, possibly with lines or boxes.
2. A display that contains blank fields, which can be filled in using a keyboard and

mouse. The data can then be transferred automatically to a database.

3. A WINDOW whose content is so laid out as to display a particular type of information to the user, and possibly to receive input. DIALOG BOXES are a type of form.

format **1.** The structure that is used in the arrangement of data. It is determined before use, and in some cases is set by industrial or international standards. There are various kinds. *Instruction format* defines the arrangement of the parts of a MACHINE INSTRUCTION. *Printer format* defines the layout on paper of the output from a PRINTER, for example the areas of a page where printing will occur and the spacing between words and between lines. *Tape format* and *disk format* define the arrangements in which data can be recorded on MAGNETIC TAPE or MAGNETIC DISK.
2. To prepare a blank storage medium, such as a disk, to accept data.
3. To put data into a predetermined structure.
4. To organize or change the appearance and arrangement of text and images on a screen or printed page.

form feed A special character (ASCII decimal 12), sometimes abbreviated to FF, that by convention causes a printer to advance one page length or to the top of the next page.

forms tractor *See* tractor feed.

Forth /forth/ A PROGRAMMING LANGUAGE designed originally for the control of scientific instruments from microcomputers with very small memories. It is a LOW-LEVEL LANGUAGE that makes great use of the STACK, and is very compact. Although the current memory sizes of microcomputers have made the small size of Forth largely irrelevant, it is still used as a substitute for ASSEMBLER language among some microcomputer users.

Fortran /**for**-tran/ A PROGRAMMING LANGUAGE whose name is derived from **for**mula **tran**slation. It was first released by IBM in the mid-1950s. Fortran was one of the earliest HIGH-LEVEL LANGUAGES generally available for scientific and technical applications, and has been under continuous development. Fortran is still widely used for scientific work, even though its CONTROL STRUCTURES are relatively crude compared with more modern languages. Each new release has more sophisticated features.

forum An on-line discussion group. Some Web sites, especially Web PORTALS, provide a large selection of forums in which participants can exchange open messages. Newsgroups and conferences are also types of forums.

forward compatibility *See* upward compatibility.

fourth generation computers Computers currently in use and designed in the period from about 1970. Since the design of computers is a continuous process by different groups in several countries, it is difficult to establish when a generation of computers starts and finishes. Conceptually the most important criterion that can be used to separate fourth generation from THIRD GENERATION COMPUTERS is that they have been designed to work efficiently with the current generation of HIGH-LEVEL LANGUAGES and are intended to be easier to program by their end-user. From a hardware point of view they are characterized by being constructed largely from INTEGRATED CIRCUITS and have very large (multi-megabyte) main store fabricated from SEMICONDUCTOR MEMORY.

The availability of an ever-increasing range of integrated circuits has produced a rapid fall in the cost of hardware while software costs have continuously increased. This has led to the existence of many computers designed for special tasks such as communications, automatic control, and military systems, which in the past would have been general-purpose machines adapted to their task by software. It has also led to a dramatic increase in home computing and Internet use.

See also first, second, fifth generation computers.

fourth generation language A programming language designed to be closer to a natural language. Generally, a first generation language is machine code, a second generation is assembler, and a third generation is a high-level programming language such as C or Java. A fourth generation language, such as SQL, allows users to specify the desired commands in more abstract terms than previously, allowing the language processor to choose the exact method used to achieve them. *See also* fifth generation language.

fractal /frak-tăl/ A curve or surface that has a fractional dimension and is formed by the limit of a series of successive operations. A typical example of a fractal curve is the *snowflake curve* (also known as the

Koch curve). This is generated by starting with an equilateral triangle and dividing each side into three equal parts. The center part of each of these sides is then used as the base of three smaller equilateral triangles erected on the original sides. If the center parts are removed, the result is a star-shaped figure with 12 sides. The next stage is to divide each of the 12 sides into three and generate more triangles. The process is continued indefinitely with the resulting generation of a snowflake-shaped curve. A curve of this type in the limit has a dimension that lies between 1 (a line) and 2 (a surface). The snowflake curve actually has a dimension of 1.26.

One important aspect of fractals is that they are generated by an iterative process and that a small part of the figure contains

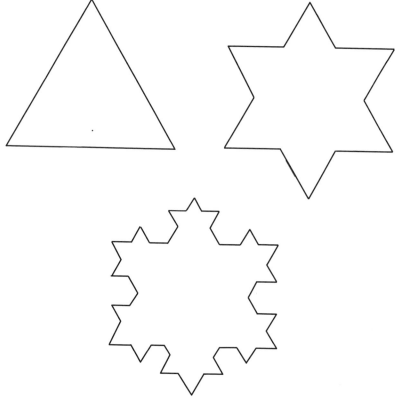

Generation of a snowflake curve

the information that could produce the whole figure. In this sense, fractals are said to be 'self-similar'. The study of fractals has applications in chaos theory and in certain scientific fields (e.g. the growth of crystals). It is also important in computer graphics, both as a method of generating striking abstract images and, because of the self-similarity property, as a method of compressing large graphics files.

See also Mandelbrot set.

fragmentation /frag-men-**tay**-shŏn/ The breaking up of disk FILES into a number of different sections scattered across a disk. It occurs when files of different sizes are frequently deleted and written. Each new file is written in the first available gap left by the deletion of a previous file; if the new file is too big for the gap then the rest is written in the next gap, and so on. The result is that a number of movements of the read/write heads of the DISK DRIVE might be required to access each fragmented file, thus reducing the efficiency of the file system. Programs exist for defragmenting disks.

frame 1. In data transmission, a unit of data complete with addressing and the necessary protocol control information that is to be transmitted between network points. The transmission is usually serial and contains header and trailer fields that 'frame' the data.
2. A separate section of a Web page that acts as an independent unit (e.g. text can be independently scrolled). Frames can be defined by tags in HTML.

free format Data arranged so that, although the items are in a specified order, the actual positions of the items are not defined. In free format it is necessary to have a separator such as some spaces and/or a comma between the items so that it is possible to tell when one ends and the next begins, for example
>   JONES,10 3
>   SMITH,8 7
>   BROWN,12 3
>   ROBINSON,25, 1

The format here could be 'A name followed by an age followed by the number in the family.' *Compare* fixed format.

free space 1. Space not being used on a storage device.
2. Space provided free of charge for a user to set up a Web page.

freeware /**free**-wair/ Copyrighted software that is offered at no cost but may have limitations on its use, most commonly that it cannot be incorporated in any program or package that is sold for profit and that the author's copyright must always be acknowledged. *See* shareware.

frequency division multiplexing (FDM) *See* multiplexer.

friction feed A technique for feeding paper into a PRINTER. Each sheet is fed in by hand and is gripped by a rotating roller, as in a typewriter. The paper cannot be positioned as accurately as it is with TRACTOR FEED. *See also* cut sheet feed.

front end The part of an application that interacts with the user, contrasted with the ENGINE, which manages and controls the application's data. In client/server applications the front end usually runs on the client while the engine runs on the server.

front-end processor A small computer that is used to relieve a mainframe computer of some of the tasks associated with input/output. It receives data from a number of input/output devices (such as the terminals in a MULTIACCESS SYSTEM), organizes it, and transmits it to the more powerful computer for processing. It also handles the output from the mainframe to the terminals. In a NETWORK, a front-end processor may handle the control of communication lines, code conversion, error control, etc. It is more powerful than a MULTIPLEXER in that it can perform simple editing, echoing, routing, etc.

FTP *See* file transfer protocol.

full-duplex transmission /**dew**-pleks/ *See* duplex transmission.

full-screen In Windows systems, the display mode in which the main window is maximized and the toolbar and the menu, title, and status bars are hidden.

full-text index *See* searching.

function A section of a program whose main purpose is to calculate a single value, such as the square root of a number, the sine or cosine of a number, or the length of a character STRING. The item of data on which the calculation is performed, or its address in memory, is known as the ARGUMENT; the calculation may be performed on more than one item of data and hence there may be more than one argument. To perform the calculation the function must be called.

A collection of different functions are provided as part of a programming language, and any of these can be used by the programmer without an understanding of the details of its internal working. Most languages also permit user-defined functions. These are usually like PROCEDURES, except that they return a calculated value. *See also* parameter.

functional programming A style of programming that emphasizes the evaluation of expressions rather than the execution of commands.

function key *See* keyboard.

function part *See* operation part.

fuzzy logic A type of logic that deals with ideas of possibility and probability. Conventional logic deals with values 'true' and 'false', whereas fuzzy logic could have a range of values such as 'true', 'very true', 'not very true', etc. It is used in EXPERT SYSTEMS and other applications of artificial intelligence.

FYI (**for your information**) Publications that convey general information about topics related to the Internet. They are a sub-series of RFCs that are not technical standards or descriptions of protocols.

G The symbol used for GIGA-, or sometimes for gigabyte.

game port The port on a PC into which devices such as joysticks or games controllers are connected.

games console An interface designed especially for computer games. Movement and speed are manipulated using buttons or levers on the console. Some controllers are specialized, such as those with steering wheels and accelerator and brake pedals for racing games. Force feedback joysticks are now available to add realism.

garbage Stored data that is no longer valid or no longer wanted. Removal of this superfluous data from store is known as *garbage collection. See also* GIGO.

garbage in garbage out *See* GIGO.

gas panel display *See* display.

gate *See* logic gate.

gateway A network connection device similar to a BRIDGE except that it is usually visible to the user. It translates data between one type of network and another.

gender bender A component for joining two male or two female connectors together.

general-purpose computer A computer that can be used for any function for which it can be conveniently programmed and can thus perform a series of unrelated tasks.

generations of computers An informal way of classifying computer systems on the basis of advances in technology, especially electronic technology, and latterly also on advances in software. The generation of computers currently in use is the FOURTH GENERATION (which succeeded the THIRD, SECOND, and FIRST GENERATIONS), with FIFTH GENERATION COMPUTERS currently under development. Since the design of computers has been a continuous process for decades – by a wide variety of people in different countries, faced with different problems – it is difficult (and not very profitable) to establish when generations start and finish.

generator A program that accepts the definition of an operation that is to be accomplished and automatically constructs a program for the purpose. An example is the *report generator*, which can be used in commercial data processing to extract information from files. The input to such a program is a description of the file structure and a specification of the information required and the way in which it should be presented to the user. The report generator can then construct a program to read the file, extract the desired information, and output it in the desired format.

GHz (gigahertz) A measure of frequency equal to 10^9 hertz.

GIF /jiff, giff/ (**graphics interchange format**) A graphics file format developed by CompuServe, which is widely used for transmitting raster image files on the Internet. The files are encoded in binary and use LZW compression. GIF files are often smaller than the same file would be if stored in JPEG format but are limited to 256 different colors and so photographic images are not stored so well. GIF files can contain simple animations. These files have the file extension gif. *See also* PNG.

giga- /gĭg-ă-/ A prefix indicating a multiple of a thousand million
    (i.e. 10^9)
or, loosely, a multiple of 2^{30}
    (i.e. 1 073 741 824).
In science and technology decimal notation is usually used, and powers of 10 are thus encountered – 10^3 10^6, 10^9, etc. The symbol G is used for giga-, as in GV for gigavolt. Binary notation is generally used in computing, and so the power of 2 nearest to 10^9 has assumed the meaning of giga-.

The prefix is most frequently encountered in computing in the context of storage CAPACITY. With magnetic disks, magnetic tape, and main store, the capacity is normally reckoned in terms of the number of BYTES that can be stored, and the *gigabyte* is used to mean 2^{30} bytes; gigabyte is usually abbreviated to G byte, GB, or just to G; thus a 1.2 G byte disk can hold 1.2885×10^9 bytes. Giga- is part of a sequence
    kilo-, mega-, giga-, tera-, peta-, ...
of increasing powers of 10^3 (or of 2^{10}).

GIGO /gȳ-goh/   Garbage in garbage out, signifying that a program working on incorrect data produces incorrect results.

global   Applicable to a whole document, file, or program rather than a particular section. *See* local.

global variable   *See* scope.

golfball printer   An obsolete type of impact character PRINTER whose printing mechanism was first used in electric typewriters. The characters were moulded on a spherical print head. The head was rotated (in two directions) in order to bring the required character opposite the printing position, and was then struck against the inked ribbon and paper. The head was interchangeable, enabling different character sets to be used.

Google   *Trademark* An Internet SEARCH ENGINE operated by Google, Inc. First developed in 1996–98 by Larry Page and Sergey Brin while they were graduate students at Stanford University, it is based on their theory that the relative importance of Web pages containing a match of the search term can best be determined by analyzing the number of links to each page from the other matching pages; the results are then presented in descending order of this importance. This theory was confirmed in practice and Google acquired a reputation for quickly and reliably returning relevant results – a notorious problem with other search engines at the time. By early 2004 the Google engine handled over 80% of all Web searches. The verb 'to Google', meaning to perform a Web search using Google, has entered the English language.

Gopher   A distributed information service (named after the mascot of the University of Minnesota) that makes hierarchical collections of information available across the Internet. Each node is either a menu, a search node, or a leaf node. A simple protocol is used that allows a single Gopher client access to information from any accessible Gopher service. However, its use declined with the spread of the more flexible World Wide Web.

GOSUB /goh-sub/   *See* Basic.

GOTO statement /goh-too/   An unconditional JUMP INSTRUCTION in a high-level language. It causes the normal flow of control to be broken by specifying explicitly the next statement to be executed, usually by means of a statement label, as in
    goto error_handler
Excessive use of GOTOs can make programs difficult to follow and their use is therefore discouraged. An extended form, the *computed GOTO*, appears in Fortran and is used to go to one of several statements, the choice dependent on the value of an integer variable. Many versions of Basic have a similar facility; however, the same effect can be achieved more elegantly by the CASE or SWITCH STATEMENTS of other languages.

GPU *See* graphics processing unit.

grandfather file *See* master file.

graph A network of lines and vertices, or a data structure that emulates this. A TREE is a kind of graph.

graphical user interface (GUI) The type of user interface provided with Windows and Macintosh operating systems (and also others), which provides the user with an intuitive graphical menu-driven system incorporating such things as windows, icons, pull-down menus, scroll bars, and pointing devices.

graphic character A symbol that can be produced by printing, typing, or handwriting. It may be an ALPHANUMERIC CHARACTER, a SPACE CHARACTER, a SPECIAL CHARACTER, or any other symbol. *Compare* control character.

graphics *See* computer graphics.

graphics accelerator A chipset attached to a video board that contains a GRAPHICS PROCESSING UNIT. This enables it to update the video display much more quickly than the CPU can and also compute special effects common to 2-D and 3-D images. Another benefit is that the CPU is freed up to execute other commands while the graphics accelerator is handling graphics information. Graphics acceleration has become a necessity since the advent of multimedia and sophisticated games, in which very large images need to change quickly in response to user input. The older graphics accelerators tend to run using the PCI bus but more power can be obtained with accelerators designed for an accelerated graphics port (*see* AGP).

graphics card (**graphics board**) *See* expansion card.

graphics display A VDU with graphics capability, i.e. one that can display pictures, graphs, and charts as well as text.

graphics processing unit (GPU; graphics coprocessor) A CHIP or CHIPSET that performs data manipulations related to graphical output on a VDU. Such operations could be accomplished in the computer's CPU by appropriate software, but using specialized hardware is much faster and modern uses of computer graphics – in particular 3-D animations – require this extra speed. A GPU is the core component of a GRAPHICS ACCELERATOR.

graphics tablet *See* digitizing pad.

graph plotter *See* plotter.

gray scale The representation of colors in a graphics file or on a monochrome output by shades of gray ranging from black to white. The number of bits per pixel determines the number of shades of gray that can be achieved.

greeking The substitution of meaningless line patterns or bars of gray to represent characters or images whose size makes them too small to be displayed at the current screen resolution. The spacing and arrangement of the text is visible but the text itself is not legible.

groupware Applications designed to be useful to a team of people. Groupware services cover such activities as the sharing of calendars, e-mail handling, shared information access, collective writing, and electronic meetings. Examples of groupware applications are Microsoft Exchange and Lotus Notes.

guest A log-in account on a network that can be accessed without a password. For example, on some subscription Web sites it is possible to sample the content by entering as a guest. The ACCESS PRIVILEGES of a guest account are generally very limited.

GUI /goo-ee/ *See* graphical user interface.

H

hacker 1. A computer enthusiast who is particularly interested in the minutiae of how PROGRAMS work and in techniques for stretching their capabilities. Hackers are usually talented programmers and their interest often extends to writing their own programs or to altering existing software to 'improve' it, sometimes without authorization. However, they tend to assume that the users of their software will not only possess their level of computing skill but also share their cast of mind. Their products are therefore often badly documented and difficult for normal users – whom hackers regard with contempt – to understand and operate.
2. (**cracker**) A person whose aim is to breach the SECURITY of remote computer systems, such as those owned by banks or government agencies. The motive is often merely personal satisfaction but occasionally malicious damage or fraud occurs. The hacker's equipment need only be a personal computer and a telephone modem.

half-duplex transmission /dew-pleks/ (HDX) Transmission of data between two endpoints, the data traveling in either direction but only in one direction at a time. It cannot flow simultaneously in both directions. Only one signal wire is therefore required in half-duplex working. The time taken to reverse the direction of transmission from send to receive or from receive to send is called the *turnaround time*. Half-duplex transmission can be used for terminals that perform their own echoing, and for any situation in which it is unnecessary for both ends to transmit simultaneously. *Compare* duplex transmission; simplex transmission.

halftone /haf-tohn, hahf-/ A method of reproducing a photograph or illustration in print by using areas of regularly spaced small dots of a color whose differing diameters allow more or less of the white background to show through and thus create the illusion of varying shades of the color. The larger the spot's diameter, the darker the shade at that point on the image. GRAY SCALE pictures use halftoning with black ink to achieve shades of gray when printed, and in color printing the intensity of each primary color (*see* CMYK system) is varied using halftoning to create other colors. The frequency of the halftone dots is measured in lines per inch (LPI).

halt instruction A MACHINE INSTRUCTION that halts the execution of a program and usually turns control of the computer to the OPERATING SYSTEM.

hand-held computer A computer that is small enough to be fitted in a handbag or pocket and used while it is being held in the hand. Special operating systems, such as Windows CE, and special applications running under the operating system are needed. Some hand-held computers accept handwriting as input; others use a small keyboard. Voice and data telephone services are sometimes combined with the hand-held computer in a single device. *See* PDA.

hand-held scanner A small scanner that is moved manually across the document. Often the scanning head is not as wide as that in a FLATBED SCANNER and most hand-scanner software automatically combines two half-page scans into a single image. Hand scanners are relatively inexpensive and convenient for capturing small amounts of text, but unsuitable for reproducing high-quality images.

handle 1. An outline box that appears in many applications when a graphical object is selected. It can be used to resize the image. The same mechanism can be used to alter the boundaries of the area selected without changing its contents. 2. A name chosen to identify a user on an on-line service. It may be the user's real name, or a nickname.

handshake An exchange of signals that establishes communication between two or more devices (by synchronizing them) and allows data to be transferred. The signals have various meanings such as
  'I am waiting to transmit.'
  'I am ready to receive.'
  'I am not ready to receive.'
  'I am switched on.'
  'The data is available.'
  'Data has been read successfully.'

hands on Describing work in which a person controls and interacts with a computer directly.

hang A program, device, or system is said to 'hang' when it has stopped responding to input, although everything appears to be otherwise normal. Sometimes this is caused by the system waiting for some event to happen and, when this occurs, normal running will resume automatically. However, it is possible that the program has crashed and operator intervention is needed to abort it.

hard copy A permanent copy, printed on paper for example, of information from a computer.

hard disk A rigid MAGNETIC DISK. A computer's ON-LINE BACKING STORE is nearly always formed by one or more hard disks. Formerly there were several types of hard disk, e.g. DISK PACK and DISK CARTRIDGE, but modern ones all use WINCHESTER DISK DRIVE technology, where the disk itself and its DISK DRIVE form a single hermetically sealed unit. Technological advances have greatly increased the recording DENSITY of hard disks and storage capacities of many gigabytes on a 3.5 inch disk are now common.

hard return The character that forces the following text to be printed on a new line in a document. In word-processing documents a hard return indicates the end of a paragraph. See also soft return.

hard-sectored disk See sector.

hardware /**hard**-wair/ The actual equipment used in a computer system, including not only the major devices – such as VDUs, disk drives, and printers – and the electronic circuitry making up semiconductor memory, logic circuits, etc., but also cables, cabinets, and so on. Compare software.

hardware character generation A technique whereby a VDU, printer, or plotter can be directed (by receipt of a particular signal from a computer) to display or draw a specified CHARACTER, the style and size of which is determined by the device's internal circuitry. This drastically reduces the complexity of the information transmitted. Instead of having to describe each character in terms of a number of dots or strokes, a single code of typically 7 or 8 bits represents each character. By means of special control codes, the size, slant, brightness, etc., of the hardware characters may often be altered. However, the variety of characters and styles is still very limited and inadequate for modern GUI-based operating systems and applications where many different fonts are used. Most VDUs, etc., now use SOFT FONTS rather than hardware character generation.

hardware interrupt See interrupt.

hardwired /**hard**-wÿrd/ Describing functions that are built in to a system rather than being added by software. Hardwired logic is logic circuitry that forms all or part of an INTEGRATED CIRCUIT and is thus permanently interconnected. Its function cannot therefore be changed.

hashing /**hash**-ing/ A technique used for organizing TABLES or FILES of information to permit rapid SEARCHING or TABLE LOOK-UP. It is particularly useful for random-access FILES and for tables to which items are added in an unpredictable manner, such as the SYMBOL TABLE of a compiler. Each item of information to be placed in the table or file has a unique *key*. A special *hashing algorithm* uses the key to allocate a position to each item, the collection of items being distributed fairly evenly over the table or file. The same algorithm is used to yield the starting point for a search for the key.

hash total /hash/ *See* control total.

Hayes compatible /hayz/ Describing a modem or system that has the same set of commands as a Hayes Microcomputer Products modem. Although the company no longer exists, Hayes compatibility remains a de facto standard for microprocessor modems.

HDTV (**high-definition television**) A television with higher resolution than one based on the NTSC standard. HDTV sets have a wider screen and roughly twice the normal resolution. The picture quality is similar to 35-mm movies with sound quality similar to that from a CD. HDTV signal transmissions are digital rather than analog, and need to be compressed, using the MPEG-2 file format and compression standard, before transmission through the narrow TV lines. The signals are then decompressed in the television receiver.

HDX *See* half-duplex transmission.

head The part of a peripheral device that is in contact with the data medium, or very close to it, and that is responsible for writing data on the medium or for reading or erasing data. For example, DISK DRIVES have *magnetic heads*; a magnetic head may be a *read head*, a *write head*, an *erase head*, or a *read/write head*, depending on its function.

head crash (**disk crash**) The accidental and disastrous contact of a read/write head with the surface of a hard disk as it rotates in a DISK DRIVE. (Normally the head flies just above the surface.) A crash is often caused by the head passing over a particle of dust on the surface. The contact destroys the track so affected – and the data stored in that track. Particles of surface material produced by the contact rapidly spread around and cause other tracks to be destroyed. The damaged disk usually has to be thrown away. Disks should be BACKED UP at regular intervals so that in the event of a crash a duplicate is available.

header 1. Text or an image that is repeated at the top of pages.
2. In digital communications, information at the beginning of a packet of data containing details about the sender, the receiver, the length of the packet, and, sometimes, the protocol being used.
3. The part of an e-mail that precedes the main message and contains the name of the sender, the addressee and other recipients, the date and time, etc.

help On-line documentation containing information about a particular application or operating system.

helper application A PLUG-IN that runs in a browser's window so that streaming video, sound, and other processes can be used in the Web page. The helper application is launched when the browser downloads a file it cannot process. Many modern Web browsers no longer need helper applications and can cope with the more common multimedia file formats themselves.

hertz /herts/ The SI unit of frequency. A frequency of 1 hertz is obtained if one cycle of an event is repeated every second.

heuristic /hyû-**riss**-tik/ A method of solving a problem that, unlike an ALGORITHM, cannot be guaranteed always to find a good solution or to run in an acceptable time. Heuristics are used where there is no satisfactory algorithm available and with the acceptance that, while they will usually work better than an algorithm, in some cases they will work very badly or even fail. Steps must be taken to ensure that such cases cannot arise in practice, or can be detected if they do and an alternative heuristic used. Systems based on heuristics often monitor their own performance and adjust their heuristics accordingly. Heuristic methods have been employed, for example, to control the routing of data through computer NETWORKS: the actual performance of a network is used to influence subsequent decisions on routing. Heuristic techniques have also been applied to problems in ARTIFICIAL INTELLIGENCE and to the detection of VIRUSES and SPAM.

hex /heks/   *See* hexadecimal notation.

hexadecimal notation /heks-ă-**dess**-ă-măl/ (**hex**)   A number system that uses 16 digits and thus has BASE 16. The 16 digits are represented by

0, 1, 2, ... 9, A, B, C, D, E, F

It is a positional notation (*see* number system), positional values increasing from right to left by powers of 16. Hex is a convenient shorthand by which people (rather than computers) can handle binary numbers. Each hex digit corresponds to a group of 4 binary digits, or bits, as shown in the table. Conversion of binary to hex is done by marking off groups of 4 bits in the binary number (starting from the right) and replacing each group by its hex equivalent. Conversion of hex to binary is done by replacing each hex digit by its equivalent binary group. *See also* binary notation.

HEXADECIMAL AND BINARY EQUIVALENTS			
Binary	*Hex*	*Binary*	*Hex*
0000	0	1000	8
0001	1	1001	9
0010	2	1010	A
0011	3	1011	B
0100	4	1100	C
0101	5	1101	D
0110	6	1110	E
0111	7	1111	F

hidden file   A file that has the hidden attribute set and is not normally visible to users; i.e. is not shown on directory listings or as an ICON when its FOLDER is opened in a window. Files are usually hidden to protect them from deletion or modification and tend to contain critical system code or data. Most file-management utilities allow the user the option to view hidden files.

hierarchical database   A type of database that is structured as a TREE: all records, or nodes, have one parent node (except for the root node) and possibly one or more child nodes. No other kind of link between nodes is allowed. Hierarchical databases are less powerful than RELATIONAL DATABASES. *See also* network database.

hierarchical file system   A FILESTORE where files are organized in directories (folders) that contain other directories or folders as well as files. The highest level directory is called the *root* and the directories passed through to reach a file is the *path*. Introduced by the Unix operating system in the 1970s, hierarchical file systems are now the norm.

hierarchy of operators   The ORDER OF PRECEDENCE of OPERATORS in a particular programming language.

high-definition television   *See* HDTV.

high-level language   A type of PROGRAMMING LANGUAGE whose features reflect the requirements of the programmer rather than those of the computer. It achieves this by being designed for the solution of problems in one or more areas of application. High-level languages are thus described as *applications-orientated* (or *problem-orientated*). They are easier for the programmer to use than LOW-LEVEL LANGUAGES, being closer to natural language and to the language of mathematics.

The programmer needs little knowledge of the computer on which the program is to run: unlike a low-level language a high-level language does not reflect the facilities provided by computer hardware. Before being run on a particular computer, however, programs written in a high-level language must be translated into a form that can be accepted by that computer, i.e. they must be converted into MACHINE CODE. This is achieved by means of a special program – either a COMPILER or an INTERPRETER. Each statement in the original program – called the *source program* – is translated into many machine instructions. A source program can be translated into different machine codes using different compilers or interpreters. As a result, programs written in high-level languages can be moved from one computer to another.

The first high-level languages were released in the late 1950s. The number and variety now available is large and still growing. Examples include BASIC, COBOL, FORTRAN, PASCAL, ADA, PROLOG, JAVA, and C++. Some high-level languages are aimed at specific areas of application. Cobol, for instance, was developed for commercial data processing while Fortran is widely used for scientific computation. Java was originally designed for programs downloaded from the Internet. Other languages, such as Basic, Pascal, C++, and Ada, are designed to be of more general use.

highlight   To change the appearance of displayed text, giving it emphasis. Data that has been SELECTED is usually highlighted, by displaying it in reverse video.

history   A list of actions recently performed by a user. For example, many applications have a feature that shows the last few files opened, so that the user is able to open a previously used file quickly. Web browsers also display a history of the previous Web sites visited.

hit   A success in matching data.

HMD (**head mounted display**)   The headgear worn in VIRTUAL REALITY applications.

home computer   A computer specifically aimed at the home computing market.

home key   The keyboard key that moves

the pointer to the beginning of the line or to the top of a menu, or to some recognized home position in an application.

home page   1. The entry page of a hypertext system, usually acting as the introductory page on a Web site. The address for a Web site is usually the home page address.
2. An individual's personal Web page.

host (**host computer**)   1. A computer running a SERVER, or that provides services in some other way to computers on its network. Hosts can range from microcomputers to mainframes. For example, the host for a Web site is the computer whose Web server software manages the site and makes it available on the WWW.
2. In mainframe computing, the host is the mainframe and may have intelligent or dumb workstations attached to it that use it as a host provider of services.

host computer   *See* host.

hot key   A user- or application-defined key combination that executes a command or causes the operating system to switch to another program. For example, in Windows applications Ctrl-S often saves the file currently being edited.

hotlist   A list of frequently accessed Web sites. *See also* bookmark.

hot swapping (**hot plugging**)   The replacement of a hard drive, CD-ROM drive, or other device without the need for switching off the computer or reconfiguration or resetting of system parameters.

housekeeping   Actions performed within a program or within a computer system in order to maintain internal orderliness. In the case of a program it may, for example, involve the management of storage space by freeing areas no longer required (*see* garbage). In the context of an entire computer system, housekeeping involves backing up the filestore, deleting files that are no longer required or whose expiry dates have passed, running appropriate programs to reduce FRAGMENTATION, and many other mundane but essential tasks.

HTML (hypertext markup language) The script programming code used to generate pages on Web sites. HTML is an application of SGML that uses tags or markup symbols to identify elements such as formatting, links, and other special handling of images, text, and objects. A Web browser then uses these tags to display the page. Files in HTML format usually have the extension htm or html. The earliest version of HTML had a fairly simple set of tags. However, with the increasingly sophisticated design of Web pages, more features have been added allowing designers to produce more advanced typographic and layout effects. HTML is being superseded by *XHTML*, a reformulation of the language as an application of XML. *See also* cascading style sheets; hypertext.

HTTP (hypertext transfer protocol) The protocol used to connect Web client software with Web servers and download Web pages. It also contains facilities for sending simple data to the Web server and also for uploading and managing Web pages. The letters 'http' are used in URLs, as in http://www.google.com.

HTTPS A variant of the HTTP protocol that uses encryption to transfer sensitive information securely over the World Wide Web. The URL of Web sites that use HTTPS begins with 'https://...' rather than 'http://...'

hub A device that allows an ETHERNET network to use a similar physical layout to a star network. Ethernet networks normally require all NODES to be connected to a single cable in a bus layout, but can use a more convenient starlike layout if each node is connected to a central device – the hub – that emulates this by broadcasting all PACKETS it receives to all connected nodes. *See also* network switch.

human-computer interface (HCI) *See* man-machine interface.

human-system interface (HSI) *See* man-machine interface.

hybrid computer *See* analog computer.

hyperlink *See* hypertext.

hypermedia /hÿ-per-**mee**-dee-ă/ *See* hypertext.

hypertext /**hÿ**-per-tekst/ Text that contains links, accessible by computer, to other parts of the document or to other documents (i.e. text with automatic cross-reference lookup). Hypertext is most usually held in HTML format, with certain words or phrases tagged to point to other places in the document or to places in other files. These words or phrases are *hyperlinks*. They are usually displayed in a different color and are often underlined. When the pointer moves over the hyperlink, it changes appearance (e.g. from an arrow to a finger), and clicking on the hyperlink takes the display to the destination it specifies. This destination can be another file on the user's machine or network, but it can also be any Web page on the World Wide Web. (Indeed, the hyperlinks in Web pages are the glue that binds the WWW together.) Consequently the user can jump from place to place, without having to read the whole document. Electronic publications now are usually not simply text but may also contain graphics, audio, video, and animations, and are generally referred to as *hypermedia*.

hypertext markup language *See* HTML.

hypertext transfer protocol *See* HTTP.

hyphenation /hÿ-fĕ-**nay**-shŏn/ The splitting of words using a hyphen at the ends of lines in text. Word-processing and DTP applications insert hyphens automatically. These are called *discretionary hyphens* or *soft hyphens*, whereas hyphens that are added explicitly by entering the dash character are called *hard hyphens*.

IA-64 A 64-bit processor and instruction set based on explicitly parallel instruction computing (EPIC) design philosophy, produced by Intel. In 64-bit mode these processors can calculate two bundles on up to three instructions at a time. However in 32-bit mode they are slower because decoders must first translate 32-bit instruction sets into 64-bit instruction sets. The ITANIUM was the first in Intel's line of these processors.

IAP (**Internet access provider**) *See* ISP.

IAR *See* instruction address register.

IAS (**immediate access store**) *See* main store.

IBG (**interblock gap**) *See* block.

IBM International Business Machines; formerly the largest computer manufacturer in the world, formed in 1911. The company's main business was originally as a producer of punched-card tabulating machines, and then later in mainframe computing and office equipment manufacture. Its first microcomputer, called the IBM PC, was built in the early 1980s and set standards that were widely adopted.

IBM-compatible (**PC-compatible**) Describing a personal computer that was a clone of the IBM PC and its successors. The PC's open architecture was further developed and improved and was subsequently known as industry-standard architecture (ISA).

IC *See* integrated circuit.

ICANN (**Internet Corporation for As-** signed **Names and Numbers**) A private non-profitmaking corporation that since 1998 has administered Internet IP ADDRESSES, DOMAIN NAMES, and some technical matters.

icon A pictorial symbol or small picture used in a GRAPHICAL USER INTERFACE to represent an available resource (a file, a printer, etc.). Double-clicking an icon or dragging it onto another icon will initiate an appropriate action.

IDE (**integrated drive electronics; intelligent drive electronics**) An interface for mass storage devices that is built into the disk or CD-ROM. This replaced the separate drive-specific disk controller card that had formerly been required. The result was greater flexibility in the choice of disk drives because all makes of IDE disk drive communicated with the rest of the computer in a standard way that concealed the details of their internal workings. Developed in the 1980s, IDE has been superseded by EIDE.

identifier A STRING of one or more letters, digits, or other characters selected by a programmer to identify some element in a program. An identifier can be a NAME or a LABEL. The kind of element that can be identified depends on the programming language. It could, for example, be a VARIABLE, an ARRAY, a PROCEDURE, or a STATEMENT. When the program is compiled or assembled, prior to execution, identifiers are converted to machine addresses using a SYMBOL TABLE. *See also* literal; reserved word.

idle time Time during which the central processor of a computer system is performing no useful function. It usually occurs when the workload on the system is insufficient to keep the processor fully occupied, or when the demands on the input/output components of the system are such that the processor runs out of work before outstanding input/output is completed.

IDSL (Internet digital subscriber line) A connection that provides DSL (digital subscriber line) technology over existing ISDN lines. IDSL has been superseded by other DSL technologies that offer higher transfer rates.

IE *See* Internet Explorer.

IEEE (Institute of Electrical and Electronic Engineers) A US society of professional engineers that defines standards. The British equivalent is the IEE (Institution of Electrical Engineers).

IEEE 1394 *See* FireWire.

IETF (Internet engineering task force) An international open community of people interested in Internet networking, including designers, operators, vendors, and researchers. They are responsible to the Internet Engineering Steering Group (IESG) and propose solutions to various technical problems occurring on the Internet. The TCP/IP protocol was formulated from their specifications.

IF, IF THEN *See* Basic.

if then else statement A simple conditional CONTROL STRUCTURE that appears in most programming languages and allows selection between two alternatives; the choice is dependent on whether a given condition is true or not. An example is
　if it rains this afternoon
　then we go to the cinema
　else we go for a walk
Most languages also provide an *if then statement* to allow conditional execution of a single statement or group of statements, for example
　if you have a dog with you
　then put it on a lead

See also case statement.

if then statement *See* if then else statement.

illegal character A character, possibly in a binary representation, that is not in the CHARACTER SET of a particular computer or of a particular programming language.

illegal instruction A MACHINE INSTRUCTION whose OPERATION CODE cannot be recognized by a particular computer with the result that the instruction cannot be executed by the computer. Since neither a COMPILER nor an ASSEMBLER ought to be able to generate an illegal instruction, a message from the OPERATING SYSTEM claiming that one has been encountered is usually caused by accidentally using part of the main store containing instructions to store data, and subsequently trying to execute the data.

iMac /ȳ-mak/ A group of Apple Macintosh computers, introduced in 1998, in which the monitor, the CPU, and the CD-ROM drive were all in one unit. A distinguishing feature was the translucent colored case. iMacs were designed for home, school, and small office use and promoted by Apple as an easy-to-use and stylish computer that outperformed other low-cost options. The iMac range has been updated regularly and is now based around a FLAT PANEL MONITOR.

image Any picture or diagram stored either as a set of pixels (*raster image*) or as a series of instructions for reproducing the picture (*vector image*).

image compression The use of data compression techniques on graphics files to reduce the amount of storage required. *See also* compression.

image map An image on a Web page that contains links to other parts of the page, other pages, or to files. On older Web browsers these maps were executed through CGI scripts. However, with newer browsers the maps are executed by the browser itself.

image processing (picture processing) The analysis and manipulation of information contained in images. The original subject can be an actual object or scene, or a photograph, drawing, etc. A numerical version of this subject is obtained electronically, for example by scanning: the original is effectively divided into a large number of tiny individual portions, and a set of numbers is produced, corresponding to the measured brightness and color of these portions. The original is thus converted into a two-dimensional ARRAY of data. The individual portions are known as *pixels*; the greater the number of pixels per image, the greater the detail available. The original is said to have been *digitized*.

This numerical version of the original is usually stored in a computer and can be manipulated in various ways to highlight different aspects of the original. For example, a specified range of brightness can be increased, or two slightly different images can be compared or superimposed. The final form of the image can be produced on a screen, printer, etc., and information derived from the image can appear in graphs or tables and can be further analyzed.

imagesetter /im-ij-set-er/ A typesetting device that produces a very high resolution output (paper or film). Imagesetters are expensive and are used by specialized companies or bureaus. Most users provide PostScript or PDF files, which have been proofed by printing to PostScript laser printers, and are then processed by the imagesetter.

IMAP *See* POP3.

IMHO In my humble opinion; one of many abbreviations in common use on the Internet, for example in newsgroups and e-mails. It is used to soften a criticism in order to avoid a FLAME.

immediate access store (IAS) *See* main store.

immediate addressing The process by which the data needed as the OPERAND for an operation is actually held in the associated MACHINE INSTRUCTION. Although the data contained there is usually restricted in size, immediate addressing provides a convenient and quick way of loading small numbers into a REGISTER or ACCUMULATOR. *See also* addressing mode.

implementation 1. The working version of a given design of a system.
2. The processes involved in developing a working version of a system from a given design.
3. The way in which some part of a system is made to fulfil its function.

import To bring some external object, such as a picture or another file, into the current system or application. The format of the information being imported must be supported by the system receiving the data.

inactive window In a system capable of displaying multiple windows, a loaded window that is not currently being used. It may be totally or partially hidden by other windows and becomes active (receives the focus) when selected by the user.

inclusive OR A Boolean operator that returns the value of true if either or both of the operands is true.

incremental backup A backup system in which the only data backed up is that changed since the last backup.

index 1. A data structure consisting of a set of POINTERS that can be used to locate records in another data structure (often a DATA FILE) based on the value of a KEY within the records. An index allows records where the key has a particular value or range of values to be accessed directly, without having to search for them sequentially through the entire structure. A small index might be a simple list of key values, with pointers to the records in which they occur, arranged in ascending order. However, large indexes have more complex structures to minimize lookup times. Common uses of indexes are in databases and in INDEXED FILES and INDEXED SEQUENTIAL FILES.

2. The value held in an INDEX REGISTER.

3. (subscript) A value, usually an integer or integer expression, that is used in selecting a particular element in an ARRAY.

indexed addressing (indexing) An ADDRESSING MODE in which the address given in a MACHINE INSTRUCTION is modified by the contents of one or more INDEX REGISTERS; the contents of the index register(s) are usually added to the address. The result identifies a storage location. The address specified in the machine instruction is called an *indexed address*. It may be an actual address or an address of an address (*see* direct addressing; indirect addressing).

indexed file A DATA FILE in which records can be accessed by means of an INDEX. This is stored in a separate portion of the file. An indexed file has characteristics of a random-access file. *See also* file; indexed sequential file.

indexed sequential file A FILE whose records are organized in such a way that they are written sequentially and can be read sequentially or by means of an index (*see* file; indexed file). The index can be used to read data randomly or to skip over unwanted records in a sequential read. The records are accessed by means of an *indexed sequential access method (ISAM)*. An indexed sequential file is similar to a textbook: It can be read in chunks or straight through, or specific information can be found using an alphabetical index.

indexing 1. *See* indexed addressing.
2. The process of selecting an element from an ARRAY by means of its INDEX.
3. The process of creating an index for a data file or other data structure by analyzing its contents.

index register A REGISTER that can be specified by machine instructions using INDEXED ADDRESSING. Its contents are used to modify the address given in the instruction. Data can be loaded into the index register

by a variety of methods, depending on the type of computer; it may for example be loaded from the ACCUMULATOR.

indirect addressing An ADDRESSING MODE in which the address specified in a MACHINE INSTRUCTION identifies a storage location that itself holds an address. The address part of the instruction is thus an address of an address. It is called an *indirect address*. The second location identified generally holds the required item of data, but may however hold a further address.

infection The corruption of a computer system by a VIRUS.

infinite loop A LOOP in a program from which, under certain circumstances, there is no exit except by terminating the program. This can be done accidentally, perhaps by a while loop whose condition can never be false, or deliberately when a program is to repeat a sequence of actions until stopped by some external event such as an interrupt or termination by the operating system.

infix notation /**in**-fiks/ A form of notation in which an OPERATOR is placed between its OPERANDS, as in the arithmetic expression

$$a + b * c$$

If no brackets are used the expression is evaluated in a computer according to the ORDER OF PRECEDENCE of the programming language. *See also* reverse Polish.

information The meaning that is carried by DATA. Computers process data without any understanding of its meaning; however, it is the information contained in data that is important to people.

information processing The organization and manipulation of pieces of INFORMATION in order to derive additional information. Information processing is the principal means of increasing the amount and variety of information. The computer is an information-processing machine in which information is enhanced or extracted by appropriate manipulations of the underlying data.

information retrieval The process of recovering information from a computer system. The aim is to access only the information required, and to do so in the minimum time with a minimum of simple instructions to the system. *See also* information storage.

information storage The recording of information, usually on magnetic disk, so that it may be used at some later time. There are various methods of organizing and labeling the information so that it may be readily accessed. *See also* information retrieval.

information superhighway A phrase coined by Al Gore, when Vice President of the USA, in one of his speeches. It points toward a worldwide system of unlimited bandwidth connections direct to businesses and homes, bringing unlimited amounts of data and information to educational institutions, businesses, and homes. The term is often used as a synonym for the INTERNET.

information technology (**IT**) Any form of technology, i.e. any equipment or technique, used by people to handle information. Although the abacus and the printing press are examples of IT, the term usually refers to modern technology based on electronics. It thus incorporates both the technology of computing and of telephony, television, and other means of telecommunication. It has applications in industry, commerce, education, science, medicine, and in the home. The development of IT is recognized worldwide as being of major importance, especially since cost reductions have made large-scale IT systems economically possible.

information theory The branch of probability theory that deals with uncertainty, accuracy, and information content in the transmission of messages. It can be applied to any system of communication, including electrical signals and human speech. Random signals (noise) are often added to a message during the transmission process, altering the signal received from that sent. Information theory is used to work out the probability that a particular signal received is the same as the signal sent. Redundancy, for example simply repeating a message, is needed to overcome the limitations of the system. Redundancy can also take the form of a more complex checking process. In transmitting a sequence of numbers, their sum might also be transmitted so that the receiver will know that there is an error when the sum does not correspond to the rest of the message. The sum itself gives no extra information since, if the other numbers are correctly received, the sum can easily be calculated. The statistics of choosing a message out of all possible messages (letters in the alphabet or binary digits for example) determines the amount of information that is contained in it. Information is measured in bits (binary digits). If one out of two possible signals are sent then the information content is one bit. A choice of one out of four possible signals contains more information, although the signal itself might be the same.

infrastructure Generally the actual hardware and software used to interconnect computers and users on a large network, including such items as telephone lines, cable television lines, satellites, and antennas together with devices that control transmission paths. The term is used especially of the INTERNET, where infrastructure companies have been significant in its growth as they affect decisions as to where interconnections are placed and made accessible and how much information is carried.

Ingres /**ang**-grĕ/ An early relational database management system that was developed from a research project at the University of California in the 1970s. It influenced several later systems.

initialize /i-**nish**-ă-lÿz/ To set COUNTERS, contents of storage LOCATIONS, VARIABLES, etc., to zero or to some other specific value before the start of some operation, usually at the beginning of a program or program unit. The value set is known as the *initial value*. Many programming languages provide a facility for specifying initial values of variables when the variable is first declared.

ink-jet printer A type of PRINTER in which fine drops of quick-drying ink are projected on to the paper to form the characters. It is thus a nonimpact matrix printer and is much quieter in operation than an impact printer. In one design the print head contains a column of nozzles, each capable of ejecting a single drop of ink at a time. As the head moves along the line to be printed, the appropriate characters are built up. In another design the head has only a single nozzle that emits a jet of ink. The jet is made to break up into droplets, and each droplet is electrically charged; this charge enables a droplet to be deflected vertically as it flies towards the paper. As the head moves along the line to be printed, the appropriate characters are built up. This latter design is slower than the design with the column of nozzles, which can print up to 400 characters per second. Color ink-jets can produce a large number of colors by mixing the inks as they print.

input /in-pût/ **1.** The data or programs entered into a computer system by means of an INPUT DEVICE.
2. The signal – voltage, current, or power – fed into an electric circuit.
3. To enter data, programs, or signals into a system.

INPUT *See* Basic.

input device Any device that transfers data, programs, or signals into a computer system. Examples include KEYBOARDS, graphical devices such as a MOUSE or DIGITIZING PAD, VOICE INPUT DEVICES, SCANNERS and DOCUMENT READERS, SAMPLING devices that convert analog data, such as video and sound, to digital form, and independent digital devices, such as digital cameras and camcorders, whose data can easily be transferred to a computer. An input device need not be operated by a person – it could

for example be a temperature sensor. There may be many input devices in a computer system, each device providing some means of communication with the computer; input devices used by people are said to provide a MAN-MACHINE INTERFACE. *Compare* output device.

input medium *See* media.

input/output (I/O) Components or processes in a computer system that are concerned with the passing of information into and out of the system, and thus link the system with the outside world.

input/output bus (I/O bus) *See* bus.

input/output device (I/O device) Any device in a computer system that can function both as an INPUT DEVICE and an OUTPUT DEVICE, and can thus be used both as an entry point and as an exit point for information. TERMINALS are I/O devices.

input/output file (I/O file) A FILE used to hold information immediately after input from or immediately before output to an INPUT/OUTPUT DEVICE.

input/output processor (I/O processor) *See* direct memory access.

insert key A keyboard key, labeled Ins or Insert, that toggles between overwrite mode (i.e. characters typed on the keyboard replace those at the current position on the screen) and insert mode (i.e. typed characters are inserted before those at the current position) when editing.

install /in-stawl/ **1.** To store and configure software permanently so that it can be run on a particular computer and make use of the various peripherals attached to that computer. Software and games can be obtained on CDs or DVDs and installed from these media, or they can be downloaded and installed from the Internet.
2. To add hardware to a system.

installation program (setup program) A program provided with an operating system or application program that facilitates the process of installation. The user is guided through a question-and-answer session that will enable the software to be set up and run effectively.

instruction   *See* machine instruction.

instruction address register (**IAR**)   A REGISTER, i.e. a temporary location, in the CONTROL UNIT from whose contents the ADDRESS of the next machine instruction is derived. It thus stores in turn the addresses of the instructions that the computer has to carry out. Each time the control unit fetches an instruction, it automatically increases the contents of the IAR by one. If the sequence of the instructions is altered by an unconditional JUMP INSTRUCTION, the contents of the IAR are changed to the address indicated by the jump instruction; in the case of a CONDITIONAL JUMP instruction, the contents of the IAR are changed only if the condition is met.

Since the contents of the IAR are incremented so as to point to the start of the next instruction, the IAR is a COUNTER and hence is also called an *instruction counter* or a *program counter*.

instruction counter   *See* instruction address register.

instruction cycle   *See* fetch-execute cycle.

instruction format   The layout of a MACHINE INSTRUCTION, showing its constituent parts.

instruction overlap   *See* control unit.

instruction prefetch   /**pree**-fech/   *See* control unit.

instruction register (**IR; current instruction register, CIR**)   A REGISTER (i.e. a temporary location) usually in the CONTROL UNIT, used for holding the MACHINE INSTRUCTION that is currently being performed or is about to be performed. Once the control unit has finished execution of one instruction, it obtains the address of

the next instruction from the INSTRUCTION ADDRESS REGISTER, fetches this instruction from main store (via the MEMORY DATA REGISTER), and places it in the instruction register. The control unit can then interpret the instruction in the instruction register and cause the instruction to be executed. *See also* microprogram.

instruction set   All the instructions that are available in a particular MACHINE CODE or ASSEMBLY LANGUAGE, i.e. all the instructions that a particular computer is capable of performing. *See* machine instruction.

INT /int/   *See* Basic.

integer /**in**-tĕ-jer/   Any positive or negative whole number, or zero. Integers belong to the set ... $-4, -3, -2, -1, 0, +1, +2, +3, +4, ...$
Negative integers carry a minus sign to distinguish them from positive integers. Positive integers do not necessarily carry a plus sign. Any integer without a sign is said to be *unsigned*.

Unsigned integers are stored in a computer in BINARY NOTATION, i.e. as a sequence of bits making up a WORD; the sequence is the binary equivalent of the integer. In a machine with a 16-bit word, 2^{16} (i.e. 65 536) possible values can be stored, ranging from 0000000000000000 to 1111111111111111. These represent the integers from 0 to $2^{16} - 1$, i.e. from 0 to 65 535. Signed integers are usually stored in two's complement (*see* complement), and a 16-bit word then stores values from -2^{15} to $+(2^{15} - 1)$, i.e. from $-32\ 768$ to $+32\ 767$. *See also* integer arithmetic; real number.

integer arithmetic   Arithmetic involving only INTEGERS. The addition, subtraction, and multiplication of integers always produces an integer as a result. In integer arithmetic, an *integer division* also always produces an integer result, unlike normal arithmetic. This is achieved by ignoring the *remainder*. Hence in integer division 7 divided by 2 equals 3 (rather than 3.5). Other examples are shown in the table overleaf.

INTEGER DIVISION			
Operation	Integer result	Remainder	Normal result
9 ÷ 4	2	1	2.250
47 ÷ 3	15	2	15.666
98 ÷ 5	19	3	19.60
1 ÷ 2	0	1	0.50

In a *modulo* operation the result is the remainder after one integer is divided by another. This is usually written i modulo j for the division of integer i by integer j. Hence 47 modulo 3 is equal to 2.

Many programming languages have special operators and functions to perform integer arithmetic. Pascal has div and mod operators for integer division and remaindering (i div j and i mod j where i and j are integers); Fortran assumes integer division if both operands are integers (I/J gives an integer result if I and J are integers), but has a special function for the remainder (MOD(I,J)); C and C++ use / for division in a similar fashion, and % for remaindering. The original version of Basic had no concept of integers as such, but most modern forms remedy this in one way or another.

integer division   *See* integer arithmetic.

integrated circuit (IC)   A complete electronic circuit that is manufactured as a single package: all the individual devices required to realize the function of the circuit are fabricated on a single CHIP of semiconductor, usually silicon. Components (mainly transistors and diodes) can be combined to make a wide variety of circuits, including LOGIC CIRCUITS and SEMICONDUCTOR MEMORY.

Integrated circuits can be fabricated using either *MOS technology* or *bipolar technology*. MOS devices in general use less power and are more densely packed with components than bipolar devices, but cannot operate as quickly as bipolar devices. Bipolar technology is used to produce the components of central processors in large computers. As these two fabrica-tion technologies have advanced, ICs have become cheaper, with improved performance and reliability. In addition, the number of components that can be fabricated on a single chip is continuously and rapidly increasing; the greatest density is now over one hundred million per chip.

IC technology can be classified according to the number of elements per chip, ranging through *ULSI* (ultra large scale integration), *VLSI* (very large scale integration), *LSI* (large scale integration), *MSI* (medium scale integration), and *SSI* (small scale integration). The range goes from more than 1 000 000 in ULSI down to 10 or more in SSI.

integrated office system   A program that performs several functions concerned with office tasks. It combines some of the functions previously performed by single-purpose programs, such as WORD PROCESSORS, SPREADSHEETS, DATABASE MANAGEMENT SYSTEMS, or programs to create graphs and charts. The results of the various sections can usually be loaded into other sections where appropriate and merged to form a final document containing textual, pictorial, and tabular material.

integrity   A measure of the correctness of data following processing. For data to have integrity, it should not have been accidentally altered or destroyed during processing by errors arising in the hardware or software of the system. (Neither should it have been deliberately altered or destroyed by an unauthorized person.) System errors do occur from time to time. Normally, therefore, protective DUMPS are organized on a regular basis. This ensures the existence of a valid copy of a recent version of every file stored in the system. *See also* security; privacy.

Intel   /in-tel/   *Trademark* One of the largest microprocessor manufacturers, Intel Corporation, founded in 1968. Intel developed the 4004, which was the first successful microprocessor, and the 8088 used in the first IBM PC. Earlier designs have now been replaced by Intel's PENTIUM series of chips.

intelligent   Describing a device that is controlled in part or totally by one or more integral processors and so does not require every detail of its operation to be specified by the computer.

intelligent terminal   *See* terminal.

interactive /in-ter-**ak**-tiv/   Denoting a device, system, application, etc., in which there is an immediate response by the computer to instructions as they are input by the user. The instructions can be presented by various means – for example, a keyboard, a mouse, a light pen, or a digitizing pad. The response usually occurs sufficiently rapidly that the user can work almost continuously; this mode of operation, involving apparently continuous communication between user and computer, is thus often called *conversational mode. See also* multiaccess system.

interblock gap /in-ter-blok/ (IBG)   *See* block.

interface /in-ter-fayss/   In general, a common boundary between two devices, two systems, or two software components. The word is often used to describe the electronic circuitry plus associated software that is required to connect two computer systems or two components of a computer system such as a peripheral device and the central processor. The interface compensates for differences in the speed of working of the interconnected units and, if different CHARACTER ENCODINGS are used in the two units, will translate from one code to the other.

Several interfaces are classified as *standard interfaces*; these may be developed by, say, a national electronics association or an international body, or they may be industry standards. All the characteristics of a standard interface are in accordance with a set of predetermined values. Use of a standard interface allows computer systems and components from different manufacturers to be interconnected. An example of a standard interface is the USB interface,

MAIN INTEL MICROPROCESSORS

Name	Date first introduced	Number of transistors on the chip	Clock speed	Data width	MIPS
4004	1971	2300	108 KHz	4 bits	0.06
8008	1972	3500	200 Khz	8bits	0.06
8080	1974	6000	2 MHz	8 bits	0.64
8088	1979	29 000	5 MHz	16 bits 8-bit bus	0.33
80286	1982	134 000	6 MHz	16 bits	1
80386	1985	275 000	16 MHz	32 bits	5
80486	1989	1 200 000	25 MHz	32 bits	20
Pentium	1993	3 100 000	60 MHz	32 bits 64 bit bus	100
Pentium Pro	1995	5 500 000	150 MHz	32 bits 64-bit bus	100+
Pentium II	1997	7 500 000	233 MHz	32 bits 64-bit bus	~300
Pentium III	1999	9 500 000	450 MHz	32 bits 64-bit bus	~510
Pentium 4	2000	42 000 000	1.4 GHz	32 bits	~1700

In 2005 Pentium chips were produced with clock speeds of over 3.5 GHz, with about 170 million transistors, and a 64 bit data width.

which was developed jointly by several computing companies. *See also* RS232C interface.

In software, an interface of a module is a set of FUNCTIONS, PROCEDURES, and PROPERTIES that can be used by the rest of the program. Interfaces are important in software development, for they allow different modules of a program to be developed by different teams, or even different companies, that do not need to know the details of each other's work. If all interaction between the modules is via agreed interfaces, the modules can be treated as BLACK BOXES. Standards that include systems for specifying interfaces, such as COM, CORBA, and .NET, allow pre-built third-party components to be integrated into a program.

interface card (**interface board**)   *See* expansion card.

interlacing /**in**-ter-layss-ing/   **1.** A technique for raster scanning in which all odd numbered scan lines are updated in one sweep of the electron beam and all even numbered in the next sweep. This means that the number of lines displayed and the information carried by the display signal are halved. Images produced by interlacing are not as clear as those produced by progressive scanning and, since the reaction time is slow, images such as animation and video may flicker or appear streaked. **2.** The technique of displaying a graphics image so that alternating rows are displayed in separate passes. The effect is that the image is displayed quickly and then the details are filled in gradually. It is often used for images on Web pages.

internal modem   A MODEM that resides on an expansion board that fits into one of the expansion slots on a personal computer. *Compare* external modem.

Internet /**in**-ter-net/ (**the Net**)   The complete worldwide system of networks and gateways that communicate using the TCP/IP group of protocols. It was set up as a decentralized network called ARPANET by the US Department of Defense in 1969. Later other networks were connected and now the Internet offers such services as e-mail, World Wide Web, IRC, Telnet, FTP, and news channels and is used for the exchange of data, news, conversation, and commerce. It is still decentralized in that no one person, organization, or country controls it or what is available through it.

Internet access   The capability of a user to connect to the Internet. A home user and many smaller organizations generally connect through a MODEM or ADSL TRANSCEIVER connected to a personal computer and, via a telephone line, to an ISP. Larger users may connect through a dedicated line. Communications can also be established using TV set-top boxes and cellphones, although these tend to offer a restricted service.

Internet address   *See* domain name.

Internet Engineering Task Force   *See* IETF.

Internet Explorer (**IE**)   A Web browser produced by Microsoft and introduced in 1995. It is, controversially, integrated into WINDOWS.

Internet presence provider (**IPP**)   A company that provides disk space, high speed Internet connection, and possibly Web design and other services for companies or individuals requiring a presence on the Internet.

Internet protocol   *See* IP.

Internet protocol security (**IPSec**)   A standard for security at the network or packet-processing layer of a TCP/IP network system. (Previously security has been applied at the application layer of the communications model.) IPSec is useful for virtual private networks (VPNs) and for remote access through dial-up connections to a private network.

Internet relay chat (**IRC**)   An Internet service provided by an IRC server, enabling a user to CHAT with other users.

Internet service provider (ISP)   *See* ISP.

Internet Voice   *See* VoIP.

internetworking   The process and science of connecting individual LANs to create WANs and connecting WANs to form even larger WANs. This is an extremely complex technology and is accomplished using routers, bridges, and gateways to deal with the differing protocols used across the various networks.

InterNIC /**in**-ter-nik/ (**Internet network information center**)   The organization formerly responsible for registering and maintaining top-level Internet domain names. Its functions have been taken over by ICANN.

interoperability   /in-ter-op-er-ă-**bil**-ă-tee/   The ability of software and hardware from different vendors, and using different operating systems, to communicate effectively.

interpreter   A means of converting a program written in a HIGH-LEVEL LANGUAGE into a form that can be accepted and executed by a computer. An interpreter is a program that analyzes each line of code in the high-level language program – the *source program* – and then carries out the specified actions. In contrast to a COMPILER, an interpreter does not translate the whole program prior to execution. Program run times are much slower than when a compiler is used. An interpreter does, however, simplify the process of loading and running source programs, and allows source programs to be changed even while running. Some languages, such as BASIC, are found in both interpreter and compiler versions; others, such as FORTRAN and C++, are not suitable for interpreters and are always compiled, while others, such as POSTSCRIPT, are always interpreted.

interrupt   A signal that is sent to a processor, making it suspend its current task and start another. The main features of interrupts are that (a) they include information as to what caused the interrupt, (b)

there are different procedures – *interrupt service routines* – to be executed for each kind of interrupt, and (c) when the correct action has been taken, control is returned to the task that was being processed when the interrupt occurred.

Interrupts can be caused by an external event, such as a key being depressed on a keyboard. They can also be caused by internal conditions in the processor, such as the detection of an attempt to divide by zero. This kind of interrupt is called a *hardware interrupt*. A *software interrupt* works in the same way but is triggered by a MACHINE INSTRUCTION in a program. Each interrupt has a priority: higher-priority interrupts may interrupt the servicing of lower-priority interrupts but not the other way round.

interrupt request   *See* IRQ.

intranet /**in**-tră-net/   A private internal company network sharing management information and computer resources among employees. It behaves like a small private WWW and is based on similar technologies. Sometimes an intranet may be connected to the Internet; usually this is done through a FIREWALL to prevent unauthorized access to information. *See also* extranet.

inverse video   *See* reverse video.

inverter   *See* NOT gate.

I/O   *See* input/output.

IP (**Internet protocol**)   The standard protocol defined in STD 5, RFC 791, which is the network layer for the TCP/IP protocol suite. It is a connectionless best-effort packet-switching protocol. The Internet is currently based on version 4 of IP (IPv4), but use of version 6 (IPV6) is spreading slowly. *See also* IP address.

IP address   The unique 32-bit number that identifies a host to other hosts on the Internet. It is used when communicating by the transfer of packets using IPv4 (*see* IP). About 4 billion hosts can be identified by this means. However, the growth of the Internet is making this number inadequate. This problem is confronted by IPV6, which supports 128-bit IP addresses.

Ipng (Internet protocol next generation) An old name for IPV6.

iPod *Trademark* A range of MP3 PLAYERS manufactured by Apple Computer, Inc. First released in 2001, the iPod has come to dominate the MP3-player market.

IPTV The delivery of television over the Internet. This is not yet practical because it requires a higher BANDWIDTH than is provided by current BROADBAND connections, but it is expected to become important in the next few years.

IPv6 (Internet protocol version 6) The latest level of the Internet protocol. Future growth of the Internet is provided for by the lengthening of IP addresses from 32 bits to 128 bits, enabling many more addresses to be assigned. However, the adoption of IPv6 has so far been slow because of the problems and expense of integration with the existing, IPv4-based, Internet. *See also* IP; IP address.

IR *See* instruction register.

IRC *See* Internet relay chat.

IRQ (interrupt request) A hardware line over which a device can send an interrupt signal to a computer. Sometimes DIP switches are used to set a device's IRQ number before attaching to a computer. Conflicts can occur when adding EXPANSION CARDS, but PLUG-AND-PLAY technology has largely resolved this problem.

ISA /ȳ-să/ **(industry standard architecture)** A standard bus architecture that was originally associated with the IBM AT motherboard. It allowed for 16 bits to flow between the motherboard and an expansion card and its associated devices. It is now obsolete. *See also* EISA.

ISAM /ȳ-sam/ **(indexed sequential access method)** *See* indexed sequential file.

ISDN (integrated services digital network) A telephone service that combines voice and digital network services in a single medium. Most ISDN lines offered by telephone companies give two 64 kbps lines at once, where one can be used for voice and one for data, or both for data, which doubles the transmission rate. For computing purposes, such as Internet access, ISDN is being superseded by ADSL.

ISO The International Organization for Standardization, the body that establishes international standards in industry, etc., including those used in computing. Examples include standards for many HIGH-LEVEL LANGUAGES, CHARACTER SETS and CHARACTER ENCODINGS, aspects of networking, and data formats for CDs and other storage devices. While compliance with such standards encourages interoperability, many manufacturers' products deviate from the relevant standard(s) for commercial reasons. *See also* ANSI.

ISO-7 A CHARACTER ENCODING scheme developed by ISO and hence internationally agreed. ISO-7 uses 7 bits for each character, and so provides 2^7, i.e. 128, different bit patterns. Certain bit patterns are reserved for national use, allowing countries to include the symbols used for their currency, the accents used above and below letters, etc. The US version is ASCII, which is a code widely used in computing.

ISP (Internet service provider) A company that provides access to the Internet, usually for a charge. A software package, username, password, and (if appropriate) access phone number are provided, and access can be obtained through a modem, ADSL transceiver, or by direct connection from a company's network. An ISP is also called an *Internet access provider (IAP)*.

IT /ȳ-tee/ *See* information technology.

Itanium /ȳ-tay-nee-ŭm/ The first member of Intel's IA-64 family of processors. It is a 64-bit microprocessor and its main use is driving large applications requiring more than 4 GB memory, such as databases, ERP, and future Internet applications. It has been succeeded by the Itanium 2.

iteration /it-ĕ-**ray**-shŏn/ (**iterative process**) Any process in which a sequence of operations is performed either a predetermined number of times or until some condition is satisfied. The repetition of the sequence can be executed by means of a LOOP in a program. Iteration is used, for example, to obtain a numerical result: as the sequence of operations is repeated, the result of each cycle comes closer and closer to the desired value.

An iteration (in Basic) for calculating the sum of the squares of numbers from 1 to 100 is as follows:

```
FOR I = 1 to 100
S = S + X(I) ↑ 2
120     NEXT I
```

Iteration is also a method of solving a problem by successive approximations, each using the result of the preceding approximation as a starting point to obtain a more accurate estimate. For example, the square root of 3 can be calculated by writing the equation $x^2 = 3$ in the form $2x^2 = x^2 + 3$, or $x = \frac{1}{2}(x + 3/x)$. To obtain a solution for x by iteration, we might start with a first estimate, $x_1 = 1.5$. Substituting this in the equation gives the second estimate, $x_2 = \frac{1}{2}(1.5 + 2) = 1.750\ 00$. Continuing in this way, we obtain:

$$x_3 = \frac{1}{2}(1.75 + 3/1.75) = 1.732\ 14$$

iterative /it-ĕ-ră-tiv/ Involving repetition. *See* iteration; control structure; loop.

J

jabber An error in which corrupted or meaningless data is continually transmitted onto a network by a faulty device, often a NIC. This may result in a network jam as other devices will 'see' the network as busy.

jaggies /jag-eez/ The display or printing of smooth straight lines or curves in stairlike steps, often because the monitor or printer does not have a high enough resolution to show a smooth line. Jaggies can also occur when a bit-mapped image is converted to a different resolution. Antialiasing and smoothing can reduce the effect.

JANET /jan-ĕt/ (Joint Academic Network) An important WAN in the United Kingdom that primarily links all university and higher education networks. It is part of the Internet.

Java /jah-vă, jav-ă/ An object-oriented programming language created by Sun Microsystems. It was specifically designed for writing programs that can be safely downloaded and immediately run on a variety of platforms without fear of viruses or other harm to the user's computer files. Java programs are compiled to a machine-independent code called *bytecode*, which is run by a JAVA VIRTUAL MACHINE. In addition, small Java programs called *applets* may be embedded in Web pages to include functions such as animations and calculations. Java was mainly inspired by C++. *See also* JDK.

JavaBeans Objects conforming to the standard developed by Sun Microsystems that defines how Java objects interact. A JavaBean is similar to an ACTIVEX control. It can be used by any application that understands the JavaBeans format and can be run on any platform. However, JavaBeans must be developed in Java.

JavaScript A programming language developed by Netscape Communications Corporation and Sun Microsystems. It allows HTML authors to script limited Java functionality directly into the text of an HTML file and is generally considered to be easier to write than JAVA. JavaScript runs in and manipulates Web browser environments, such as INTERNET EXPLORER, MOZILLA FIREFOX, and NETSCAPE NAVIGATOR. When JavaScript is combined with CASCADING STYLE SHEETS (CSS), and versions of HTML 4.0 and later, the result is often called DYNAMIC HTML.

Java virtual machine (JVM) An INTERPRETER or JUST-IN-TIME COMPILER used to translate compiled Java code, known as *bytecode*, into the machine code of whatever computer the Java application or applet is running on. This is the way in which Java achieves its multiplatform ability.

JCL *See* job control language.

JDK (Java development kit) A software development package from Sun Microsystems that provides the basic set of tools needed to write, test, and debug Java applications and applets.

Jini Sun Microsystems software that aims to simplify the connecting and sharing of devices on a network. A device that incorporates Jini announces itself to the network, providing details about itself, and immediately becomes available to other devices on the network. The software works by passing applets back and forth among devices and any computer that can run Java will be able to access this code.

JIT compiler *See* just-in-time compiler.

job A set of programs and the data to be manipulated by these programs, regarded by a computer system as a unit of work. The complete description of a job, required for its execution, is written in a JOB CONTROL LANGUAGE.

job control language (JCL) A special kind of programming language by means of which a user can communicate with the OPERATING SYSTEM of a computer. It is employed most often in connection with BATCH PROCESSING

A JCL is used to write the sequence of commands that will control the running of a JOB on a particular computer: an initial command identifies the job while subsequent commands are each an instruction for the operating system to take some action on behalf of the user. For example, there are usually commands to run a specified program, find input data at a specified source, and output results at some specified destination. The objects manipulated in a JCL are thus not VARIABLES as in a normal programming language but can, for example, be complete programs or the input or output streams for these programs. A JCL command can be obeyed as soon as it is keyed. Alternatively a complete set of JCL commands can be stored and executed later. *See also* command language.

job queue/stream The lining up of jobs for a computer device such as a printer.

Josephson junction /**joh**-zĕf-sŏn/ The junction between a thin layer of insulating material and a superconducting material across which a superconducting current can flow in the absence of an applied voltage. Josephson memory, a cryogenic memory, uses Josephson cells that contain Josephson junctions, and is inherently extremely fast but expensive to run.

journal A log that is computer-pro-

duced and often used as a trail to aid recovery in the event of problems occurring. It is used to contain details of network transactions, such as file usage or message transfer information.

joystick A device for generating signals that can control movement on a display. It is a shaft a few inches in height that is mounted in an upright position in a base and can be pushed or pulled by the operator in any direction. Generally, the device also has control buttons. Joysticks are most commonly used in computer games.

JPEG /**jay**-peg/ (**joint photographic experts group**) A compression standard common format for image files. It is especially used for photographic images. Files in this format often have the extension jpg.

Julia set A set of points in the complex plane defined by iteration of a complex number. One takes the expression $z^2 + c$, where z and c are complex numbers, and calculates it for a given value of z and takes the result as a new starting value of z. This process can be repeated indefinitely and three possibilities occur, depending on the initial value for z. One is that the value tends to zero with successive iterations. Another is that the value diverges to infinity. There is, however, a set of initial values of z for which successive iterations give values that stay in the set. This set of values is a Julia set and it can be represented by points in the complex plane. The Julia set is the boundary between values that have an attractor at zero and values that have an attractor at infinity. The actual form of the Julia set depends on the value of the constant complex number c. Thus, if $c = 0$, the iteration is $z \rightarrow z^2$. In this case the Julia set is a circle with radius 1. There is an infinite number of Julia sets showing a wide range of complex patterns. *See also* Mandelbrot set.

jump (**branch**) A departure from the normal sequential execution of program instructions. It is used when writing a program in ASSEMBLY LANGUAGE or MACHINE CODE. Instead of performing instruction 1, then instructions 2, 3, 4, and so on, there could be a jump from, say, instruction 3 to instruction 19. This departure from normal program flow is specified by using a special instruction in a program, known as a *jump instruction* (or *branch instruction*).

A jump can be either *unconditional* or *conditional*. An unconditional jump takes place whenever the instruction specifying it is executed (*see* instruction address register). A conditional jump takes place only if specified conditions are met, for example if the contents of the accumulator equal zero or if the value of some variable exceeds a stated number.

Jump instructions offer the only way of controlling the flow of events when using assembly language or machine code. This can lead to programming errors. Most high-level languages offer improved facilities in the form of a variety of CONTROL STRUCTURES. See also GOTO statement.

jumper A small piece of plastic-encased metal that closes a circuit so that electricity can flow through. The jumper is placed across two metal pins sticking out of a circuit board. Sometimes jumpers are used to configure expansion boards; by placing them over different sets of pins the board's parameters can be changed.

jump instruction (branch instruction) *See* jump.

justify 1. To control the positions of printed or typed characters so that regular vertical margins to a page or column of text are obtained; the lines are then said to be *justified*. In some cases only one margin is aligned; the lines are then said to be *left-justified* when the left-hand margin is regular or *right-justified* when the right-hand margin is regular. When both margins are to be aligned, the spaces between words (and sometimes between letters) is adjusted until each line fills the distance between the two specified margins. Hyphenation of a

word can be used if the space in a line is too great. WORD PROCESSING systems allow the user to specify both the right and the left margin, between which text is automatically justified as it is typed in; hyphenation is offered as an option.

2. To shift the contents of a REGISTER, if necessary, so that the bit at a specified end (right-hand or left-hand) of the data entered in the register is at a specified position (usually the right- or left-hand end) of the register. The contents are then said to be *right-justified* or *left-justified*, as the case may be.

just-in-time compiler (JIT compiler) A compiler that operates when a program is run. Such programs are normally compiled in two stages. The source code is first processed by a normal compiler, which checks for syntax errors, performs optimizations, etc.; however, this compiler does not generate MACHINE CODE but rather a machine-independent code, known as *intermediate language* or *bytecode*. When the program is run, the JIT compiler translates this code into machine code for each program section when it is first used. (This differs from an INTERPRETER, which translates each program section every time it is used.) Unlike a traditional compiler a JIT compiler is aware of its operating environment and can vary the machine code it generates accordingly; for example it can often determine the exact version of the computer's central processor and take advantage of any special MACHINE INSTRUCTIONS that version provides to produce optimal code. The slight overhead of compiling during program execution has been shown not to affect the speed of many applications to any significant degree; nevertheless there remain programs where this overhead is significant, and for these traditional compilation remains the appropriate method.

JVM *See* Java virtual machine.

K

K (**k**) A symbol used in computing for KILO-, or sometimes for kilobyte or kilobit.

Kb A symbol for kilobit.

KB A symbol for kilobyte.

Kerberos /ker-bĕ-ross/ A secure method of computer network user authenticating developed by Massachusetts Institute of Technology. Kerberos is used in network applications and relies on trusted third-party servers for authentication. A user requests an encrypted ticket from the authentication process that can then be used to access a particular service from a server. A version of Kerberos can be downloaded from MIT and it is used in commercial products.

Kermit A popular file transfer protocol used in asynchronous communications between computers (not the same as FTP). It runs in most operating environments and provides an easy method of file transfer. Kermit was developed at Columbia University in the USA.

kernel /ker-nĕl/ The central module of an operating system, typically responsible for hardware resource allocation and management. In order to share the hardware between multiple processes, it implements the basic facilities for MULTITASKING. The kernel is the part of the operating system that loads first, and it remains in memory until the computer is shut down. It is part of the SUPERVISOR.

key 1. A labeled button or marked area on a KEYBOARD.
2. A value held in one of the fields of a RECORD and used to identify that record in a collection of records (usually a FILE or TABLE). *See also* index; searching; sorting.
3. In CRYPTOGRAPHY, a pattern of bits, usually expressed as a STRING or a large number, that governs the cipher code produced by an encryption algorithm for a particular piece of plain text, and allows the equivalent decryption algorithm to recover the plain text from the cipher.

keyboard A manually operated input device for converting letters, digits, and other characters into coded form. It is the most commonly used means by which people can communicate with a computer. It

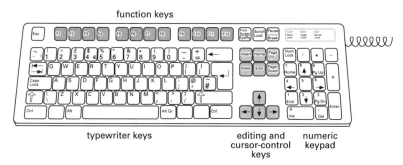

function keys

typewriter keys

editing and cursor-control keys

numeric keypad

A typical keyboard

consists of an array of labeled *keys* that are operated by pressure applied by the fingers. In current devices the operation of a particular key generates a coded electrical signal that can be fed directly into a computer. The characters that have been keyed may be displayed on a VDU screen.

A computer keyboard consists of the standard typewriter layout – the QWERTY keyboard – plus some additional keys (see diagram). These can include a *control key, function keys, cursor keys,* and a *numeric keypad.* The control key operates in the same way as a shift key but allows CONTROL CHARACTERS to be sent to the computer; the function keys send not one but a whole sequence of characters to the computer at a time, and can often be programmed by the user to send commonly used sequences; the cursor keys are used to move the screen CURSOR to a new position; and the numeric KEYPAD duplicates the normal typewriter number keys and speeds up the entry of numerical data by allowing one-handed operation. *See also* alt key.

key frame An animation technique in which the start and end frames of a sequence are defined but the intervening frames are interpolated by a computer to give a smooth animation.

keypad A small KEYBOARD with only a few keys, often 12 or 16, that can be operated with one hand. It is usually used for encoding a particular sort of information. For example, a *numeric keypad* is used to feed numerical data into a computer, and can be operated much faster than an all-purpose keyboard; there are 10 keys for the digits 0–9 plus additional keys for, say, the decimal point and minus sign. A keypad may be hand-held or may be part of a larger keyboard.

keystroke The action of pressing a keyboard key to enter a character or initiate a command in a computer system. A keyboarder's efficiency can be measured in keystrokes per minute, or ease of use of an application can be assessed by how many keystrokes it takes to perform common operations.

key to disk An obsolete system of data entry in which the data entered by a number of keyboard operators was recorded on a magnetic disk under the control of a small computer. The data from each keyboard was routed by the computer to the appropriate file on the disk. The data encoded on the disk was verified, often by comparing it with data entered by a second operator working from the same source. The disk could then be transferred to the computer on which the data was to be processed.

key to tape An obsolete system of data entry in which the data entered by a keyboard operator was recorded on a magnetic tape. The data encoded on the tape was then verified, often by comparing it with data entered by a second operator working from the same document. The tape could then be transferred to a computer so that the data could be processed.

kHz *See* kilohertz.

kill To stop a process in a program or operating system.

kill file A feature of some e-mail clients or newsgroup readers, or an independent Internet utility, that is used to block out or filter unwanted messages or newsgroup articles from particular individuals. *See also* spam.

kilo- /kil-ŏ-/ A prefix indicating a multiple of 1000 (i.e. 10^3) or, loosely, a multiple of 2^{10} (i.e. 1024). In science and technology decimal notation is usually used and powers of 10 are thus encountered – 10^3, 10^6, etc. The symbol k is used for kilo-, as in kV for kilovolt. In communications, kilo- (as in kilobaud) means 1000. Binary notation is generally used in computing and so the power of 2 nearest to 10^3 (i.e. 2^{10} or 1024) has assumed the meaning of kilo-.

The prefix is most frequently encountered in computing in the context of storage CAPACITY. With magnetic disks and main store, the capacity is normally reckoned in terms of the number of BYTES that can be stored, and the word *kilobyte* (KB) is used to mean 2^{10} bytes; kilobyte is usually abbreviated to K byte or just to K, thus a 256 K byte file is 262 144 bytes in size. SEMICONDUCTOR MEMORY – RAM or ROM – is sometimes considered in terms of the number of bits that can be stored in the device, and the word *kilobit* (kb) is used to mean 2^{10} bits.

Kilo- is part of a sequence

kilo-, mega-, giga-, tera-, peta-, ...

of increasing powers of 10^3 (or of 2^{10}).

kilohertz /**kil**-ŏ-herts/ (kHz)   A unit of frequency equal to 1000 HERTZ.

kiosk   A booth providing a computer-related service such as a theater booking service or tourist information service. A kiosk requires a very simple, physically robust interface, such as a TOUCH SCREEN, that anyone can use without training or documentation.

knowledge acquisition   The translation of human knowledge into a form suitable for use on a computer. The term is used in EXPERT SYSTEMS.

knowledge base   A database of accumulated human knowledge on a particular subject used in EXPERT SYSTEMS. For example, a knowledge base of symptoms for a particular disease would be used in medical diagnosis.

knowledge-based system   *See* expert system.

knowledge domain   The area of knowledge in which a particular EXPERT SYSTEM is involved.

L1 cache (primary cache) A small cache memory in microprocessors. *Compare* L2 cache.

L2 cache (secondary cache) Cache memory that was originally external to the microprocessor. Modern micr-oprocessors now include L2 caches in their architectures. *Compare* L1 cache.

label An IDENTIFIER associated with a STATEMENT in a program and used in other parts of the program to refer to that statement.

LAN /lan/ *See* local area network.

landscape The arrangement of printed matter on a page so that the top of the page is on the wider side. *Compare* portrait.

language *See* programming language.

laptop /lap-top/ A small portable personal computer. Laptops can be run from battery packs or from the mains. Flat panel display screens are used and a TRACKERBALL is often embedded in the keyboard. In other respects the memory capacity, hard drives, and CPUs are the same as those used in a personal computer.

large-scale integration *See* integrated circuit.

laser printer A type of PRINTER in which a page-sized image is produced in the form of a pattern of very fine dots by the action of a laser beam. The principle is similar to that used in many office copiers. The laser beam writes the image on the surface of a drum or band in the form of a pattern of electric charge. This charge pattern is coated with varying amounts of toner (dry particles of pigment) according to the strength of the charge. The toner is rolled onto the paper and fused in place by heat, producing a permanent image. Laser printers work quietly and at high speed, generating high-quality print. They readily produce graphs and diagrams and a wide variety of typestyles. Color laser printers are also available.

latency 1. In general, the period of time that one component in a system is waiting for another component. For example, in accessing data on a disk, latency is the time it takes to position the sector under the read/write head. *See* access time.
2. In networking, the amount of time it takes a packet to travel from source to destination. Together latency and BANDWIDTH determine the speed and capacity of the network.

LaTeX /lay-tekh/ A typesetting system based on the TeX programming language for typesetting. It provides for higher level macros than TeX, making it easier to format documents, but in doing so it loses some flexibility.

launch To start a program.

layer A level in a program or system that performs a particular function or functions and has a defined interface for passing results to/from the layers next to it. OSI is an example of a layered system.

layout In word processing and desktop publishing, the arrangement of the text and graphics on a page.

LCD (**liquid crystal display**) A device used in many FLAT PANEL MONITORS. The screen is composed of many dots, each of which contains a liquid that is usually transparent. Individual dots can be darkened however (by applying an electric field) and so an image can be built up. Note that LCDs do not emit light, and can only be seen because light shone from behind the display makes the dark dots stand out. Color images are achieved by each 'dot' in fact being a triplet of red, green, and blue dots. A variant of LCD technology called ACTIVE MATRIX DISPLAY is generally used in computer monitors. LCDs are also used in digital watches, calculators, cellphones, etc. *See also* display.

LDAP /el-dap/ (**lightweight directory access protocol**) A significantly simpler version of the X.500 standard for a set of protocols for accessing information directories. Unlike X.500 it supports TCP/IP and is an open protocol, which means that applications do not need to know about the type of server hosting the directory.

leaf An item at the very bottom of a hierarchical TREE structure. In hierarchical file systems, files are leaves whereas directories are NODES.

leased line A telephone line that is rented for exclusive use by the customer from one location to another. Typically leased lines are used by businesses to connect geographically distant sites for high-speed data transfer.

least significant digit (**lsd**) The digit in the least significant, i.e. rightmost, position in the representation of a number, and thus making the smallest contribution to the value of the number. With a binary number, this digit is referred to as the *least significant bit (lsb)*.

LED display A device used in some calculators, digital clocks, etc., to display numbers and letters. The characters are formed from groups of segments. The segments are small electronic components called LEDs, i.e. *light-emitting diodes*. In-

dividual LEDs can be made to emit light (usually red) so as to form the shape of a particular character. A simple SEVEN-SEGMENT DISPLAY can display the numbers 0–9 and some letters. LED displays require more electrical power than LCDS.

left-justified *See* justify.

legacy system A computer, software, operating system, network, or other computer equipment that exists before a new system is installed, and is still required afterward. BACKWARD COMPATIBILITY is an important part of the design and implementation of new systems, but cannot always be maintained. The question of when to engage in the expensive operation of upgrading legacy systems is therefore important.

LEO /lee-oh/ Lyons Electronic Office. *See* first generation computers.

LET /let/ *See* Basic.

letter quality (**LQ**) A quality of printing that is as good as a top-quality electric typewriter. The term is mainly used of dot matrix printers and is becoming less common with the increased use of laser printers, which generally give better quality than electric typewriters.

lexical analysis The initial phase in the COMPILATION of a program during which the source program is split up into meaningful units. These units could, for example, be NAMES, CONSTANTS, RESERVED WORDS, or OPERATORS. The part of the compiler program that does this analysis is called a *lexical analyzer*. The units recognized by the analyzer are known as *tokens*. They are output in some conveniently coded form for further processing by the compiler. *See also* syntax analysis.

library *See* program library; subroutine library.

library program *See* program library.

license agreement An agreement defin-

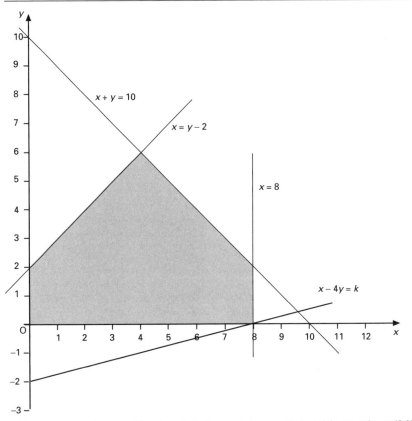

Linear programming: possible values lie in the shaded area with the minimum value at (8,0)

ing terms for the use of hardware or software by a user.

LIFO /lȳ-foh/ (**last in first out**)   *See* stack.

light-emitting diode   *See* LED display.

light pen   An obsolete hand-held penlike instrument used together with a DISPLAY device for the input of data to a computer graphics system. (The display had to be a refreshed cathode-ray tube.) The main function of the light pen was to point to small areas of the screen, such as a single character or a small graphical object, and it could for example indicate a selection from a displayed list (a MENU). It could also be used to draw shapes on the screen.

line   1. A hardware circuit connecting two devices.

2. A single line in a computer program.

3. A single data entry in a cache. For example in an L2 cache each line is 32 bits wide and can contain data from one or more addresses.

linear programming   The process of finding maximum or minimum values of a linear function under limiting conditions or constraints. For example, the function $x - 4y$ might be minimized subject to the constraints that $x + y \leq 10$, $x \leq y - 2$, $x \leq 8$, $x \geq 0$, and $y \geq 0$. The constraints can be shown as the area on a Cartesian coordinate graph bounded by the lines $x + y = 10$, $x = y - 2$, $x = 8$, $x = 0$, and $y = 0$ (see diagram). The minimum value for $x - 4y$ is chosen from points within this area. A series of parallel lines $x - 4y = k$ are drawn for different values of k. The line $k = -2$ just reaches the constraint area at the point (8,0). Higher values are outside it, and so $x = 8$, $y = 0$ gives the maximum value of $x - 4y$ within the constraints. Linear programming is used to find the best possible combination of two or more variable quantities that determine the value of another quantity. In most applications, for example, finding the best combination of quantities of each product from a factory to give the maximum profit, there are many variables and constraints. Linear functions with large numbers of variables and constraints are maximized or minimized by computer techniques that are similar in principle to this graphical technique for two variables.

line editor *See* text editor.

line feed **1.** A control mechanism on a PRINTER, or a programmed instruction, that causes the paper to move up a specified distance until the printing mechanism is opposite the next line to be printed. *See also* paper throw.
2. A means by which the CURSOR on a screen is moved down one line. If that would result in the cursor dropping off the bottom, the whole display moves up one line and the cursor is placed on a new blank bottom line.
3. (LF) The ASCII control character, with the decimal value 10, that is often used to invoke either of these operations. Sending a line-feed character to a printer or screen may or may not also generate a CARRIAGE RETURN.

line printer A PRINTER that produces a complete line of characters at a time, or more precisely produces a complete line during one cycle of its operation. Printing rates range from 150 to 3000 lines per minute (lpm). There are normally between 80 and 160 character positions on a line. Each line is first assembled in the printer's memory and then printed. Once the line has been printed the paper is moved so that the next line is opposite the printing mech-

anism. Line printers are usually found on large computers rather than personal computers, but their use is declining as they are replaced by high-volume LASER PRINTERS.

Line printers may be impact or nonimpact printers (*see* printer). Modern line printers use the same technology as DOT MATRIX PRINTERS except that they have enough needles to print a line at a time. However, there are two older types still in use, *barrel printers* (also known as *drum printers*) and *band printers* (see illustration overleaf).

A barrel printer involves a horizontal cylinder with rows of characters embossed along its length. The cylinder rotates continuously at high speed, and one line is printed during each rotation. The paper is positioned between the cylinder and a set of print hammers; there is one hammer for each printing position along a line, i.e. for every character in a row on the cylinder. The complete character set that can be printed is provided at each printing position. During one rotation of the cylinder all the characters of one sort are printed at the same time; for example, all the As in a line are printed, then all the Bs, then all the Cs, and so on.

A band printer involves a steel band with characters embossed along its length. The band forms a loop and is rotated continuously at high speed. The band moves horizontally, spanning the line to be printed. The paper is positioned between the band and a set of print hammers; there is one hammer for every printing position along a line. As the appropriate character on the band passes beneath a particular hammer, the hammer strikes and prints it in the required position on the line. Bands can be replaced when necessary. Mistiming of the hammer actions in a band printer produces slight variations in character spacing; these are less obvious than the vertical displacements resulting from mistiming in a barrel printer.

Chain printers and *train printers* are earlier versions of the band printer. The former involves small metal plates linked into a continuous chain rather than a metal band; the latter involves metal slugs guided around a track.

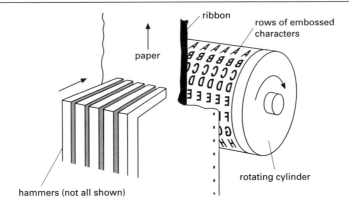

barrel printer

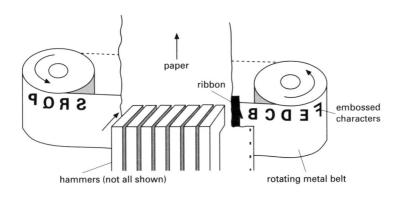

band printer

Line printers

link 1. (linkage) The part of a computer program, possibly a single instruction or address, that passes control and PARAMETERS between separate portions of the program.

2. (pointer) A character or group of characters that indicates the LOCATION of an item of data in memory.

3. A path for communications. *See* data link.

4. *See* hypertext.

5. To join together program modules that have been compiled separately to form a program that can be executed. This operation is performed by a LINK EDITOR.

linkage *See* link.

link editor (linkage editor; linker) A UTILITY PROGRAM that combines a number of user-written or library routines, which have already been compiled or assembled but are individually incomplete, into a single executable program. This program is then either stored on disk or placed in main store for immediate execution. A link editor that performs the latter function is sometimes called a *link loader*. *See also* program library.

linked list *See* list.

linker   *See* link editor.

link loader   *See* link editor.

linkrot /**link**-rot/ *Informal* The condition in which HYPERTEXT leads nowhere because Web sites have removed or reorganized their Web pages and the links are outdated.

Linux /**lin**-ŭks/   A freely distributable operating system that runs on several hardware platforms including Intel and Motorola microprocessors, and thus on PCs and Macintoshes. Linux is a form of UNIX written by Linus Torvalds. Its use has increased rapidly in recent years, especially for SERVERS.

liquid crystal display   *See* LCD.

Lisp   A high-level programming language whose name is derived from **list** processing. It was developed in the early 1960s. Lisp was designed specifically for the manipulation of LIST and TREE structures of various kinds and has an unusual SYNTAX. It was later taken up by those working in the emergent field of ARTIFICIAL INTELLIGENCE. There are now a number of dialects of Lisp, extended in various directions. Lisp programs are generally translated by an INTERPRETER.

list   One form in which a collection of data items can be held in computer memory. The items themselves are in a particular order. The items could all be integers, for example, or real numbers, or letters, or they could be a mixture of, say, integers and real numbers. If all the items are of the same type then the list is a one-dimensional ARRAY.

There are two commonly used forms in which this collection of data items is held in memory. In a *sequentially allocated list* the items are stored in their correct order in adjacent LOCATIONS. In a *linked list* each location contains a data item and a LINK containing the address of the next item in the list; the last item has a special link indicating that there are no more items in the list. Various operations can be performed on the list components without moving the data items; only the links are modified. For example, a new item can be inserted in its correct place by having the preceding item link to the new item and the new item link to the succeeding item.

list box   A type of CONTROL on a GUI FORM. The user selects its content from a list of available options. *Compare* text box; *see also* combo box.

listing   An output of a computer system, often diagnostic, that can be printed. The word is usually qualified in some way to give a more precise idea of the contents, as in *program listing* or *error listing*. A listing may be created in a file rather than printed, which allows it to be examined on line and perhaps analyzed by program.

literal   A character or group of characters that stands for itself, i.e. is to be taken literally, rather than being a NAME for something else. In nearly all programming languages numbers are always taken as literals. When letters are used as literals they must be placed within some form of quotation marks to distinguish them from letters used as names. For example, X may be the name given to a VARIABLE in a program written in Basic.
PRINT X
would then mean print the value assigned to that variable.
PRINT 'X'
would mean print the letter X.

load   1. To enter data into the appropriate REGISTERS in a processor or into the appropriate storage LOCATIONS in main store. 2. To copy a program from backing store into main store so that it may be executed.

loader   A UTILITY PROGRAM that LOADS a program into main store ready for execution. The program loaded is an OBJECT PROGRAM that is read from backing store. *See also* program library.

local   1. In communications, describing a device, file, or other resource that does not need a communications line for access; in other words, it is either part of the user's computer or directly attached to it. *See* remote.

2. In programming, describing a variable that is only defined for one part (subroutine, procedure, or function) of the program and does not have GLOBAL scope.

local area network (LAN) A communication system linking a number of computers within a defined and small locality. This locality may be, say, an office building, an industrial site, or a university. Linking the computers together allows resources, such as HARD DISKS, PRINTERS, or an Internet connection, to be shared between them, allows FILES to be shared, gives all users access to specialized application SERVERS (e.g. database servers that manage shared databases), allows an INTRANET to be established, and permits messages to be sent between the computers by E-MAIL.

The computers in the network are directly connected to the transmission medium by electrical INTERFACES; normally the medium is in the form of electric cables, optical fibers, or wireless. There are various ways in which the interconnections between machines can be organized (*see* network). The system is usually controlled overall by a server that validates all users when they LOG ON and ensures they do not access resources or perform operations forbidden to them. Often, and especially on small networks, this function is combined with that of FILE and PRINT SERVERS on a single machine, known as the server.

Local area networks generally provide high-speed data communications and have very low ERROR RATES. It is possible to interconnect two or more local networks, and to connect a local network to a larger network, i.e. to a WIDE AREA NETWORK.

local bus A PC architecture design that gives high performance by allowing expansion boards to communicate directly with the microprocessor.

LocalTalk /**loh**-kăl-tawk/ An obsolete cabling system supported by the APPLETALK network protocol for Macintosh computers.

local variable *See* scope.

location (storage location) In general, any place in which information can be stored in a computer – in MAIN STORE or BACKING STORE. A storage location is an area within a computer memory capable of storing a single unit of information in binary form. Each location can be identified by an ADDRESS, allowing an item of information to be stored there or retrieved from there. In most cases it is possible to change the stored value during execution of the program. The words location and address are often used as synonyms, as in ordinary speech.

Main store is divided into either WORDS or BYTES: in some computers each location holds a word; in other machines each location holds a byte. The memory is said to be *word-addressable* or *byte-addressable*. A word is now generally 32 bits and is usually large enough to contain a MACHINE INSTRUCTION. A byte is always 8 bits and thus is the storage space required to hold a small number or a common CHARACTER.

Locations on disk and tape hold a number of bytes and are addressed by TRACK and SECTOR or by BLOCK, respectively.

lock 1. A mechanical device on some removable storage media that prevents the user writing to the media.
2. A software security feature that ensures it cannot be used unless unlocked with a key or DONGLE.

locked up The condition of a system or an application that seems unable to respond to the user. *See also* hang.

log /log/ A record containing details of past activities or actions on a computer system, e.g. details of problems found (or not) and actions taken when scanning a disk.

logic /**loj**-ik/ *See* binary logic; digital logic.

logical device /**loj**-i-kăl/ A device that is given a logical name or number and is treated by the operating system as a distinct device regardless of its physical relationship to a system. For example, a hard drive of a computer can be partitioned into a number of logical drives E, F, G, etc.

logical error *See* error.

logical expression *See* Boolean expression.

logical function *See* Boolean expression.

logical instruction *See* logic instruction.

logical operation *See* logic operation.

logical operator *See* logic operator.

logical record *See* record.

logical shift *See* shift.

logical value *See* truth value.

logic bomb **1.** A catastrophic program error that only occurs under certain conditions when least expected.

2. A piece of code that deliberately introduces provision for such an error into a system, often as the PAYLOAD of a virus.

logic circuit (**logic network**) An electronic circuit consisting of interconnected LOGIC GATES and possibly other components (such as FLIP-FLOPS). These elements are connected by paths that carry electrical signals from one to another. A circuit consisting only of logic gates is shown in the diagram, LOGIC SYMBOLS being used to represent the different gates. The inputs to one gate in a circuit are usually the outputs of other gates, or signals from, say, a peripheral device. The output of a gate can be fed to another gate or to some other unit in the computer system.

Logic circuits are found in all types of computer devices, including the CENTRAL PROCESSOR where they perform various arithmetic, logic, and control operations, and in memory devices (such as ROM and RAM) and COUNTERS.

A TRUTH TABLE can be drawn up to determine the output of a COMBINATIONAL LOGIC circuit for all possible combinations of inputs (see diagram). It can also be used to obtain an expression of the logical func-

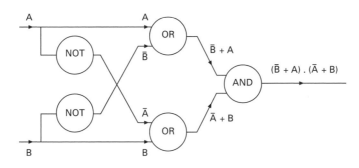

inputs						output
A	B	$\bar{A}$	$\bar{B}$	$\bar{A}$ + B	$\bar{A}$ + B	$(\bar{B} + A).(\bar{A} + B)$
0	0	1	1	1	1	1
0	1	1	0	0	1	0
1	0	0	1	1	0	0
1	1	0	0	1	1	1

Logic circuit and truth table of this circuit

tion of the circuit. In the circuit and truth table shown, +, ., and ⁻ are the OR, AND, and NOT OPERATORS. The expression derived for the circuit output indicates that it is an EQUIVALENCE GATE.

logic diagram A diagram in which LOGIC SYMBOLS are used to show how the inputs and outputs of components are connected together in a LOGIC CIRCUIT. It thus displays the design of the circuit but gives no indication of the electronics involved.

logic element A LOGIC GATE or combination of gates.

logic error An error in the logic of a program that causes incorrect results but does not stop the program running.

logic gate A device, almost always electronic, that is used to control the flow of signals in a computer. It does this by performing LOGIC OPERATIONS on its inputs. Normally there are between two and eight inputs to a gate and one output. An input signal is either at a high level or at a low level, the level changing with time. (It is actually the voltage level of the signal that changes in an electronic gate.) The signal thus has a binary nature; the two values can be represented as 1 (high) and 0 (low). As the input signals switch between their high and low levels, the output level changes, the value depending on the logic operation performed by the gate. For example, one type of gate has a high output only when all its inputs are high at the same time. In any logic gate the output changes extremely rapidly in response to a change in input.

There are several simple gates that each perform a basic logic operation. For example, the AND GATE and OR GATE perform the AND and OR OPERATIONS, respectively, on their inputs. The NOT GATE performs the NOT OPERATION on its single input. The other simple logic gates are the NAND GATE, NOR GATE, NONEQUIVALENCE GATE, and EQUIVALENCE GATE. Selected gates can be connected together to form a LOGIC CIRCUIT.

TRUTH TABLES are used to describe the output of a logic gate for all possible combinations of inputs. A simple gate with 2 inputs has 2^2, i.e. 4, possible combinations of inputs and hence 4 possible outputs; with 3 or with 4 inputs there are 2^3 or 2^4, i.e. 8 or 16, possible outputs. Truth tables are given at individual entries for gates in this dictionary. *See also* logic symbols; logic diagram.

logic instruction A MACHINE INSTRUCTION specifying a LOGIC OPERATION and the OPERAND or operands on which the logic operation is to be performed. An example, expressed in ASSEMBLY LANGUAGE, might be

OR A,B

This is an instruction to perform the OR operation between the contents of registers A and B, placing the result in A. *See also* arithmetic instruction.

logic network *See* logic circuit.

logic operation An operation performed on quantities (OPERANDS) that can be assigned a truth value; this value is either *true* or *false*. The result of the operation is in accordance with the rules of BOOLEAN ALGEBRA, and again has a value *true* or *false*. The operands may be statements or formulae, such as

It is July
the wire is less than 10 cm long
x + 2 = 16

The operands of a logic operation can be represented by letters. For example, in

P AND Q, P OR Q, NOT (P OR Q)

P and Q are the operands and AND, OR, and NOT are LOGIC OPERATORS that represent three commonly used logic operations, namely the AND, OR, and NOT OPERATIONS. Logic operations can be described by means of TRUTH TABLES. LOGIC GATES are constructed so as to perform logic operations on their input signals.

logic operator A symbol representing a LOGIC OPERATION such as the AND or OR operation. A logic operator has logical quantities as its OPERANDS and delivers a logical result. For example, the expression 'I have no money' and 'it is Tuesday' is true if both the operands are true, otherwise it is false. In programming languages logic operators are either spelled out in full or various symbols are used; some of these are shown in the table.

An example of how logic operators are used, in Basic, is as follows:

IF (A=3) AND (X<10) THEN A=0

See also order of precedence.

LOGIC OPERATORS	
Operation	*Operators*
AND	AND .AND. & .
OR	OR .OR. ! | +
NOT	NOT .NOT. ¬ ! ¯ ´

logic symbols   The symbols used in a LOGIC DIAGRAM to represent the various LOGIC GATES and the other possible components in a LOGIC CIRCUIT. One set that is used to represent logic gates is shown in the diagram: the logic operation performed by the gate is printed inside a circle, and arrowed lines labeled by letters represent the inputs and output. In another set (widely used in technical literature) the shape of the symbol is used to indicate its function.

log in   *See* log on.

Logo /**loh**-goh/   A PROGRAMMING LANGUAGE developed in the 1960s by Seymour Papert for use in teaching young children. Logo is a simple but powerful language. It incorporates the concepts of PROCEDURES and LISTS, and helps children to think algorithmically. Logo includes TURTLE GRAPHICS: the turtle is a simple pen-plotter that can be steered around a large piece of paper on the floor, under program control, allowing the children to create complex patterns. In some microcomputer implementations the turtle pictures may be drawn on the screen. This is cheaper but perhaps not so effective.

log off   (**log out**; **sign off**)   To inform a computer system or NETWORK from a TERMINAL or WORKSTATION that one's work is finished for the moment. This log off or *log out* process is required in any MULTIACCESS SYSTEM for security and accounting purposes. After a user has logged off, the computer system or network will not be available on that terminal or workstation until someone has logged on there. *See also* log on.

log on   (**log in**; **sign on**)   To identify oneself to a computer system or NETWORK when wishing to enter the system from a TERMINAL or WORKSTATION. This log on or *log in* process is required in any MULTIACCESS SYSTEM. The log on sequence involves first saying who one is by means of one's name or account number and then entering one's PASSWORD. In some cases where, say, security is very strict, there may be more to do. *See also* log off.

log out   *See* log off.

long filename   In Windows, a FILENAME

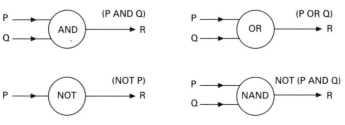

Logic symbols for gates

that can be up to 255 characters long and include spaces. The term is used to contrast such names with the *short filename* (or *8.3 filename*) system used by MS-DOS, where filenames were limited to 8 characters plus a 3-character EXTENSION and spaces were not allowed. Windows still maintains short filenames for BACKWARD COMPATIBILITY.

look-up table   A TABLE used in table look-up.

loop   A statement or group of statements whose execution is repeated either a specified number of times or while or until some condition is met. Both types of loop are iterative CONTROL STRUCTURES, and appear in various forms in almost every programming language.

The *for loop* appears in several languages and permits a specified number of repetitions to occur. An example, in Pascal, is

for i := 1 **to** 10 **do** statement

This causes the statement to be executed 10 times. In Basic this would be written

FOR I = 1 TO 10

statement

NEXT I

There are two main forms of the second type of loop, i.e. the conditional-type loop. In one form the condition is given at the beginning of the loop and execution continues while this condition is satisfied. This is commonly referred to as a *while loop* or a *while do loop*. For example, Pascal has

while condition **do** statement

The statement or statements will not be executed at all if the condition does not hold the first time round. The other form is the *repeat until loop*, which is executed at least once since the condition is given at the end of the loop, causing it to terminate when necessary. For example, Pascal has

repeat statement **until** condition

It is possible for one or more loops to be included within a loop; the incorporated loop or loops are referred to as *nested loop(s)*. The nesting must be such that each inner loop is entirely contained within the outer loop or loops.

loopback /**loop**-bak/   In telephone sys-tems, a test signal sent to a network destination that is then returned as received by the sender. Loopback is used in fault diagnosis. On the Internet a comparable utility is called PING.

lossless compression /**loss**-liss/   Any data compression accomplished by a technique that reduces file size by eliminating redundancies in the data. Data compressed by these methods can be restored exactly to its original value. Lossless compression is used for data and programs. *Compare* lossy compression. *See also* PKZIP.

lossy compression /**loss**-ee/   Any data compression accomplished by a technique that reduces file size by discarding some data forever, such as small variations in color, whose loss won't be noticed. Lossy files cannot be decompressed to their original state. Only certain types of data, such as graphics, audio, and video, can tolerate lossy compression. JPEG and MPEG, for example, use lossy compression techniques. *Compare* lossless compression.

lost cluster   A CLUSTER on a disk storage device that the FILE SYSTEM thinks is in use but is not part of any file. Lost clusters are usually caused when a system or program is closed down without going through the proper procedures, for example in the case of a power cut. Some file systems (e.g. FAT) are more vulnerable to lost clusters than others (e.g. NTFS).

Lotus 1-2-3 /**loh**-tŭs/   *Trademark* An electronic spreadsheet application introduced in 1982. It is widely credited with establishing the popularity of the IBM PC.

Lotus Notes   A GROUPWARE application developed by Lotus, now part of IBM. Notes was one of the first applications to support a distributed database of documents that could be accessed by users across a LAN or WAN. Its sophisticated REPLICATION features enable users to work with local copies of documents and have their modifications propagated throughout the entire Notes network. With the growing popularity of the WWW, IBM integrated Web support into Notes.

lower case Describing a small alphabetic character as opposed to a capital letter.

low-level language A kind of PROGRAMMING LANGUAGE whose features directly reflect the facilities provided by a particular computer or class of computers. It is designed mainly around the set of instructions (i.e. the INSTRUCTION SET) that a particular computer or computer class is able to perform: low-level languages are thus described as *machine-orientated* (or *computer-orientated*). It is possible from the program source code either to control exactly what MACHINE CODE is generated when the program is compiled or to predict what it will be. ASSEMBLY LANGUAGES are the most common types of low-level language. *See also* high-level language.

lpm (**lines per minute**) A measure of the rate of output of LINE PRINTERS.

LPT A name frequently used by operating systems to identify a printer. The printers attached to a computer are called LPT1, LPT2, LPT3, etc. It was originally an abbreviation of 'line printer'.

LQ *See* letter quality.

lsb (**least significant bit**) *See* least significant digit.

lsd *See* least significant digit.

LSI (**large scale integration**) *See* integrated circuit.

LTO (**Linear Tape-Open**) A type of TAPE CARTRIDGE used to back up computers and networks. The first version, introduced in 1999, could hold 100 Gb of uncompressed data; usually data compression allowed up to 200 Gb to be stored. The current version, LTO3, can store up to 400 Gb (800 Gb compressed) with future capacity increases planned. LTO is an open format developed by Hewlett-Packard, IBM, and Certance. *See also* DLT.

Lycos /lȳ-kos/ An Internet SEARCH ENGINE.

LZW compression (**Lempel-Ziv-Welch compression**) A data-compression technique developed in 1977 by J. Ziv and A. Lempel and later refined by Terry Welch. It is the compression algorithm used in GIF files and other graphics-file formats.

M

M The symbol used for MEGA-, or sometimes for megabyte.

Mac The popular abbreviation for the Apple MACINTOSH computer.

MacBinary /mak-**bȳ**-nă-ree, -nair-ee/ A file transfer protocol that preserves Macintosh file formats when stored on non-Macintosh computers.

machine An actual computer, or sometimes a model of a computer used in theoretical studies.

machine address A unique number that specifies a unique storage LOCATION (or in some computers a particular REGISTER or INPUT/OUTPUT DEVICE). A machine address contains a fixed number of bits, n, and hence can identify 2^n different locations. The number of distinct locations to which a machine address can refer is known as the *address space*. A machine address is determined by the computer hardware and when specified in a MACHINE INSTRUCTION can be used directly by the hardware. A machine address is thus an ABSOLUTE ADDRESS. *See also* address.

machine code The code used to represent the instructions in the INSTRUCTION SET of a particular computer. It is therefore specific to a particular computer. It is the form in which programs must be recorded on, say, magnetic disk and entered into main store, and is the form subsequently used for processing. Machine code consists of a sequence of MACHINE INSTRUCTIONS. In order for the computer hardware to handle it, machine code is expressed in binary form: every operation, every address, every item of data consists of a particular sequence of bits.

It is very slow and tedious for a person to write a program in machine code, and very difficult to read and understand such a program. Instead a programmer normally uses a HIGH-LEVEL LANGUAGE (such as Basic or C++), which is translated into machine code by a program – either a COMPILER or an INTERPRETER. A large number of machine code instructions are required for each high-level language statement. Another less easy alternative is for the programmer to use ASSEMBLY LANGUAGE, which gives a 'readable' and hence more convenient representation of machine code; a program written in assembly language is converted into machine code by a program known as an ASSEMBLER.

machine instruction (**computer instruction**) An instruction that can be recognized by the processing unit of a computer. It is written in the MACHINE CODE designed for that particular computer, and can be interpreted and executed directly by the CONTROL UNIT and the ARITHMETIC AND LOGIC UNIT of that computer. A machine instruction consists of a statement of the OPERATION to be performed, and some method of specifying the object upon which the operation is to be performed (i.e. the OPERAND) together with an indication of where the result is to go. There may be more than one operand involved.

The portion of the machine instruction in which the operation is specified is known as the *operation part*. The operand or operands are generally specified in the instruction by stating their actual ADDRESSES in main store, or by some other scheme such as INDIRECT ADDRESSING, INDEXED ADDRESSING, or RELATIVE ADDRESS-

ING; in IMMEDIATE ADDRESSING the operand itself is stated explicitly in the instruction. This information about the operand or operands is given in the *address part* or *parts* of the instruction.

In addition to the operation and address parts, special bit positions are required in the machine instruction to indicate whether a particular addressing mode is to be used. There are thus bit positions to specify indirect addressing, the use of index registers for indexed addressing, or the use of base registers in relative addressing. Additional bits may be used for other purposes.

The layout of a machine instruction, showing its constituent parts, is known as the *instruction format*. It is laid down by the computer manufacturer, and specifies the number of bits allocated to each part – operation part, address part or parts, etc. – and the relative positions of each of these parts.

machine-orientated language *See* low-level language.

machine-readable Available in a form that can be input to a computer system or application. Digital data in a computer file is machine-readable, but the term is also applied to such things as bar codes and magnetic inks.

machine word *See* word.

Macintosh /**mak**-in-tosh/ A series of personal computers introduced in 1984 by Apple Computer Corporation. They were one of the first personal computers to use GUIs (graphical user interfaces) and the 3½″ floppy, and are noted for their user-friendly features and stylish design. Most personal computers are Windows-based PCs rather than Macintoshes, but the Macintosh is used extensively in certain areas such as graphic design and publishing. The Windows interface drew many features from the Macintosh. *See also* iMac.

macro /**mak**-roh/ An instruction that represents a sequence of other instructions. Many application programs allow the user to record a series of keystrokes and commands, which can be given a name or assigned to a combination of keys. The macro can then be used over and over again to repeat the recorded actions.

In programming languages or applications that provide a macro facility, which may include a macro language, a set of instructions can be given a name. When the program is compiled or assembled, whenever the macro name is used the set of instructions is substituted by a process called *macro expansion.*

macro virus A virus written in the macro language associated with an application. The document file that uses the macro carries the virus which is usually triggered when the document is opened or sometimes when other actions are performed. A macro virus is often spread as an e-mail virus. A well-known example was the Melissa virus of 1999.

magnetic disk A storage medium in the form of a flat circular plate that can be rotated by some means and that is covered on one or both sides with a magnetic film. Data is recorded in concentric TRACKS in the film (see below). Disks can be either rigid or flexible, leading to the two categories of HARD DISK and FLOPPY DISK. The storage CAPACITY depends on a number of factors including type and size of disk, the number of tracks per disk, and the recording density along the tracks.

Magnetic disks are RANDOM ACCESS devices: items of data can be located directly, independently, and within a very short period of time (the ACCESS TIME) of some tens of milliseconds. It follows that FILES of data can be stored on disk either in an ordered sequence or in a random manner. In addition, data within a file can be accessed either in sequential order or randomly. The access time to disks is brief enough for ONLINE processing and magnetic disks are used as BACKING STORE in the whole range of computers.

Data is recorded and retrieved when a disk is rotating rapidly in a peripheral device known as a DISK DRIVE, and this is achieved by means of read/write heads.

When the head is to write data on the disk it receives a coded electrical signal. This signal causes magnetization of tiny regions in the magnetic film immediately below the head. The regions are magnetized in either of two directions; magnetization in one direction represents a binary 1 and in the other direction a binary 0. A letter, digit, or other character can therefore be encoded as a particular sequence of magnetized spots (*see* binary representation). This pattern of spots is written on the disk in different SECTORS of the tracks. The sectors may be in a continuous sequence or scattered in different tracks. When the head is to read data from a specified track and sector, it senses the magnetized patterns there and converts them into a corresponding electrical signal. Items of data can be deleted from a disk and new data added – a magnetic disk is a reusable resource.

magnetic drum An obsolete type of BACKING STORE that has now been displaced by the MAGNETIC DISK. It was a cylindrical storage device whose curved outer surface was coated with a magnetic material in which data could be recorded.

magnetic ink character recognition *See* MICR.

magnetic media Material on which data can be recorded in the form of a magnetic pattern on the surface of a magnetizable film. The data is represented in binary form by the direction in which points on the surface are magnetized. Magnetic media form the recording basis of MAGNETIC DISKS, MAGNETIC TAPE, and MAGNETIC STRIPES.

magnetic storage Any noninternal data storage medium that uses magnetic properties to effect the representation of data, for example disk.

magnetic stripe (**magnetic strip**) A strip of magnetic film on a plastic card on which machine-readable data is encoded. The most common application of magnetic stripes is for identification and security purposes; they are found for example on credit cards and swipe cards.

magnetic tape A cheap robust storage medium consisting of a flexible plastic tape with a magnetic coating on one side. Formerly held on large reels, magnetic tape is now found mainly in small CASSETTES. It is essentially the same as that found in domestic video cassettes. SERIAL ACCESS must be used for storage and retrieval from tape.

When magnetic tape was first used in computer systems in the 1950s it was the only form of high-speed BACKING STORE. It is still a cheap way of storing very large amounts of information, especially when the information is accessed rarely, as in FILE ARCHIVES or BACKUP copies of information held on MAGNETIC DISK.

Items of data are recorded on and retrieved from magnetic tape by read and write heads operating in a TAPE DRIVE. The write head receives a coded electrical signal. This signal causes magnetization of tiny regions in the magnetic film immediately below the head. Each spot represents one BIT: a spot magnetized in one direction represents a binary 0 while a spot magnetized in the opposite direction represents a binary 1. A letter, digit, or other character is encoded as a pattern of magnetized spots.

The data is written on the tape in BLOCKS; a sequence of blocks makes up a FILE. When data is to be read from a block, the read head senses the magnetized patterns there and converts them into a corresponding electrical signal. Blocks of data can be erased from tape and new data added – magnetic tape is a reusable resource.

See also digital audio tape.

magnetic tape unit (**MTU**) *See* tape unit.

magneto-optic recording A technology used with optical disks in which small areas of the surface are heated by a laser so that they can be easily magnetized and the regions of magnetization can be confined and precisely located. A lower power laser is used to read the data from the disk. The disks are erasable and have a capacity similar to a CD-ROM.

mailbox   A network user's area for storing incoming e-mail messages. Mail systems usually allow mail to be read, archived, copied to a file, deleted, printed, or forwarded to another user.

mailing list   1. A system, usually automated, that allows users to send e-mail to one address, whereon their message is copied and sent to all of the other subscribers to the list.
2. A named list of e-mail addresses. Entering the name of the list in the address field of an e-mail causes the e-mail to be sent to all addresses on the list.

mail merge   A technique used in word-processing applications for the merging of names and addresses and sometimes other facts from a list (perhaps derived from a database) into a form letter or other basic document for mass mailing purposes.

main body   The program code that is executed when the program is started and that then calls subroutines and functions as necessary.

mainframe /**mayn**-fraym/   Any large general-purpose computer system, as opposed to the smaller MINICOMPUTERS and MICROCOMPUTERS. Mainframes are typically expensive computers used by large organizations and are capable of supporting a large number of users.

main memory   *See* main store.

main store (**main storage; main memory; RAM; immediate access store, IAS; primary store**)   The storage closely associated with the PROCESSOR (or processors) of a computer and from which program instructions and data can be retrieved extremely rapidly and copied directly into the processor registers, and in which the resulting data is stored prior to being transferred to BACKING STORE or OUTPUT DEVICES. Main store thus stores instructions waiting to be obeyed or currently being obeyed,

and data awaiting processing, being processed, or resulting from processing.

Programs can only be executed when they are in main store. Programs and associated data are not, however, retained permanently in main store. They are kept in backing store until required by the processor. The backing store in a particular computer (large or small) has a larger storage CAPACITY and is also less expensive (in terms of cost per bit stored) than the main store in that computer.

Main store consists of SEMICONDUCTOR MEMORY. There is RANDOM ACCESS to data and the average ACCESS TIME is less than a millionth of a second, typical values being a few nanoseconds. Semiconductor memory is very reliable and is highly miniaturized.

A single item of data is stored in a particular LOCATION in main store. In most computers each location holds a fixed number of BITS; this number is the same for all locations in the main store of a particular computer, and is usually 8, 16, or 32 bits. Each location can be identified by its ADDRESS.

maintenance   1. Any activity, including tests, measurements, adjustments, replacements, and repairs, that aims to prevent the occurrence of faults in hardware or corrects an existing hardware fault. *See* preventative maintenance; corrective maintenance.
2. (**program maintenance**) *See* program.

malfunction   The occurrence of a FAULT, usually a hardware fault.

malware (**malicious software**)   An overall term denoting any software intended to damage a computer system or subvert its operation. Malware includes ADWARE, SPYWARE, TROJANS, VIRUSES, and WORMS.

MAN (**metropolitan area network**)   Usually characterized by very high speed connections using fiber optic cables or other digital media, a network that can consist of several LANS as well as satellite and microwave relay stations. It is smaller, but generally speaking faster, than a WAN.

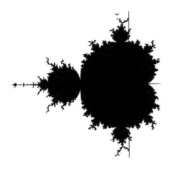 

Mandelbrot set (left) with a magnified detail (right)

Manchester Mark I *See* first generation computers.

Mandelbrot set /**man**-dĕl-brot/ A set of points in the complex plane generated by considering the iterations of the form $z \rightarrow z^2 + c$, where z and c are complex numbers. This process can produce an infinite number of patterns depending on the value chosen for the constant complex number c (*see* Julia set). These patterns are of two types: they are either connected, so that there is a single area within a boundary, or they are disconnected and broken into distinct parts. The Mandelbrot set is the set of all Julia sets that are connected. It can be represented by points on the complex plane and has a characteristic shape.

It is not necessary to generate Julia sets for the Mandelbrot set to be produced. It can be shown that a given value of c is in the Mandelbrot set if the starting value $z = 0$ is bounded for the iteration. Thus, if the sequence
$$c, c^2 + c, (c^2 + c)^2 + c, \ldots$$
remains bounded, then the point c is in the Mandelbrot set.

The set has been the subject of much interest. The figure is a FRACTAL and can be examined on the computer screen in high 'magnification' (i.e. by calculating the set over a short range of values of c). It shows an amazingly complex self-similar structure characterized by the presence of smaller and smaller copies of the set at increasing levels of detail.

man-machine interface (MMI; human-computer interface, HCI; human-system interface, HSI) The means of communication between a human user and a computer system, referring in particular to the use of input/output devices with supporting software.

mantissa /man-**tiss**-ă/ *See* floating-point notation.

map 1. In programming, a file showing the structure of a program after it has been compiled. It lists variables and memory addresses and is useful for debugging purposes.
2. To make logical connections between two entities.
3. To reproduce an object in a different format while preserving the object's internal organization. For example a graphics image can be mapped onto a display screen.

mapped drives 1. In a Windows system, NETWORK DRIVES or DIRECTORIES that have been given local drive letters.
2. In a UNIX system, defined drives that can be activated.

MAR *See* memory address register.

marker In communications, a marker is used to define a part of a communications signal that specifies the structure of the message.

mark reading Optical mark reading (*see* OMR). *See also* document reader.

mark sensing An obsolete process of data entry to a computer in which a machine sensed pencil marks made by a person in his or her choice of indicated positions on a document. The pencil marks were electrically conductive and could thus be sensed by electrical methods. Mark sensing has been displaced by optical mark reading, i.e. OMR, which employs more reliable (photoelectric) sensing methods.

markup language A set of formatting codes and indexing and linking information included in a text file, indicating how the file should look when it is printed or displayed or describing the document's logical structure. The markup indicators are often called *tags*. Some markup languages are specific to an application, but others are designed to enable documents to be platform independent, offering great portability between applications and machines. *See also* HTML; SGML; XML.

mask 1. A pattern of bits that is used to modify or identify part of a byte, word, etc., with the same number of bits. An operation, such as an AND OPERATION or an OR OPERATION, is performed on corresponding pairs of bits in the two patterns; this operation is called *masking*. For example, with an AND operation a 1 in the mask will retain the value of the corresponding bit in the input byte, word, etc., for subsequent processing, while a 0 in the mask will set or keep the corresponding bit as a 0. Thus an AND on a byte containing an 8-bit number with a mask of 00000011 will clear all bits except the two least significant, i.e. it will return a number between 0 and 3 that is the modulo 4 of the input. Masking is often used to store more than one piece of information in a single byte or word. For example, the 8 bits of a byte can be made to store 5 TRUTH VALUES in one bit each and also a 3-bit number, i.e. between 0 and 7.
2. A chemical shield used to determine the pattern of interconnections in an INTEGRATED CIRCUIT.

mass storage device Any backing storage device that contains one or more high-capacity disk drives used primarily as FILESTORE. *See also* capacity.

master file A DATA FILE that is subject to frequent requests for data and frequent updating of the values stored. It can therefore be considered an authoritative source of information. A master file can be updated by being completely rewritten by the system. The version prior to a particular update is called the *father file*, while the one before that is the *grandfather file*. The father file, and often the grandfather file, are retained as BACKUP copies of the master file. *See also* file updating.

math coprocessor A coprocessor designed to enable quick processing of floating-point numbers. This speeds up the processing of math and graphs considerably if the application is designed for its use. Formerly important in microcomputers, the functionality of math coprocessors is now built into most microprocessors.

matrix A two-dimensional ARRAY.

matrix printer *See* printer; dot matrix printer; ink-jet printer.

maximize To make a window take up the full screen.

Mb A symbol for megabit.

MB A symbol for megabyte.

MBR *See* memory buffer register.

MDR *See* memory data register.

mean In general, a number representing the average or typical value of a group of numerical values, such as measurements of a quantity. If the values are represented by
$$x_1, x_2, x_3, \dots x_n$$
so that there are n values in all, then their mean (symbol: $\bar{x}$) is given by the formula
$$\bar{x} = (x_1 + x_2 + x_3 + \dots + x_n)/n$$
The mean of 8 numbers is thus the sum of those numbers divided by 8. For example, with the numbers
$$2, 5, 3, 7, 5, 1, 3, 2$$

the mean is 3.5. The amount of variation within the group is indicated by its STANDARD DEVIATION.

media (*singular:* **medium**) Materials used in the entry or storage of data in a computer, or in the recording of results from a computer. The latter group are known as *output media*, the former as *input media* and *storage media*. Input and storage media include MAGNETIC MEDIA, CDs, and DVDs. Data is represented in coded form on these media. Output media include the sheets of paper or film or the CONTINUOUS STATIONERY used in PRINTERS, the paper or film used in PLOTTERS, and the microfiche used in COM.

mega- /meg-ă-/ A prefix indicating a multiple of a million (i.e. 1 000 000 or 10^6) or, loosely, a multiple of 2^{20} (i.e. 1 048 576). In science and technology decimal notation is usually used and powers of 10 are thus encountered – 10^3, 10^6, etc. The symbol M is used for mega-, as in MV for megavolt. In communications mega- (as in megabaud) means a million. Binary notation is generally used in computing and so the power of 2 nearest to 10^6 has assumed the meaning of mega-.

The prefix is most frequently encountered in computing in the context of storage CAPACITY. With magnetic disks, magnetic tape, and main store, the capacity is normally reckoned in terms of the number of BYTES that can be stored, and the word *megabyte* (MB) is used to mean 2^{20} bytes; megabyte is usually abbreviated to M byte or just to M, thus a 256 M byte flash drive can hold 268 435 456 bytes.

Mega- is part of a sequence

kilo-, mega-, giga-, tera-, peta-, ...

of increasing powers of 10^3 (or 2^{10}).

membrane keyboard A keyboard in which the keys are covered by a transparent plastic shell. It is difficult to type accurately and quickly on such a keyboard but it has the advantage that the components can be kept very clean.

memory A device or medium in which data or program instructions can be held

for subsequent use by a computer. The word is synonymous with *store* and *storage* but is most frequently used in the terms random-access memory (RAM) and read-only memory (ROM), i.e. semiconductor memory used for MAIN STORE. The word 'memory' by itself is generally used to mean main store. *See also* storage device; media.

memory address A number representing a location in MAIN STORE.

memory address register (MAR) A REGISTER (i.e. a temporary location) in the central processor, used for holding the address of the next location in main store to be accessed. Whenever the main store is to be used, the required address must be sent to the MAR. The corresponding storage location can then be selected automatically, and data read from there or stored there. *See also* memory data register.

memory buffer register (MBR) *See* memory data register.

memory card (**memory board**) *See* expansion card.

memory data register (MDR) A REGISTER (i.e. a temporary location) used for holding all instructions and data items as they are transferred between main store and central processor. It thus acts as a BUFFER, and is therefore also called *memory buffer register* (MBR). *See also* memory address register.

memory mapping 1. A technique allowing a processor to access more memory than it is ordinarily capable of addressing. Different banks of memory are electronically switched so that they appear to the processor to have the same address. 2. In microcomputers, a way of making input/output devices appear to be memory locations so that instructions that would normally store or retrieve information from main store perform output or input to an external device.

memory-resident Describing programs

that are permanently in memory. The operating system is not permitted to swap such programs out to a storage device.

memory size   The total number of bytes of MAIN STORE available on a computer.

menu   A list of options, usually displayed on a screen, from which a choice can be made by the user. The list may be displayed with a different letter or number in each option underlined or shown separately opposite each option, and the selection made by typing in the appropriate letter or number using a keyboard. The selection can also be made by positioning the cursor on the selected option using, for instance, a mouse or cursor keys on a keyboard, or by touching a selected area of a touch-sensitive screen. The selected item will often be highlighted in some way; REVERSE VIDEO or blinking are commonly used.

Any program that obtains input from a user by means of a menu is said to be *menu-driven*.

menu bar   A horizontal menu that appears at the top of a window. Usually each option in the menu is associated with a PULL-DOWN MENU.

menu-driven   *See* menu.

merge   To combine together in logical sequence.

message switching   A process whereby information (in digital form) can be sent automatically from one computer on a NETWORK to any other, the information being switched from one communication line to another until it reaches its specified destination. The information to be communicated is known as a *message*. A message may be of any length, and is transmitted as a whole rather than being split up. At each switching point along its route the message is checked for errors, stored temporarily, and forwarded to the next point when the necessary resources become available. *Compare* packet switching.

metafile /met-ă-fÿl/   1. A file that contains information about another file, e.g. a format for graphics files. The commonest is the Windows metafile (which has the file extension wmf).

metalanguage /met-ă-lang-gwij/   A language used to describe other languages in the theory of computer languages. SGML and XML are metalanguages.

meta tag   A tag used to describe the contents of a Web page. Meta tags do not affect how the page is displayed but provide information such as authorship, what the page is about, and keywords that represent the page's content. Many search engines use this information when building their indexes.

metatext /met-ă-tekst/   Text contained in a document that does not appear on the screen but is included to trick Web search indexing software into giving the site a higher relevance rating in a search. For example, by repeating the word aeroplane 100 times, the document is more likely to turn up in a search that includes aeroplane as a key word. Search engines are now aware of this trick and compensate for it.

method   *See* class.

MFLOPS /em-flops/ (**mega floating-point operations per second**)   A measure of the speed of computers used to perform floating-point calculations. *See also* MIPS.

MICR (**magnetic ink character recognition**)   A process in which a machine recognizes characters printed on a document in *magnetic ink* and converts them into a code that can be fed straight into a computer. The machine is called a *magnetic ink character reader*. The characters can be read not only by the machine but also by people: they may be digits, letters, or other symbols. The ink contains a magnetic substance. When the document is fed into the machine the ink is magnetized, enabling each character to be sensed by magnetic read heads. For the machine to recognize and encode the characters, they must be of

0 1 2 3 4 5 6 7 8 9

Digits that can be used in MICR

an appropriate shape and size and have a high print quality; one standardized and widely used set of type is rectangular in appearance with thickened vertical sections. The most common application of MICR is in the handling of cheques by banks; cheque number, branch number, and account number are printed in magnetic ink characters along the bottom of the cheque, the amount of the cheque being added manually. *See also* OCR; OMR.

micro /mӯ-kroh/ *See* microcomputer.

micro- 1. a prefix to a unit, indicating a submultiple of one millionth (i.e. 10⁻⁶) of that unit, as in *micrometer, microsecond,* or *microvolt.* The Greek letter μ (mu) is used as the symbol for micro-; micrometer, microsecond, and microvolt are thus often written μm, μs, μV. Micro- is part of a sequence milli-, micro-, nano-, pico-, ... of decreasing powers of 1000.
2. A prefix indicating smallness or comparative smallness, or something concerned with small objects, quantities, etc.

microchip /mӯ-kroh-chip/ A CHIP (integrated circuit) used in a microprocessor.

microcode /mӯ-kroh-kohd/ 1. The code in which microinstructions of a MICROPROGRAM are written. It is used to create the microprogram that defines exactly what each MACHINE INSTRUCTION does.
2. A sequence of microinstructions.

microcomputer /mӯ-kroh-kŏm-**pyoo**-ter/ (**micro**) A computer system that uses a MICROPROCESSOR as its processing unit. The central control and arithmetic element is thus fabricated on (usually) one CHIP of semiconductor material. In addition a microcomputer contains storage and input/output facilities for data and programs, possibly all on the same chip as the microprocessor.

Microcomputers were originally rather limited in what they could do. With the emergence of more powerful, more sophisticated, but cheaper systems they are now regarded as serious problem-solving tools in many fields, including business, science, and engineering, and are a standard feature in many homes. The capability of the system depends not only on the characteristics of the microprocessor employed but also on the amount of storage provided, the types of peripheral devices that can be used, the possibility of expanding the system with add-on peripherals and additional storage, and other possible options.

microelectronics /mӯ-kroh-i-lek-**tron**-iks/ Electronics involving the use of very small electronic parts to build circuits and systems.

microfiche, microfilm /mӯ-kroh-feesh, mӯ-kroh-film/ *See* COM.

microinstruction /mӯ-kroh-in-**struk**-shŏn/ *See* microprogram.

microprocessor /mӯ-kroh-**pross**-ess-er/ The physical realization of the CENTRAL PROCESSOR of a computer on a single chip of semiconductor, or on a small number of chips. It forms the basis of the MICROCOMPUTER. The INTEGRATED CIRCUIT manufactured on the chip or chips implements – at the very least – the functions of an arithmetic and logic unit and a control unit. Normally there are other sections such as semiconductor memory in the form of RAM (random-access memory) for data and programs and ROM (read-only memory) for in-built programs, plus special REGISTERS and input/output circuits.

Microprocessors are characterized by a combination of their speed (i.e. CYCLE time), their WORD length, their INSTRUCTION SET, and their ARCHITECTURE (i.e. storage organization and addressing methods, I/O operation, etc.). They are used not just in microcomputers but are embedded in a wide variety of devices.

The first microprocessor – the Intel 4004, produced by the US company Intel – appeared in 1971. It had a 4-bit word length, 4.5 kilobytes of memory, and 45 instructions. Its 8-bit counterpart, the Intel 8008, was introduced in 1974 and its improved derivative, the Zilog Z80, in 1976. The number of microprocessors, and their sophistication, has increased rapidly – alongside developments in the fabrication technology of integrated circuits. The Intel PENTIUM 4 series of microprocessors, introduced in 2000 and now the usual chip in computers designed for SOHO use, has a 32-bit word length and can address at least 4 gigabytes of memory; other microprocessors, such as the ITANIUM, are more powerful still.

microprogram /mȳ-kroh-**proh**-gram/ A sequence of fundamental instructions, known as *microinstructions*, that describe all the steps involved in a particular computer operation. It is the means by which an operation is accomplished. A microprogram is set into action once a MACHINE INSTRUCTION is loaded into the INSTRUCTION REGISTER of a computer. One part of the instruction – the operation part – specifies the operation to be performed, and the microprogram for that operation can then be executed. Almost all processors in both large and small computers are now *microprogrammed*, i.e. controlled by microprograms. (It is often thought, mistakenly, that only *micro*computers use *micro*programs.)

The memory in which microprograms are stored forms part of the CONTROL UNIT or ARITHMETIC AND LOGIC UNIT of the computer. It is sometimes referred to as *control store*. It must be very fast (i.e. have a very short storage cycle) and not prone to errors. Semiconductor ROM (read-only memory) is almost always used. The contents are therefore fixed, and the circuitry itself is usually permanently connected. In larger computers semiconductor RAM (random-access memory) can be used to store microprograms, either instead of or in addition to ROM. Use of RAM allows modifications to be made and specialized microinstructions to be introduced.

Microsoft /**mȳ**-kroh-soft/ *Trademark* Microsoft Corporation is the largest producer of PC operating systems and applications, founded in 1975 by Paul Allen and Bill Gates. *See also* Windows.

MIDI /**mid**-ee/ (**musical instrument device interface**) A standard INTERFACE that allows electronic instruments such as musical keyboards and synthesizers to be attached to and used on a computer. The MIDI standard is supported by most synthesizers and sounds created by one synthesizer can be played and manipulated by another. Applications are available for composing and editing music on computers to MIDI standards.

migration The process of moving from the use of one operating environment to another. Migration can be small, for example taking an existing application and making it work on another computer or operating system, or large scale involving many systems, new applications, or a redesigned network. Data can be migrated from one kind of database to another, usually involving the conversion of data to some intermediate common form.

millennium bug A problem that faced some computers when year 2000 was reached. There were two main problem areas. The first was that some software, mainly accounting and database applications, relied on 2 digits to represent the year and produced incorrect follow-on dates. The second involved the BIOS giving incorrect dates after Dec 31, 1999. Most of the problems were resolved either by new issues of software or patches to existing software and very little disruption was caused.

milli- /**mil**-ee/ A prefix to a unit, indicating a submultiple of one thousandth (i.e. 10^{-3}) of that unit, as in *millimeter, millisecond*, or *millivolt*. The symbol m is used for milli-; millimeter, millisecond, and milli-

volt are thus often written mm, ms, mV. Milli- is part of a sequence
> milli-, micro-, nano-, pico-, ...

of decreasing powers of 1000.

MIMD (**multiple instruction multiple data stream processing**) A type of computer architecture used in parallel processing in which CPUs are used to process data and fetch instructions independently. *See* CPU; parallel processing.

MIME /mȳm/ (**multipurpose Internet mail extensions**) A method for attaching files to an e-mail that allows 8-bit binary data to be included with the e-mail rather than encoding the binary data as 7-bit ASCII text. An e-mail program is said to be MIME compliant if it can both send and receive files using the MIME standard.

mini /**min**-ee/ *See* minicomputer.

minicomputer /min-ee-kŏm-**pyoo**-ter/ (**mini**) Loosely, a medium-size computer, i.e. one that is usually considered to be smaller and less capable in terms of performance than a contemporary MAINFRAME computer, and hence cheaper. A modern minicomputer might well operate faster than many of the mainframes used in the l970s. Again, it might be outperformed by a sophisticated MICROCOMPUTER. Minicomputers came on the scene before microcomputers and the boundary between the two is no longer clear.

Minicomputers are often to be found in laboratories built into complex experiments, or controlling industrial processes. They are also used for small multiuser systems, although networks of PERSONAL COMPUTERS attached to a server are now commonly used for this purpose. Mainframe computers often have a number of minicomputers to perform such tasks as input/output or communications, thus freeing the power of the mainframe CPU for more demanding tasks.

minimize To shrink a window down so that it is displayed only on the TASK BAR. A minimized application remains active and can be brought back by clicking on the task bar button with the application's name on it.

minitower /**min**-ee-tow-er/ A microcomputer in which the power supply, motherboard, and mass storage devices are stacked on top of one another in a small vertical cabinet.

MIP mapping A technique used in 3D graphics for correcting pixelation and speeding up rendering times. Different prebuilt texture maps are applied when the object on the screen is close up or far away. MIP is an abbreviation for the Latin *multum in parvo* (many in few).

MIPS Millions of instructions per second; a measure of processor speed.

mirror An arrangement where data is written to more than one hard disk simultaneously, so that if one disk fails, the system can instantly switch to the other disk without any loss of data or service. Mirrored disks are a type of RAID.

mirror site A Web or FTP site that maintains exact copies of material originated at another location, usually in order to provide more widespread access to the information or to improve the performance of a heavily used site.

mod /mod/ Modulo. *See* integer arithmetic.

model A mathematical or graphical representation of something. Mathematical models can usually be changed so that 'What if' analyses can be carried out. *See also* simulation.

modem /**moh**-dem/ A device that converts a DIGITAL SIGNAL into an ANALOG SIGNAL suitable for transmission over an analog communication channel, and that converts an incoming analog signal back into a digital signal. It does this by MODULATION of the digital signal and demodulation of the incoming analog signal. (The word 'modem' is a contraction of 'modulator/demodulator'.) A pair of modems can thus be used to connect two units – two computers, say, or a computer and a terminal – across a telephone line; the streams of bits from each unit are digital signals that are converted by the modems into continuously varying signals suitable for transmission.

moderator /**mod**-ĕ-ray-ter/ A person or group who has the authority to intervene in a newsgroup or other on-line discussion forum and to block messages where appropriate. A moderated forum has fewer FLAMES and less SPAM than unmoderated forums.

Modula-2 /**moj**-ŭ-lă/ A programming language developed in the 1970s to correct some of the deficiencies in PASCAL. Its emphasis is on modular programming and it is used to teach beginners good programming techniques.

modular programming /**moj**-ŭ-ler/ A style of programming in which the complete program is broken down into a set of components, called *modules*. Each module is of manageable size, has a well-defined purpose, and has a well-defined boundary composed of one or more INTERFACES. Almost all programs are modular in structure. The basis on which a program is broken down could be that, for example, each module involves one specific task to be performed by the computer. Alternatively each module could involve a single design decision. It should be possible for the modules to be developed, tested, and debugged independently. Modules can then be brought together and tested to see if they operate together correctly. Finally the whole program can be tested to ensure that it is performing in the desired way. A modular program is easy to amend or update, in many cases the modifications being necessary in just one module.

modulation /moj-ŭ-**lay**-shŏn/ The process of altering one SIGNAL (called the *carrier*) according to a pattern provided by another signal. The device used to achieve this is known as a *modulator*. The carrier is usually a continuously and regularly vary-

ing ANALOG SIGNAL. The other signal may for instance be a DIGITAL SIGNAL, e.g. a stream of bits. The pattern conveyed by this signal is superimposed on the carrier by varying one of the properties of the carrier, e.g. its frequency, amplitude, or phase. This modified analog signal – the *modulated* signal – can then be transmitted.

Demodulation is the reverse process to modulation: a signal is obtained from a modulated signal (by means of a *demodulator*) and has the same shape as the original modulating signal.

See also modem.

modulator /moj-ŭ-lay-ter/ *See* modulation.

module /moj-ŭl/ 1. A largely self-contained component of a program. *See* modular programming.
2. A unit of hardware that has a specific function or carries out a specific task and is designed for use with other modules.

modulo /moj-ŭ-loh/ *See* integer arithmetic.

modulo-*n* check (checksum; remainder check) A type of CHECK, i.e. an error-detecting process performed on an item of information. In the simplest case the individual bits making up the item are added together and divided by a number n to generate a remainder. For example, in a modulo-5 check the number n is 5 and the remainder will be 0, 1, 2, 3, or 4. This remainder is added as an additional digit – the *check digit* – to the item of information. The check digit can be recomputed after the item has been transferred or otherwise manipulated. The new value is compared with the received value, and most simple bit errors (certainly all single bit errors) will be detected.

If the number n is made equal to 2, then the process is known as a PARITY CHECK and the check digit can be a parity bit.

monadic operator /mon-**ad**-ik/ (**unary operator**) *See* operator.

monitor A video display unit in a computer system.

monochrome /**mon**-ŏ-krohm/ Using or involving only one color. For example, a monochrome graphics image uses one color, shades of that color, and no color (i.e. white). The color is usually black, in which case the image is a GRAY SCALE image.

monospaced /**mon**-ŏ-spaysst/ Describing a font in which all the characters occupy the same width. Typewriter-like fonts such as Courier are monospaced. They are often used in software programs where vertical alignment needs to be maintained. *See also* fixed format.

morphing /**mor**-fing/ The gradual transformation of one image into another. This is a common motion-picture special effect and is available in computer animation packages.

MOS /moss/ (metal oxide semiconductor) *See* integrated circuit.

Mosaic /mŏ-**zay**-ik/ Client browser software for access to WWW, Gopher, and FTP servers. It was the first WWW browser that was available for Macintosh, Windows, and Unix, all with the same interface. Development ceased in 1997. *See also* NCSA.

MOSFET /**moss**-fet/ (metal oxide silicon field-effect transistor) A field-effect transistor that uses the MOS capacitor, which is a two-terminal capacitor structure, as its basis.

most significant digit (msd) The digit in the most significant, i.e. leftmost, position in the representation of a number, and thus making the largest contribution to the value of the number. With a binary number this digit is referred to as the *most significant bit (msb)*.

motherboard /**muth**-er-bord, -bohrd/ A PRINTED CIRCUIT BOARD into which other boards can be plugged. In most microcomputer systems, the motherboard carries all the major functional elements, e.g. the processor and some of the memory; additional boards plugged into the motherboard perform specific tasks such as memory extension or disk control.

Motorola /moh-tŏ-**roh**-lă/ Formerly one of the leading manufacturers of microprocessors. Until the early 1990s, Motorola chips were used in all Macintosh computers and in many workstations. Together with Apple Computer and IBM, they launched the PowerPC architecture in 1994. In 2004 Motorola's semiconductor division became a separate company (Freescale Semiconductor), allowing Motorola to concentrate on cellphones and other communications equipment.

mouse A pointing device that is moved by hand around a flat surface: the movements on the surface are communicated to a computer and cause corresponding movements of the CURSOR on the display. The mouse has one or more buttons to indicate to the computer that the cursor has reached a desired position (by a 'click'). Usually, a click on the main mouse button is equivalent to 'enter' on the keyboard; however, on Windows systems the default is that a click selects an item while a double-click is equivalent to 'enter' and opens it. Using suitable software, other mouse buttons can be assigned different functions. A mouse is normally connected by cable to the computer; a 'tailless' mouse communicates by means of infrared rather than electrical signals (as occurs in a TV remote control system).

mousepad /**mowss**-pad/ (mouse mat) A small mat that provides better traction for a mouse ball than the smooth surface of a desk or table. It usually has a plastic surface and a thin non-slip rubber or plastic backing.

MP (multilink point-to-point protocol) An extension of the PPP protocol that allows multiple physical connections between two points to be combined into a single logical connection. The combined

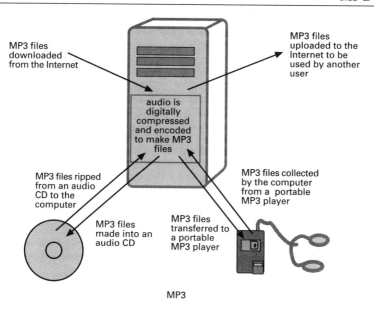

MP3 files downloaded from the Internet

MP3 files uploaded to the Internet to be used by another user

audio is digitally compressed and encoded to make MP3 files

MP3 files ripped from an audio CD to the computer

MP3 files collected by the computer from a portable MP3 player

MP3 files made into an audio CD

MP3 files transferred to a portable MP3 player

MP3

connections are called a bundle and provide greater BANDWIDTH than a single connection. MP is an open standard specified in RFC 1990.

MP3 A form of audio compression adapted from the MPEG format for compressed video. Similar to Dolby Digital, MP3 compresses the sound by filtering out the differences in the sound signal that the human ear cannot detect. Additionally, a musical CD has a sampling rate of 44 100 times a second, whereas MP3 can produce almost the same quality by sampling at 16 000 times a second for music and 10 000 times a second for voice. This means a sound track can be compressed to about one-tenth its original size.

MP3 player A portable device that can store sound tracks in MP3 format, usually music, and play them through earphones. MP3 players can be connected to computers and tracks transferred from the computer to the inbuilt memory (RAM) in the MP3 player. 32-MB flash memory can store 32 minutes of CD-quality sound using 128 kilobits of storage for each second of sound. MP3 files can be stored on

removable flash memory cards, and plugged into the player to swap collections of music. The illegal sharing of music in MP3 format, in breach of copyright, has been a cause of controversy in recent years. *See also* iPod.

MPEG /**em**-peg/ (**moving picture experts group**) The set of digital video compression standards and file formats developed by a working group of the ISO. MPEG achieves a high compression rate by only storing the changes from one frame to the next, instead of each entire frame. The video information is then encoded using discrete cosine transform. MPEG uses a type of lossy compression as some data is removed. However this is generally imperceptible to the eye. There are several MPEG standards, the latest being MPEG-7. Files in this format have the file extension mpeg or mpg.

msb (**most significant bit**) *See* most significant digit.

msd *See* most significant digit.

MS-DOS /em-ess-**doss**/ *Trademark* An

OPERATING SYSTEM produced by Microsoft Corp. It was introduced in 1981 and intended for use on microcomputers that supported a single user at any one time. It was the standard operating system on PCs before Windows was created. It provided a basic set of services for using and storing files on disk and communicating with other peripheral devices, as well as an environment to run one program at a time, a command line interface, and a simple BATCH FILE facility. Windows still supports MS-DOS programs for backward compatibility and provides an MS-DOS CLI that runs in a window. This is sometimes called a *DOS box*.

MSN Microsoft's Web PORTAL, which offers e-mail, news, information, entertainment, on-line shopping, and chat rooms.

MTBF (mean time between failures) *See* failure.

MTU (magnetic tape unit) *See* tape unit.

multiaccess system /mul-ti-**ak**-sess/ A system allowing several people to make apparently simultaneous use of one computer, usually a mainframe or minicomputer. Each user has a TERMINAL, usually a VDU plus keyboard. The terminals are connected to a MULTIPLEXER or to a FRONT-END PROCESSOR, both of which transmit the data, etc., from the various terminals to the computer in a controlled way. The OPERATING SYSTEM of the computer is responsible for sharing main store, processor, and other resources among the terminals in use, which it accomplishes by MULTITASKING or TIME SHARING. As the number of users at any one time increases, the RESPONSE TIME of the computer gets longer. A multiaccess system is an example of a MULTIPROGRAMMING SYSTEM.

multicasting /mul-ti-**kast**-ing, -kahst-/ The simultaneous sending of a message to several network destinations. Multicasting refers to sending a message to a select group whereas *broadcasting* refers to sending a message to everyone on a network.

multifunction card /mul-ti-**funk**-shŏn/ (**multifunction board**) *See* expansion card.

multimedia /mul-ti-**mee**-dee-ă/ Electronic media that includes pictures, animations, video, and sounds as well as text.

multiplate disk /**mul**-tĭ-playt/ A disk storage medium in which there is more than one plate on which data can be stored.

multiplexer /**mul**-tĭ-pleks-er/ (MUX) A device that merges information from several input channels so that it can be transmitted on a single output channel. One transmission channel can thus be shared among multiple sources of information or multiple users. This can be achieved by allocating to each source or user a specific time slot in which to use the transmission channel; this is known as *time division multiplexing (TDM)*. Alternatively the transmission channel can be divided into channels of smaller bandwidth – frequency range – and each source or user is given exclusive use of one of these channels; this is known as *frequency division multiplexing (FDM)*. Multiplexers are often used in pairs, connected by a single transmission channel and allowing contact to be made between any of a large choice of points at each end.

multipoint /**mul**-ti-point/ A communications line in which several secondary stations are sequentially connected along the line and the line is usually controlled by a primary station.

multiprocessing system /mul-ti-**pross**-ess-ing/ A computer system in which two or more PROCESSORS may be active at any particular time. Two or more programs can therefore be executed at the same time – a multiprocessing system is thus also a MULTITASKING system. The processors share some or all of main store. They must also share other resources, such as input/output devices and parts of the system software. The allocation and release of resources is under the control of the OPERATING SYSTEM.

multiprogramming system /mul-ti-**proh**-gram-ing/ A computer system in which several individual programs may be executed at the same time through MULTITASK-ING.

multitasking /mul-ti-**task**-ing/ The process of loading and running at least two PROCESSES (tasks) at the same time. A MULTIPROCESSING SYSTEM, containing several processors, is a multitasking system. It is also feasible for a multitasking system to be operated with only one processor. On any processor only one processing task can take place at a particular time. In a multitasking system there is rapid switching between the tasks to be performed for each of the processes running. For example, while one process is waiting for an input or output operation to be completed, another process can be allocated use of the processor. TIME SLICING is also used. This switching procedure is under the control of the OPERATING SYSTEM, which also controls the allocation and release of other resources in the system.

Multitasking enables optimum use to be made of the computer resources as a whole: a processor operates very much faster than its peripheral devices, and valuable processor time is wasted if the processor remains idle while a process uses a peripheral.

Most current operating systems support multitasking, which can be cooperative or preemptive. In *cooperative multitasking*, processes give up control voluntarily, whereas in *preemptive multitask-ing* the operating system takes the decision. Preemptive multitasking is generally preferred, because a single misbehaving process that does not give up control properly can bring other processes to a halt in a cooperative multitasking system.

Some operating systems allow processes themselves to be multitasked, by subdividing them into parallel units of execution called *threads*. In such systems the allocation of processor time is performed at the thread level rather than the process level. In general all a process's threads share its resources, such as memory, files, etc. (by contrast, each process has its own independent set of resources). The use of more than one thread in a process is called *multithreading*.

multithreading /mul-ti-**thred**-ing/ **1.** *See* multitasking.
2. In data manipulation, the use of extra pointers in a tree data structure to facilitate moving around the structure.

multiuser system /mul-ti-**yoo**-zer/ A computer system that is apparently serving more than one user simultaneously. A multiuser system is generally a MULTIACCESS system.

mutual exclusion A programming technique used to ensure that a resource, such as a file, can be used by only one program at a time.

MUX *See* multiplexer.

nagware /**nag**-wair/ *Informal* Shareware that reminds the user at regular intervals that payment or registration is needed.

NAK (**negative acknowledgment**) The ASCII character 21 transmitted as a control code by a receiving station to a sending station to indicate that the transmitted information has been received incorrectly and should be resent. *See also* acknowledgment.

name A means of referring to or identifying some element in a computer program, in a computer NETWORK, or in some other system. For example, VARIABLES, ARRAYS, and PROCEDURES in a program would be named, as would be the NODES in a network. In many programming languages a name must be a simple IDENTIFIER consisting of a string of one or more characters. In most of these languages names can be chosen quite freely by the programmer. The names used in a program are stated in a series of DECLARATIONS. When the program is compiled or assembled, prior to execution, names are converted to MACHINE ADDRESSES by means of a SYMBOL TABLE. *Compare* literal.

namespace A named group of IDENTIFIERS. This device, found in many modern computer languages, aids program or data organization and in particular prevents name clashes between modules. A software library, for example, might define a namespace called 'nsa' and assign to it the procedures 'x', 'y', and 'z'. Another software library might also have procedures with these names, but it assigns them to its own namespace, 'nsb'. A programmer can use both libraries without problems by referring to the functions by both their name-space and their identifier. A common syntax, used in C++, is 'nsa::x' and 'nsb::x' to distinguish the two procedures called 'x'. Namespaces are not only applicable to programs: XML uses them to prevent duplicate ELEMENT and ENTITY names. Namespaces are generally hierarchical: i.e. it is possible to qualify the contents of a namespace by dividing it into two or more sub-namespaces. There are various conventions to ensure that publicly-used namespaces are unique; for example JAVA namespaces should begin with the Internet DOMAIN NAME of the author.

NAND gate /nand/ A LOGIC GATE whose output is low only when all (two or more) inputs are high, otherwise the output is high. It can thus be regarded as an AND GATE followed by a NOT GATE: it effectively first performs the AND operation on its inputs then negates the output, i.e. performs the NOT operation on it. The TRUTH TABLE for a NAND gate with two inputs (A, B) is shown, where 0 represents a low signal level and 1 a high level. The LOGIC OPERATION implemented by the gate is known as the *NAND operation* and can be written as A NAND B for inputs A and B.

A	B	output
0	0	1
0	1	1
1	0	1
1	1	0

Truth table for NAND gate

The NAND gate is important because any logic gate can in theory be constructed from a suitable combination of NAND gates. For example, a two-input NAND gate with one input kept permanently high acts as a NOT gate. An AND gate is produced by feeding the output of a NAND gate through such a NOT gate, i.e.

NOT (A NAND B) = A AND B

This can be checked by means of the NAND and AND truth tables. Again, an OR GATE is produced if each of the two inputs to a NAND gate is first fed through a NOT gate, i.e.

(NOT A) NAND (NOT B) = A OR B

See also NOR gate.

NAND operation *See* NAND gate.

nano- /**nan**-oh-/ A prefix to a unit, indicating a submultiple of one thousand millionth (i.e. 10^{-9}) of that unit, as in *nanometer* or *nanosecond*. The symbol n is used for nano-; nanometer and nanosecond are thus often written nm and ns. Nano- is part of a sequence

milli-, micro-, nano-, pico-, ...

of decreasing powers of 1000.

NAP (**network access point**) An interchange point for network access. Information originating on a network and being accessed by another network invariably passes through at least one NAP.

NAS *See* network-attached storage.

native compiler The usual form of compiler, which runs on a computer producing an object file for that computer, as opposed to a CROSS COMPILER.

natural language Language as used and spoken by humans as opposed to a programming or machine language. One facet of ARTIFICIAL-INTELLIGENCE research is the approximating of natural language in a computer environment.

natural-language processing The study of computer systems that can react to and understand both spoken and written natural language.

natural-language query A query to a database framed in natural language. Various syntax rules must be obeyed so that the system can parse the query.

navigation 1. The control of cursor movement using the navigation keys on the keyboard or a mouse.
2. The process of moving around in a document or on the WWW using hyperlinks.

NCR paper A paper that is impregnated with a special chemical so that it darkens when pressure is applied, used in multipart forms, especially those printed on continuous stationery to be used on a computer system. NCR stands for 'no carbon required'.

NCSA The National Center for Supercomputing Applications; a research center based in the University of Illinois and formed in 1985 as part of the National Science Foundation. NCSA MOSAIC and NCSA TELNET are two of its best-known products.

NDIS (**network driver interface specification**) A Windows device driver interface developed by Microsoft and 3COM that enables a single NETWORK interface card (NIC) to support multiple protocols, for example TCP/IP and IPX connections.

near letter quality (**NLQ**) A print quality that is not quite as good as LETTER QUALITY, the quality produced by a top-quality electric typewriter, but is better than draft quality. The term is mainly used of dot matrix printers and is becoming less common with the increased use of laser printers, which generally give better quality than electric typewriters.

negation 1. In computing, a NOT OPERATION, the performance of a NOT operation, or the result of a NOT operation. The NOT operation can be performed, for example, on a binary digit or number or a logic statement. The outcome is the COMPLEMENT of that entity. (It should be noted that if a number is stored in two's complement, then to negate it is to take the two's complement.)

2. The changing of the sign of a nonzero number or quantity. The negation of +42 is –42; the negation of –42 is +42; the negation of 0 is 0.

negative acknowledgment *See* acknowledgment; NAK.

NEQ gate *See* nonequivalence gate.

nested structure A programming structure that is incorporated into another structure of the same kind. More than one structure, of the same kind, can be incorporated. For instance, one or more LOOPS can be nested in another loop.

Net *See* Internet.

NetBIOS (**network basic input output system**) An application programming interface (API) that adds special functions to the MS-DOS BIOS for local area networks.

netiquette /net-ă-ket/ Internet etiquette; guidelines for posting messages to on-line services and particularly newsgroups.

Netscape Navigator /net-skayp/ A Web BROWSER produced by Netscape Communications Corporation, which was founded in 1994 by James H. Clark and Marc Andreessen. It was based on MOSAIC and was one of the first commercial Web browsers. It runs on all major platforms, Windows, Macintoshes, and UNIX. Formerly the leading Web browser, it has been overtaken by INTERNET EXPLORER.

Netware /net-wair/ A local area network (LAN) operating system developed by Novell Corporation. Formerly the leading network operating system, it has been overtaken by WINDOWS-based systems.

network /net-werk/ Loosely, a group of computer systems that are situated at different places and are interconnected in such a way that they can exchange information by following agreed procedures (*see* protocol). The computer systems must be capable of transmitting information onto and receiving information from the

connected system. The information is sent as an encoded signal, and is transmitted along communication lines such as telephone lines, satellite channels, electric cables, and optical fibers.

It is not sensible (in terms of cost and efficiency) for each computer to be directly connected to every other computer on the network: communication lines are expensive to provide and maintain and would be unused for much of the time. Direct connections are made only at certain points in the network, called NODES. (Strictly, a network is a set of interconnected nodes.) Computers are attached to some or all of the nodes. Nodes may be at a junction of two or more communication lines or may be at an endpoint of a line. A particular piece of information has to be routed along a set of lines to reach its specified destination, being *switched* at the nodes from one line to another, often by a computer. A particular destination is specified by its ADDRESS, which is indicated by the sender and forms part of the transmitted signal.

The computers on a network may be organized in several ways (see diagram). In a *star* structure all communication lines connect directly to a single central or controlling node. The endpoints on each line cannot therefore communicate directly. In a *ring* structure there is a single loop of communication lines (usually electric cables) between the various nodes on the network. Each node is thus connected to two adjacent ones. Information can pass in only one direction around the loop. A *bus* configuration is linear rather than circular. All nodes are connected to a single electric cable acting as a bus, and information can travel in both directions.

See also local area network; wide area network; packet switching; message switching.

network administrator The person who is in control of a network.

network architecture The aspects of the design of a communications network that involve the way in which the network communicates and how its interconnections are arranged.

network-attached storage (NAS) A device, connected to a network, on which data is stored rather than on the network's servers. NAS devices consist of an *NAS Head*, which is connected to network and handles all data requests, and one or more data-storage devices, such as RAID disk arrays, controlled by the NAS Head. NAS devices perform many of the functions of FILE SERVERS but do so more efficiently: because they are special devices and not general-purpose computers running file-serving software, they can be optimized to do nothing but handle data quickly and securely. As well as providing general shared storage for the network's users, NAS devices can take the place of the local hard disks on all types of server. This consolida-

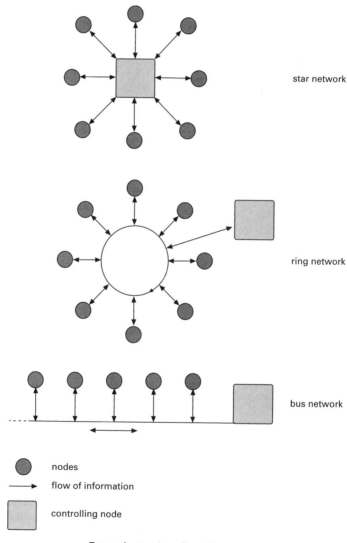

star network

ring network

bus network

nodes

flow of information

controlling node

Types of network configuration

tion of data storage and its separation from other network functions simplifies both network management and the addition of new storage capacity. NAS is not appropriate to all networks: it is unnecessarily sophisticated for small networks, while the faster STORAGE AREA NETWORKS are often preferred in large networks.

network computer *See* thin client.

network database **1.** A type of database that is similar to but less rigorously structured than a HIERARCHICAL DATABASE, in that records, or nodes, can have more than one parent. Network databases are less powerful than RELATIONAL DATABASES. **2.** A database that runs in a network. **3.** A database containing the addresses of users on a network.

network directory A DIRECTORY that can be used, if rights have been granted, by other computers on a LAN.

network drive A disk drive that can be used, if rights have been granted, by other computers on a LAN.

network interface card *See* NIC.

network layer *See* OSI.

network operating system (NOS) An operating system that includes the capability for organizing computers and devices on a local area network (LAN). Some, such as UNIX, Windows, and Mac OS, are standard operating systems, but others, such as Netware, are exclusively devoted to networking functions.

network switch A device that performs the same function as a HUB but in a more efficient fashion. In particular, as far as possible it only sends PACKETS it receives to their correct destination NODE rather than broadcasting them to all connected nodes. This strategy makes better use of available BANDWIDTH and gives improved network performance.

network topology *See* network architecture.

neural network /new-răl/ A network of interconnected components (called *neurons*), designed so that an input signal to each component is multiplied by a weighting factor and passed to two or more other components, and so on. The output signal depends on the weighting factors used, and these can be adjusted so as to optimize the final output. Neural networks are designed by analogy with the transmission of impulses in the human brain. They do not have to be physical networks; usually they are simulated by a computer. They can be 'trained' to recognize patterns in large amounts of data and are used for making predictions (e.g. in financial markets). They are also used for research in ARTIFICIAL INTELLIGENCE.

neuron /new-ron/ *See* neural network.

news aggregator *See* Web feed.

newsgroup A discussion group on USENET. Newsgroups cover every conceivable interest. A NEWSREADER is needed to view and post messages to a newsgroup.

newsreader A client application enabling the user to read messages posted to, and send messages to, Internet newsgroups. Microsoft Internet Explorer and Netscape Navigator both include newsreaders. A number of freeware, shareware, and commercial newsreaders are also available.

news server A computer or program that exchanges Internet newsgroups with newsreader customers and other servers.

NEXT *See* Basic.

nibble (quadbit) A unit of 4 bits (i.e. half a byte).

NIC (network interface card) An expansion board enabling a computer to be attached to a network. A NIC is usually designed to interface with a particular network, protocol, and media.

NLQ   *See* near letter quality.

NNTP (network news transport protocol) The protocol used by client and server software to carry USENET postings back and forth over a TCP/IP network.

node   1. A point in a computer NETWORK where communication lines, such as telephone lines, electric cables, or optical fibers, are interconnected, or the endpoint of such a line. Some nodes are used to connect client computers, SERVERS, printers, etc., to the network. Others, especially on larger networks, are devices dedicated to the operation of the network itself, for example HUBS, NETWORK SWITCHES, REPEATERS, BRIDGES, and ROUTERS.
2. *See* tree.

noise   Any unwanted electrical signals occurring in a system and producing disturbances in the output of the system. Noise can disturb a TV picture or make a telephone message inaudible. It can also introduce errors in data communications.

nonbreaking space   A character representing a space between words that must not be split between lines in text formatting.

nondestructive read   The process of reading a memory device in a way that does not alter the contents of the device. Most devices are read in this way.

nonequivalence gate (NEQ gate)   A LOGIC GATE whose output is high only when one of its (two or more) inputs is

high and the rest are low; otherwise the output is low. It effectively performs the operation of EQUIVALENCE on its inputs and then negates the output, i.e. performs the NOT OPERATION on it. The TRUTH TABLE for a gate with two inputs (A, B) is shown, where 0 represents a low signal level and 1 a high level.

The NEQ gate is also called the *exclusive-OR* (or *XOR*) *gate*: it excludes the possibility (allowed in the OR GATE) that the output is high when more than one input is high. *See also* equivalence gate.

noninterlaced /**non**-in-ter-laysst/ Describing monitors and video standards that do not use INTERLACING techniques to improve resolution.

nonlocal variable /non-**loh**-kăl/ *See* scope.

nonprocedural language   A PROGRAMMING LANGUAGE where the programmer states the problem but does not specify the exact steps (procedure) to be used in solving it. FOURTH and FIFTH GENERATION LANGUAGES are nonprocedural languages. *Compare* procedural language.

nonvolatile memory /non-**vol**-ă-tăl/   A type of memory whose contents are not lost when the power supply to the memory is switched off. ROM is nonvolatile memory. *Compare* volatile memory.

no-op instruction /**noh**-op/ (**do-nothing operation**)   An instruction whose execution causes no action to take place in a computer. The computer just proceeds to the next instruction to be executed.

NOR gate /nor/   A LOGIC GATE whose output is high only when all (two or more)

A	B	output
0	0	0
0	1	1
1	0	1
1	1	0

Truth table for nonequivalence (NEQ) gate

A	B	output
0	0	0
0	1	0
1	0	0
1	1	0

Truth table for NOR gate

inputs are low, otherwise the output is low. It can thus be regarded as an OR GATE followed by a NOT GATE: it effectively first performs the OR operation on its inputs then negates the output, i.e. performs the NOT operation on it. The TRUTH TABLE for a NOR gate with two inputs (A, B) is shown, where 0 represents a low signal level and 1 a high level. The LOGIC OPERATION implemented by the gate is known as the *NOR operation* and can be written as A NOR B for inputs A and B.

The NOR gate is important because any logic gate can in theory be constructed from a suitable combination of NOR gates. For example, a two-input NOR gate with one input kept permanently low (i.e. earthed) acts as a NOT gate. An OR gate is produced by feeding the output of a NOR gate through such a NOT gate, i.e.

NOT (A NOR B) = A OR B

This can be checked by means of the NOR and OR truth tables. Again, an AND GATE is produced if each of the two inputs to a NOR gate is first fed through a NOT gate, i.e.

(NOT A) NOR (NOT B) = A AND B

See also NAND gate.

normalization 1. In relational database design, the organization of data to minimize duplication. The ideal is that data is organized into tables and relationships between tables in such a way that field modifications, additions, and deletions need take place in only one table. Although normalization makes databases more efficient to maintain, they can become very complex.
2. In programming, changing the representation of a floating-point number so that the most significant digit in the mantissa is not zero.

NOR operation *See* NOR gate.

NOS /noss/ *See* network operating system.

NOT /not/ A Boolean operator that returns the value true when the operand is false and vice versa.

notation The set of symbols and formats used in a programming language.

notebook A small LAPTOP computer usually weighing less than 6 lbs that will fit inside a briefcase.

NOT gate (inverter) The only LOGIC GATE with a single input. When the input is high then the output is low; when the input is low the output is high. The NOT gate thus performs the NOT OPERATION on its input and has the same TRUTH TABLE. The truth table is shown in the diagram, where 0 represents a low signal level and 1 a high level.

NOT operation A LOGIC OPERATION with only one OPERAND. When applied to an operand A (a statement or formula, for instance), the outcome is false if A is true and is true if A is false. It thus negates the truth value of A. The TRUTH TABLE is shown in the diagram, where the two truth values are represented as 0 (false) and 1 (true). The operation can be written in several ways, including

NOT A $\neg A$ $\tilde{A}$ ~A   A$'$

The symbol next to the operand A (i.e. NOT, $\neg$, etc.) is a LOGIC OPERATOR for the NOT operation.

In a computer the NOT operation is used in HIGH-LEVEL LANGUAGES to negate a logical expression. It is used in LOW-LEVEL LANGUAGES on a BYTE or WORD, producing a result by performing the NOT operation on each bit in turn. If 8-bit words are used, then for example with the operand

01010001

the outcome of the NOT operation is

10101110

See also NOT gate.

NSFnet /en-ess-eff-**net**/ A wide area network that was developed by the National Science Foundation and replaced ARPANET as the main USA government network linking research establishments and universities. It was replaced in 1995 with a commercial Internet backbone. A new backbone for research and development, the *very high speed backbone network service* (vBNS), was also set up.

NT   *See* Windows NT.

NTSC (**national television standards committee**)   An organization for setting television and video standards in the USA.

null character   A character denoting nothing. When interpreted as text it can have a special meaning. For example, in database and spreadsheet applications null characters are often used as padding and are displayed as spaces. Most character sets use character number 0 as the null character.

null modem   A technique for allowing two computers to be connected together through a special null-modem cable via their communications ports. An actual modem is not required.

null string   *See* empty string.

number cruncher   *Informal* Any powerful computer designed or used mainly for numerical and mathematical work, usually of a scientific or technical nature. *See also* supercomputer.

number system   Any of various systems used to represent and manipulate numbers. A particular system is characterized by its BASE, i.e. the number of different digits (and possibly letters) used to represent the numbers. All present-day number systems are *positional notations*. In such a notation the value of a number is determined not only by the digits it contains but also by the position of the digits in the number. The po-sitional value increases from right to left by powers of the base; in BINARY NOTATION, for example, the positional value increases by powers of 2. *See also* hexadecimal notation; floating-point notation; octal notation.

numerical control   The control, by means of a computer, of a manufacturing process. It typically involves a machine tool used in metal-working, such as a milling machine, lathe, or welding machine. Numerical control systems range from very simple to quite complex. The principal variable that is controlled is the position of the machine tool during the manufacturing process. The desired positions for a particular process are calculated and stored in numerical form on, for example, a magnetic storage medium, together with other necessary information. When fed into the computer associated with the machine tool, this information is converted into signals that operate servo-mechanisms that in turn guide the tool through the desired sequence of positions.

numeric keypad   A section on the right-hand side of a keyboard that duplicates the normal typewriter number keys and which also doubles up the cursor control keys. Using the numeric keypad can speed up the entry of numerical data.

num lock key   A toggle key that turns on/off the locking of the numeric keypad into producing numeric codes or enabling cursor control.

object An item of data, such as a picture, a spreadsheet, or a CAD document, to be used in a program. *See also* object-oriented programming.

object code *See* object program.

object file A file containing object code, which has normally been generated by a compiler or assembler.

object linking and embedding (OLE) The use of a data OBJECT created in one application (the server or object application) by another (the client application). A file created by the server application is inserted at the desired point in the client document. A compound document is created and the object can be edited from the client application using the server application. For example, a spreadsheet inserted into a word-processing document will appear as a seamless part of the document when it is printed, but is updated from within the word-processing application by automatically calling the spreadsheet application. An object can be inserted in one of two ways. If it is *linked*, the compound document contains a link to the object's original file; any changes made to this file are reflected in the compound document, and vice versa. If the object is *embedded*, a copy of the original file is taken and stored within the compound document; no connection between the original file and the compound document is maintained and changes will not be replicated.

object module A unit of output from a COMPILER or ASSEMBLER, usually the translated form of one file of source code. An object module may be a self-contained program, or it may be part of a larger program

and must be LINKED to other, separately compiled, object modules to form the complete program. An object module contains *object code*, which is either MACHINE CODE or, for such languages as JAVA, a machine-independent code that is interpreted at runtime. *See also* object program.

object-oriented graphics The representation of images by the use of a combination of basic graphical shapes (polygons, ellipses, curves, lines, etc.). Because of the way they are defined such objects can be layered, rotated, or resized with relative ease. *See also* computer graphics.

object-oriented programming (OOP) Programming using a language in which all the files, inputs, and outputs are considered as discrete objects, and the programs dealing with these are part of the same objects. This is a form of modular programming in which the definition of each data structure is normally contained in one module. When a data structure's internal design is altered and the implementation details of actions carried out on it need to be changed, the program modifications are localized and easier to control.

object program The program that is in object code and is ready to be loaded into a computer for execution. An object program consists of one or more OBJECT MODULES.

OCR (optical character recognition) A process in which a machine scans a document and is able to convert the characters that it recognizes into a code that can be fed straight into a computer. The machine is called an *optical character reader*, also sometimes abbreviated to OCR. Scanners

binary	000	001	010	011	100	101	110	111
octal	0	1	2	3	4	5	6	7

Octal and binary equivalents

with appropriate software can also function as optical character readers. The characters can be read not only by the machine but also by a person: they can be letters, digits, punctuation, and other symbols. Early optical character readers could only recognize a limited number of rather stylized characters. Current devices can now recognize characters in a variety of typefaces, sizes, layouts, etc., and may even recognize handwritten characters. OCR is widely used in billing and banking, optical characters being used for example on gas and electricity bills. *See also* MICR; OMR.

octal notation /ok-tăl/ (**octal**) A number system that uses 8 digits, 0–7, and thus has BASE 8. It is a positional notation (*see* number system), positional values increasing from right to left by powers of 8. Octal is a convenient shorthand by which people (rather than machines) can handle binary numbers. Each octal digit corresponds to a group of 3 binary digits, or bits. Conversion of binary to octal is done by marking off groups of 3 bits in the binary number (starting from the right) and replacing each group by its octal equivalent. Conversion of octal to binary is done by replacing each octal digit by its equivalent binary group. The use of octal notation has declined in favour of HEXADECIMAL NOTATION. *See also* binary notation.

ODBC (**open database connectivity**) The Windows interface, part of the WOSA (Windows open system architecture), that provides a common language for applications needing network database access. The goal of ODBC is to make it possible to access any data from any application regardless of which database management system is handling the data. In order to achieve this, both the data and the application must be ODBC-compliant. Despite the subsequent release of new Microsoft database-access technologies, ODBC is still widely used.

odd parity *See* parity check.

OEM (**original equipment manufacturer**) Usually, an organization that purchases equipment, such as electronic circuitry, that is to be built into one of its products. In some cases the term is used to refer to the supplier of the equipment.

Office (**Microsoft Office**) A suite of related applications, developed by Microsoft and designed to meet the WORD-PROCESSING, SPREADSHEET, E-MAIL, etc., needs of computer users. There are several editions of Office, aimed at different types of user. All mainstream editions include WORD, EXCEL, OUTLOOK, and POWERPOINT; ACCESS and other programs are added in those aimed at business users. Although some of these applications existed prior to Office and all can be used on their own, they can easily share data and successive versions have increased common features that assist integration and management in a corporate environment. First released for the Macintosh in 1989 and Windows in 1990, the current versions are Office 2003 (Windows) and Office 2004 (Macintosh).

off-line 1. Of PERIPHERAL DEVICES or FILES: not under the control of the processing unit of a computer, i.e. not connected to a computer system or not usable. A peripheral device may be connected to a computer but is off-line if the system has been instructed not to use it.
2. Involving peripheral devices or files that are off-line (as in *off-line processing* or *off-line storage*).
   Compare on-line.

off-line browser A browser designed to read HTML files from a disk rather than on the Internet.

off-line navigator A piece of software used to download and save to disk information from the Internet so that it can be browsed off-line. This minimizes connection time and reduces costs.

OLAP *See* on-line analytical processing.

OLE /oh-el-*ee*/ *See* object linking and embedding.

OLTP *See* on-line transaction processing.

OMR (optical mark reading; optical mark recognition) A process of data entry to a computer in which a machine detects marks made by a person in his or her choice of indicated positions on a document. The marks could be put for example in one or more chosen columns of a table or in selected areas of a multiple-choice answer sheet. OMR displaced the less reliable but similar process of MARK SENSING. In OMR each mark is detected photoelectrically, usually by the reduction in intensity of a narrow light beam after reflection by the mark rather than the less absorbent paper or card. The positions of a mark on a document indicate the input data. Once a mark has been detected, the relevant data is converted into a code that can be fed straight into the computer. *See also* OCR; MICR.

one's complement *See* complement.

on-line **1.** Of PERIPHERAL DEVICES or FILES: under the control of the processing unit of a computer, i.e. connected to the computer and usable.
2. Involving peripheral devices or files that are on-line (as in *on-line processing* or *on-line storage*).
Compare off-line.

on-line analytical processing (OLAP) Describing a database whose contents are stable and are used as source data for analysis, and whose structure is designed to optimize this usage. The term is also used of the activities that typically utilize such a database. An OLAP database is up-dated periodically, often by adding a copy of the current contents of an OLTP database together with a record of the date and time the copy was taken. OLAP databases are often used in DATA WAREHOUSES. *Compare* on-line transaction processing.

on-line information service A business that provides its users with information transmitted over, for example, telecommunications lines and possibly a range of services such as e-mail, chat, on-line conferencing, forums, etc. Many on-line information services are now based on the Internet.

on-line transaction processing (OLTP) **1.** An application that facilitates and manages activities such as banking, mail order, and travel booking. Because transactions may span a network and involve more than one company, modern OLTP software uses client/server processing and brokering software.
2. Describing a database whose contents are frequently updated by many users and whose structure is designed to optimize this type of usage. The term is also used of the activities that typically utilize such a database. *Compare* on-line analytical processing.

on the fly Describing a process that is carried out when needed while normal operations are still proceeding. The term is also used to describe activities that occur or develop dynamically rather than as the result of something predefined. An example is a Web page that can be varied 'on the fly' depending maybe on the time of day, what previous pages have been looked at, and possibly user input. *See also* cookie.

OOP *See* object-oriented programming.

op code *See* operation code.

open Accessible or providing accessibility. For example, an open file is one that is ready to be used.

open architecture **1.** Any computer architecture or peripheral design whose spec-

ifications are public. Open architecture allows other manufacturers or developers to produce add-ons.
2. A design allowing for expansion slots on a motherboard.

open database connectivity *See* ODBC.

Open Group An international group of computer hardware and software manufacturers, and users, formed in 1996 by the consolidation of the Open Software Foundation and X/Open Company Limited, to promote the furthering of multivendor information systems. The Open Group owns the trademark of the UNIX operating system and so certifies whether a particular implementation is or is not permitted to describe itself as 'UNIX'.

open source A programming methodology founded on the premise that better software will result by making the SOURCE CODE available for any user or interested party to study and modify. All contributions are then fed back into the master version, which is updated frequently. Open source has had some successes, notably the LINUX operating system. It contrasts with *closed source*, or *proprietary*, software, where the source code is a closely-guarded secret of the developer and all development is strictly controlled. The merits of the two methodologies are hotly debated.

open standards Standards for the specification of hardware or software that is generally known and can be used by anyone. Open standards encourage the development and use of new technologies.

open system Any system that conforms to standards that are not specific to the particular supplier, but that are generally well defined and well known.

open systems interconnection *See* OSI.

OpenType *Trademark See* font.

operand 1. In general, a quantity or function upon which an arithmetic or LOGIC OPERATION is performed. For example, in

$$(3 + 5)/2$$

3, 5, and 2 are the operands. The symbols + and / are called OPERATORS.
2. The parts of a MACHINE INSTRUCTION that specify the objects upon which the operation is to be performed. For instance, in the instruction

$$ADD\ A,B$$

A and B are the operands and could be REGISTERS in the central processor, or actual values, or ADDRESSES of values, or even addresses of addresses of values. *See also* addressing mode.

operating system (OS) A program or collection of programs used for managing the hardware and software resources of a computer system on behalf of the user(s). Every general-purpose computer has an operating system. In general a portion of the operating system is always resident in MAIN STORE; the rest is kept on BACKING STORE and is read into main store when required.

The operating system provides the USER INTERFACE that interprets commands typed on a user's KEYBOARD and movements of the MOUSE, and provides information about the current state of the system on the VDU. In GUI systems it manages the WINDOWS that appear on the screen. It also manages queues for slower devices such as PRINTERS, keeps track of where FILES are stored, what they are named, and who they belong to, and transfers files between disks or between disks and main store when instructed. Input from and output to any connected INPUT or OUTPUT DEVICES is controlled by the operating system. In MULTITASKING systems the operating system is responsible for deciding in what order and for how long each processing task is to have the central processor, and what sections of memory they are to use. All communications are managed by the operating system – not only links between the user(s) and the computer but also links to other computers or to a NETWORK.

operation 1. In mathematics, the act of

OPERATING SYSTEMS BASIC FUNCTIONS	
Function	*Description*
CPU management	To ensure that every process and application that is running has enough processor time to funtion properly and to optimize the use of processor cycles by scheduling processor priorities. Dealing with hardware and software interrupts.
Memory management	The different types of memory in the system must be used effectively and each process must have enough secure memory in which to run.
Device management	Usually performed through device drivers.
I/O management	Largely the management of data in the form of a stream of bits coming in from or going out to a device such as a keyboard, monitor, printer, or disk.
Application interface	The provision and management of an API that lets applications use operating system functions and request input/ouput functions without having to keep track of the CPU's basic operation.
User interface	To provide a structure for the communication between a user and the computer.

doing something to an entity or entities that produces a new entity. *See* operator.
2. An action carried out during the execution of a program.

operation code (op code) The code that can be placed in the operation part of a MACHINE INSTRUCTION and that specifies the operation to be performed.

operation part (function part; operator part) The part of a MACHINE INSTRUCTION in which the operation to be performed is specified.

operator 1. A symbol representing the operation that is to be performed on one or more OPERANDS so as to yield a result. An operator taking one operand is called a *monadic operator* or *unary operator*. A monadic operator usually precedes the operand, as in –1, NOT P. An operator taking two operands is called a *dyadic operator* or *binary operator*. A dyadic operator usually appears between the operands, as in 1 + 1, 3/X; in certain notations, however, it appears before or after the operands (*see* reverse Polish). *See also* arithmetic operator; logic operator; relational operator; order of precedence.
2. A person responsible for the supervision of the hardware of a computer system.

optical character recognition *See* OCR.

optical computer A theoretical device that uses light or infrared beams instead of an electric current to perform digital computations. Because light travels about 90% faster than an electric current it may be possible to build an optical computer that is much smaller and faster than a conventional electronic one.

optical disk Any of various disk storage devices on which the data is read or written by a laser. Examples are the CD and DVD.

optical drive A disk drive that uses a laser to read or write information. *See also* CD.

optical fiber *See* fiber optics transmission.

optical font A font designed for easy use by an optical character reader.

optical mark reading *See* OMR.

optical mouse A mouse that uses light-emitting diodes (LEDs) and optical sensors to detect motion.

optical scanner *See* flatbed scanner; hand-held scanner; sheet scanner.

optomechanical mouse /op-tŏ-mĕ-**kan**-ă-kăl/ A mouse that uses a combination of optical and mechanical technology to detect movement. Because there are fewer moving parts than in a standard mouse fewer faults occur.

Oracle *Trademark* A database system from Oracle Corporation, which runs on Windows, Macintosh, and several varieties of UNIX, including Linux. Oracle was the first database to support the SQL query language, which has since become an industry standard.

ORB (object request broker) *See* CORBA.

order 1. The relative significance of a bit or byte. Usually high order will be the most significant (leftmost) and low order the least (rightmost).
2. The sequence in which arithmetic operations are performed (*see* order of precedence).

order of precedence The order in which ARITHMETIC or LOGIC OPERATIONS are

performed in an expression. Programs treat expressions in parentheses as a unit to be calculated. In a set of nested parentheses, the innermost ones are handled first. For instance, in evaluating the expression
$$((3*(6+(4-2)))/5)+7$$
the program first subtracts 2 from 4, then adds 6 to the result, multiplies this value by 3, divides the result by 5, and finally adds 7. Sets of nested parentheses can always be used in programs to make the expression unambiguous.

When no parentheses are used, arithmetic and logic operations are performed in a particular order. This means that the OPERATORS are applied in a particular order. This order of precedence is not the same for all programming languages. A typical order is as follows (the operator symbols are not included since they vary from one language to another):

unary minus
exponentiation
multiplication   division
addition   subtraction
relationals
NOT
AND
OR

Using this order of precedence, the example given above could be written
$$(3*(6+4-2)/5)+7$$
There are always several levels of precedence, some operations/operators lying on the same level and thus having the same precedence. When an expression is to be evaluated, operations of the highest precedence are performed first, then the next highest, and so on. With the order shown, UNARY MINUS would be done first, i.e. all negative numbers would be set as such before any EXPONENTIATION, multiplication, division, etc., was performed. Operations on the same level are usually performed from left to right but in some languages the order is undefined. Again, in some languages the RELATIONAL OPERATORS (less than, greater than, etc.) come after the logic operators (NOT, AND, OR, etc.); the relationals do not always have the same precedence.

OR gate /or/ A LOGIC GATE whose output is low only when all (two or more) inputs are low; otherwise the output is high. It thus performs the OR OPERATION on its inputs and has the same TRUTH TABLE. The truth table for a gate with two inputs (A, B) is shown, where 0 represents a low signal level and 1 a high level.

A	B	output/outcome
0	0	0
0	1	1
1	0	1
1	1	1

Truth table for OR gate and OR operation

OR operation   A LOGIC OPERATION combining two statements or formulae, A and B, in such a way that the outcome is false only when A and B are both false; otherwise the outcome is true. The TRUTH TABLE is shown in the diagram, where the two truth values are represented by 0 (false) and 1 (true). The operation can be written in several ways, including

A OR B   A ! B   A | B   A + B

A and B are known as the OPERANDS and the symbol between them (OR, !, |, +) is a LOGIC OPERATOR for the OR operation.

In a computer the OR operation is used in HIGH-LEVEL LANGUAGES to combine two logical expressions according to the rules stated above. It is used in LOW-LEVEL LANGUAGES to combine two BYTES or WORDS, producing a result by performing the OR operation on each corresponding pair of bits in turn. If 8-bit words are used, then for example with operands

01010001
11110000

the outcome of the OR operation is

11110001

See also OR gate.

OS /oh-**ess**/   *See* operating system.

OS/2 /oh-ess-**too**/   A multitasking GUI operating system created by Microsoft and IBM in the late 1980s. Originally intended as the successor to MS-DOS, OS/2 suffered

after the market success of Windows prompted Microsoft to withdraw from the partnership. Following a period of rivalry in the early 1990s OS/2 is no longer a major force but is still supported by IBM and is used in specialized areas.

OSI /oh-ess-ȳ/ (**open systems interconnection**)   A standard description for how messages should be transmitted between two points in a telecommunications network. Its purpose is to guide developers to produce products that will work with other products in a consistent fashion. OSI divides telecommunications into seven layers of functions.

outage /**owt**-ij/   The time when a computer system is out of use because of servicing or maintenance.

outline font   A font in which the shape of the character is defined by a set of commands. Page description languages such as PostScript treat everything on a page as a graphic and use outline fonts, which are more versatile at producing different sizes of type with different attributes or special effects. TrueType and OpenType are other kinds of outline fonts.

Outlook (**Microsoft Outlook**)   *Trademark* An E-MAIL and personal-information management application developed by Microsoft. First released with OFFICE 97, it has been upgraded several times and the current version is Outlook 2003.

output   1. The results obtained from a computer system following some processing activity. Computer output may be in a form that people can understand and use, such as words or pictures that are displayed on a screen or printed or drawn on paper. Alternatively, it may be written on, say, a magnetic disk or a CD so that it can be fed back into the computer system – or into another system. *See also* output device.
2. The signal – voltage, current, or power – obtained from an electric circuit.
3. To produce a signal or some other form of information.

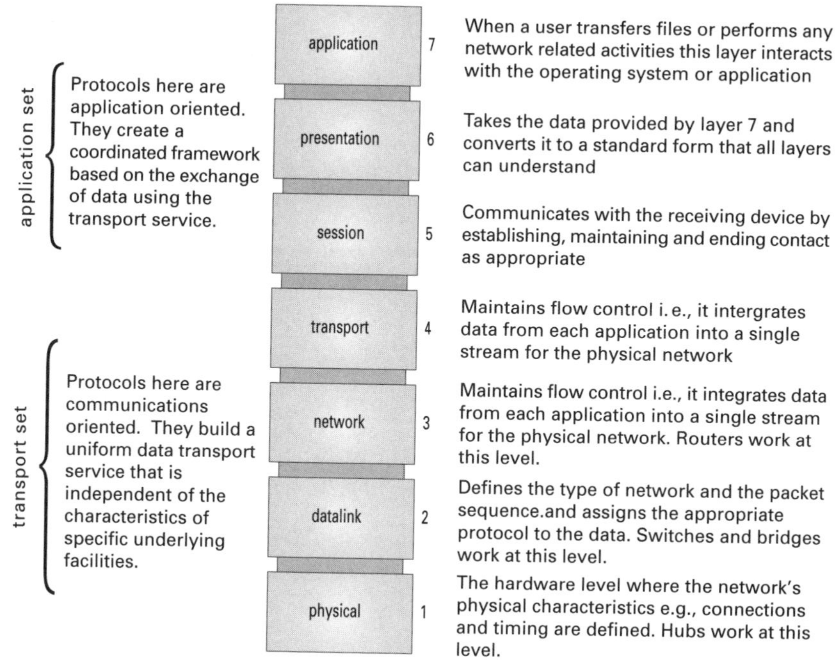

application set
Protocols here are application oriented. They create a coordinated framework based on the exchange of data using the transport service.
transport set
Protocols here are communications oriented. They build a uniform data transport service that is independent of the characteristics of specific underlying facilities.
application
7
When a user transfers files or performs any network related activities this layer interacts with the operating system or application
presentation
6
Takes the data provided by layer 7 and converts it to a standard form that all layers can understand
session
5
Communicates with the receiving device by establishing, maintaining and ending contact as appropriate
transport
4
Maintains flow control i. e., it intergrates data from each application into a single stream for the physical network
network
3
Maintains flow control i.e., it integrates data from each application into a single stream for the physical network. Routers work at this level.
datalink
2
Defines the type of network and the packet sequence.and assigns the appropriate protocol to the data. Switches and bridges work at this level.
physical
1
The hardware level where the network's physical characteristics e.g., connections and timing are defined. Hubs work at this level.

OSI 7-layer model

output device Any device that converts the electrical signals representing information within a computer system into a form that can be understood and used by people. PRINTERS, VDUS, loudspeakers, and PLOTTERS are the most common types of output device. An output device may also convert internal signals into a coded form on, say, magnetic disk that can be fed back into a computer system at a later date. *Compare* input device.

output medium *See* media.

overclocking The process of reconfiguring a computer to operate the CPU at a higher clock speed than specified for. Overclocking carries risks to the CPU including the possibility of permanent damage. Successful overclocking is more likely to be achieved if the system has a well-designed motherboard with a fast enough bus and a suitable cooling fan and an Intel microprocessor. It is illegal for an overclocked system to be sold without informing the buyer.

overflow The condition arising when the result of an arithmetic operation in a computer exceeds the size of the location allocated to it. For example, a number represented by 33 bits would not fit in a 32-bit word. If there were no means to detect overflow, results would be unreliable and might be incorrect. Facilities are therefore provided to detect overflow: a particular bit, known as an *overflow flag* or *bit*, is set to 1 when overflow occurs. The overflow can then be corrected. *See also* underflow.

overlay **1.** Part of a program, i.e. a section of code, that is loaded into MAIN STORE during the execution of that program, overwriting what was previously there. The section of code is thus not permanently resident in main store but is held in BACKING STORE until required by the processor.

The parts of the program to be used as overlays, and the other parts that these are allowed to overwrite, are determined by instructions to the LINK EDITOR at link time. This has to be done with care: if an overlay tries to use a procedure in a part of the code it has just overwritten, then the program will fail.

In general several overlays are loaded into the same area of main store at different stages of the program, each erasing the code already there but no longer needed. This process of repeatedly using the same area of store during program execution overcomes space limitations in main store. The technique, which is called *overlaying*, is used mainly in systems that do not have VIRTUAL STORAGE – principally some older mainframes.

2. A piece of card or plastic placed over or around a KEYBOARD to indicate what special functions have been assigned to particular keys.

overrun The loss of data incurred when data arrives at a port faster than it can be processed.

overtype (**overwrite**) In text entry, the mode in which characters being typed replace those under or to the right of the insertion point.

overwrite 1. To destroy the contents of a storage location by writing a new item of data to that location.
2. To write a new version of a file on top of the existing version.
3. *See* overtype.

package   *See* application package.

packet   *See* packet switching.

packet assembler/disassembler (PAD) The device that builds packets of information in order to transmit them across a PACKET-SWITCHING network, and breaks up incoming packets.

packet Internet groper   *See* ping.

packet switching   A process whereby information (in digital form) can be sent automatically from one computer on a NETWORK to any other, the information being switched from one communication line to another until it reaches its specified destination. The information is transmitted in units known as *packets*. A packet is a group of bits of fixed maximum length. If the information to be communicated exceeds the packet size it is split up into a

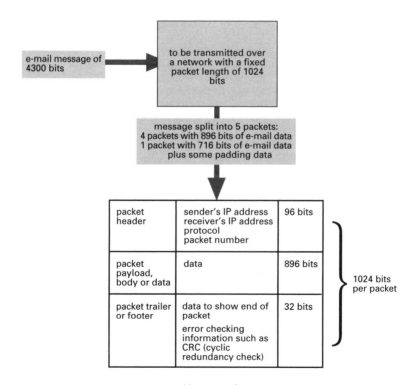

e-mail message of 4300 bits

to be transmitted over a network with a fixed packet length of 1024 bits

message split into 5 packets:
4 packets with 896 bits of e-mail data
1 packet with 716 bits of e-mail data
plus some padding data

packet header	sender's IP address receiver's IP address protocol packet number	96 bits	
packet payload, body or data	data	896 bits	1024 bits per packet
packet trailer or footer	data to show end of packet error checking information such as CRC (cyclic redundancy check)	32 bits	

Message packet

number of packets. Each packet is switched as a composite whole from one communication line to another. At each switching point along its route the packet is checked for errors, stored temporarily, and forwarded to the next point when the necessary resources become available. The packets of a partitioned message, document, etc., are reassembled when they reach their final destination (*see* packet assembler/disassembler). *Compare* message switching.

packing The storage of data in a compact form in order to reduce the amount of storage space required. The term is used as a synonym for the simpler methods of COMPRESSION – i.e. those that remove obvious redundancies in the data without using a sophisticated encoding algorithm. For example, several BYTES may be stored in one WORD, or multiple occurrences of a character or word may be replaced by a single instance of the character or word and an indication of the number of times it occurs. The original data can be recovered from its packed form by *unpacking* it.

packing density 1. The number of components, or LOGIC GATES, per unit area of an INTEGRATED CIRCUIT.
2. The amount of information in a given dimension of a storage medium, for example the number of tracks per inch of a magnetic disk.

PAD /pad, pee-ay-**dee**/ *See* packet assembler/disassembler.

padding The adding of blanks to extend a string or record to a set length.

page The unit of interchange between main store and a backing-store device used for SWAPPING. A page contains instructions or data, or both, and generally the number of words or bytes is fixed for a given computer system. *See also* virtual storage.

page break The position in a document at which text will move to the top of a new page. Word-processing systems will usually paginate documents automatically ac-cording to rules set by the user. However a 'hard' or 'manual' break can be placed in a document to force the following text to be printed or displayed on a new page.

page description language (PDL) A type of programming language in which commands are used to describe the layout of a complete page of text and graphics, including font styles and sizes and picture shapes and positions. POSTSCRIPT is a widely used PDL.

page orientation The position of a page according to whether the top of the page is along the short edge of the paper (*portrait*) or along the long edge of the paper (*landscape*).

page printer A PRINTER that produces a complete page at a time, an example being the LASER PRINTER.

paging The process of managing VIRTUAL MEMORY by dividing the space into fixed-size pages, which can then be mapped onto an available physical address. Address translation from physical to virtual is done by memory management hardware.

paint To infill a part of a drawing on the screen with color or a pattern.

paint program A simple drawing program that allows freehand drawing and also provides tools to facilitate the drawing of lines and curves. Paint programs produce raster graphics files.

palette 1. The tools provided in a graphics program, which enable different patterns, colors, brush shapes, and line widths to be selected for use.
2. The set of different colors available in a program.

palmtop /**pahm**-top/ A hand-sized computer. Often palmtops have no keyboard but the screen serves as both an input and output device. A stylus is used to enter information; text can be entered by manually writing on the screen with the stylus. A palmtop is generally used for simple applications such as personal organization and note taking. *See also* PDA.

PANTONE /**pan**-tohn/ *Trademark* A system of coding colors used in commercial color printing.

paperless office The (mistaken) idea that the introduction of computers would mean that paper would no longer be necessary in offices.

paper tape An obsolete data medium consisting of a continuous strip of paper of uniform width and thickness on which data could be encoded by punching patterns of holes; materials of greater strength were also used, such as laminates of paper and polyester. All these forms can be referred to as *punched tape*. Punched tape was once widely used for the input, output, and storage of data, especially in scientific and engineering applications, and in fact was used for data communications (telex) before its use in computing.

paper throw A rapid and continuous movement of the paper in a CHARACTER PRINTER or LINE PRINTER through a distance of several lines without any printing. A paper throw may be used, for example, to print the next line on a new page – described as *throwing a page*.

parallel port An interface for PARALLEL TRANSMISSION.

parallel processing Strictly, the execution of two or more PROCESSES at the same time in a single computer system. This implies that two or more PROCESSORS are in operation at the same time. The term is often loosely applied to MULTITASKING, a situation in which a number of processes are potentially active, but only one is actually being run at any particular instant. *See also* multiprocessing system.

parallel transmission A method of sending data between devices, typically a computer and its peripherals, in which all the BITS associated with a unit of data (e.g.

a character) are transmitted at the same time along different paths. For example, if two devices wish to communicate a 16-bit unit of data, the sending device would transmit all the 16 bits at the same time, one bit per wire. More than 16 wires are required, however, since control signals are sent simultaneously, again one bit per wire, indicating that data is available (*see* handshake). The receiving device would then signal that the data had been received. *See also* serial transmission.

parameter /pă-**ram**-ĕ-ter/ **1.** A quantity whose value is selected according to the application or the circumstances.
2. Information passed to a PROCEDURE, SUBROUTINE, or FUNCTION. Different parameters may be provided for each CALL of the procedure, subroutine, or function. The *actual parameters* (or *arguments*) are those supplied at the time of a particular call. The mechanism by which the parameters are supplied is known as *parameter passing*. Normally this involves either passing the value of the actual parameter or passing the address of the location in which the actual parameter is stored.

parameter passing *See* parameter.

parameter RAM *See* PRAM.

parity bit *See* parity check.

parity check A simple CHECK used to detect ERRORS in items of information in binary form. The item may represent a piece of data or part of a program, and may be a single CHARACTER, BYTE, or WORD. An extra BIT, called the *parity bit*, is added to the leftmost end of the item. There are two kinds of parity: *even parity* and *odd parity*. In the former, the parity bit is used to make the total number of 1s an even number; in the latter, the parity bit is used to make the total number of 1s an odd number. For example, the 7-bit ASCII code for W is 1010111. The number of 1s is 5; the parity bit for even parity is thus a 1, making the number of 1s up to 6 (an even number) and the 8-bit ASCII code with parity 11010111.

The parity bit is recomputed after the augmented item – original item plus parity bit – has been transferred or otherwise manipulated. The new value is compared with the received value, and most simple bit errors (certainly all single bit errors) will be detected.

The same principle can be used to check larger blocks of data. In most RAID disk arrays, for example, the equivalent bits on all but one of the disks generate a parity bit on the last disk.

park To lock the read/write heads in a secure position (over an unused track or beyond the surface of the disk) before shutting down the drive, so that data cannot be corrupted when the computer is physically moved. On all modern drives parking is done automatically but it can also be achieved by using a disk utility. The disk will unpark itself automatically when the power is turned back on.

parsing, parser /**par**-zing, **par**-zer/ *See* syntax analysis.

partition A logical division of a hard disk. It enables the user to have different operating systems on the same drive (in different partitions) and can also create the appearance of having separate drives, which can make it easier to organize files and provide access for multiple users.

Pascal /**pass**-kăl/ A PROGRAMMING LANGUAGE named after the French philosopher and mathematician Blaise Pascal and developed by Niklaus Wirth in the late 1960s. Pascal was intended to be used for teaching students the principles and practice of programming, and is a relatively small concise HIGH-LEVEL LANGUAGE with what were at the time a number of novel features. Pascal has certain drawbacks when it comes to large-scale programming but is nevertheless popular and has been implemented on all sizes of computer.

passive matrix display An inexpensive type of low-resolution LCD. *See also* active matrix display.

password A means of verifying the identity of someone wishing to gain access to a computer system or to certain parts of a system. A password is a particular sequence of characters (letters, numbers, etc.) that is allocated to or chosen by an authorized user and is unique to that person. A potential user must feed his or her password into the computer to establish identity. The password entered is checked against the stored version of the password. Only if the two sequences agree will the person be granted access.

Passwords form part of the LOG ON sequence in most modern computer systems. They are used, for example, in systems that hold sensitive data (i.e. for security reasons) or when charges are made for use of system resources. Passwords can also be granted to programs in order to control access to data.

Methods commonly used to make a password system more effective are:
1. the passwords are stored on the computer in code so that unauthorized people looking at the password file cannot read the passwords;
2. the password does not appear on the screen when it is typed in, thus making it harder to discover users' passwords by watching them log on;
3. users are urged, and in some cases forced by the system, to change their passwords frequently.
4. users are urged, and in some cases forced, to use passwords that are not simple words or predictable numbers (e.g. birthdays) but a mixture of letters, numbers, and other characters, sometimes intermingled in a random fashion.
5. the password must be of a minimum length, often 6 or 7 characters.

paste To insert information that has been cut or copied from a document into a different place in that document or into another document.

patch A small section of code, often provided by a software supplier, that allows a user to correct or modify a program. It is used for purposes of convenience and speed of change. The patch may be written in MACHINE CODE and introduced into the compiled (or assembled) version of a program, i.e. the OBJECT PROGRAM.

path 1. A link between two nodes in a network.
2. The route followed by an operating system through a disk's directories and subdirectories to a file.

pattern recognition The ability of a computer to recognize visual images or sound patterns that are represented as an array of numbers.

pause instruction An instruction that specifies that the execution of a program is to be suspended. After a pause, a program may be continued from where it left off or was abandoned. For example, a program might pause to allow the screen to be read, or so that the paper may be changed on a printer.

pause key A keyboard key positioned to the right of the function keys. When pressed it pauses the execution of a program.

payload The part of a VIRUS or WORM that is intended to damage the infected system, as opposed to the part concerned with spreading the program to other machines.

PC *See* personal computer.

PCB (pcb) *See* printed circuit board.

PC-compatible *See* IBM-compatible.

PC-DOS /pee-see-**doss**/ *Trademark* An OPERATING SYSTEM produced by Microsoft Corp. for IBM and intended for use on the IBM Personal Computer. (Its official name in IBM documentation was 'DOS'.) It was essentially the same as MS-DOS, but made assumptions about the underlying hardware that were only fully justified on IBM PCs. IBM took over development in the 1990s and PC-DOS diverged from MS-DOS. However, its use declined as WINDOWS became the dominant PC operating system.

PCI (**peripheral component interconnect**) A local bus developed by Intel Corporation and some other large companies to alleviate problems caused by the slow speed of ISA. Introduced in the mid-1990s, it has continued to develop and is still the standard expansion bus for PCs.

PCL (**printer control language**) *Trademark* A control language developed by Hewlett Packard for laser printers.

PDA (**personal digital assistant**) Any small portable hand-held device, possibly a PALMTOP computer, that provides a computing capability covering information input, retrieval, and storage for business, college, or personal use. A PDA is often used as a personal organizer with calendars, diaries, and address books. Increasingly PDAs are combining with cellphones and paging systems. Data can be transferred between a PDA and a personal computer.

PDF (**portable document file**) *See* Acrobat.

PDL *See* page description language.

peer-to-peer Describing the linking of computers together in a NETWORK without a dedicated SERVER. Each individual computer acts as a server to other computers (its peers) on the network and is also a client to all its peers acting as servers. Resources can be shared and messages passed between users. All versions of Windows since Windows 95 have offered limited peer-to-peer networking.

PEM *See* privacy-enhanced mail.

pen computer Any computer whose primary input device is a stylus ('pen'). *See* palmtop.

pen plotter *See* plotter.

Pentium /**pen**-tee-ŭm/ *Trademark* A family of processors first introduced in 1993 by Intel. These had speeds up to 133 MHz. The original Pentium has been succeeded by the Pentium Pro, the Pentium II, the Pentium 3, and the current version, the Pentium 4. Speeds of 3 GHz are now standard. Each model has, in its turn, been the standard microprocessor for PCs, although other companies have brought out competitive clones. Since the Pentium II each model has been marketed in three versions: the Pentium, aimed at desktop and notebook PCs; the *Xeon*, a high-end version aimed at servers and high-specification workstations; and the *Celeron*, a reduced specification aimed at the bottom end of the home and small business market.

peripheral component interconnect /pĕ-**riff**-er-ăl/ *See* PCI.

peripheral device (**peripheral**) Any device that can be connected to and controlled by a computer. It is external to the CENTRAL PROCESSOR and MAIN STORE of the computer. Peripherals may be INPUT DEVICES, OUTPUT DEVICES, or BACKING STORE.

Perl /perl/ (**practical extraction and report language**) A script programming language with a syntax similar to C and incorporating some popular UNIX features. It has good text-handling features and can also deal with binary files, making it suitable for developing common gateway interface software.

personal computer (**PC**) A general-purpose computer designed for operation and use by one person at a time. Such a device became a reality when MICROCOMPUTERS became available. All personal computers are microcomputers, but microcomputers are not necessarily personal computers. Personal computers range from cheap domestic machines to expensive and highly sophisticated systems. Often the abbreviation 'PC' is used in a more restricted sense to mean a computer that will run Windows, as opposed to an Apple Macintosh computer.

personal digital assistant *See* PDA.

personal identification device (**PID**) A device issued to an authorized user of a computer system and containing a machine-readable sequence of characters that identifies that person. It may be a card or badge. The PID must be inserted into a terminal of the computer system before access to the system is granted. In many cases a person must use a PID in conjunction with a PIN (personal identification number) to gain access.

personal identification number (**PIN**) A number allocated to an authorized user of a computer system that is unique to that person, and which he or she must feed into the computer system in order to establish identity. A PIN is a form of PASSWORD. It is often used in conjunction with a PID (PERSONAL IDENTIFICATION DEVICE). PINs and PIDs are used, for example, at cash dispensers.

personal information manager *See* PIM.

peta- /**pet**-a/ A prefix indicating a multiple of a thousand million million (i.e. 10^{15}) or, loosely, a multiple of 2^{50} (i.e. 1 125 899 906 842 624). In science and technology decimal notation is usually used, and powers of 10 are thus encountered - 10^3, 10^6, 10^9, etc. The symbol P is used for peta-, as in PV for petavolt. Binary notation is generally used in computing, and so the power of 2 nearest to 10^{15} has assumed the meaning of peta-.

The prefix is most frequently encountered in computing in the context of storage CAPACITY. With magnetic disks, magnetic tape, and main store, the capacity is normally reckoned in terms of the number of BYTES that can be stored, and the *petabyte* is used to mean 2^{50} bytes; petabyte is usually abbreviated to P byte, Pb, or just to P.

Peta- is part of a sequence
   kilo-, mega-, giga-, tera-, peta-, ...
of increasing powers of 10^3 (or of 2^{10}).

phishing An attempt to obtain personal information from Internet users through fake Web sites and fraudulent e-mails. A common tactic is to send an e-mail allegedly from a bank that links to a Web site where users are asked to 'confirm' such details as their online-banking username and password, credit-card number and expiry date, etc. These are then used to steal money from the user's account or run up bills on the credit card. Alternatively the user can be asked for personal information, which is then used to set up bank accounts, obtain credit, etc. Such frauds can vary from the crude to the highly sophisticated: the fake Web site, for example, is often identical in look and feel to the bank's genuine one. The term 'phishing' derives from the analogy with fishing: while many people will ignore the bait (the e-mail), some will bite and be reeled in. To *phish* is to engage in phishing, and the people who do so are sometimes known as *phishers*.

phono connector /**foh**-noh/   A connector used to attach a device such as a microphone or pair of headphones to a computer peripheral with audio capability.

PhotoCD /foh-toh-see-**dee**/   A Kodak system that can store 35-mm film pictures, negatives, slides, and scanned images as digitized images on CD. The images can usually be viewed on a computer with a CD drive and the appropriate software or by using a special player.

photo illustration   Computer art that starts with a digitized photograph that is subsequently manipulated by image-enhancement software to apply special effects.

PHP (**Hypertext Preprocessor**)   An OPEN SOURCE SCRIPTING LANGUAGE widely used for SERVER-SIDE SCRIPTING. First released in 1995, its syntax is closely based on C, JAVA, and PERL.

physical record   *See* block; record.

pico- /**pÿ**-koh-/   A prefix to a unit, indicating a submultiple of one million millionth (i.e. 10^{-12}) of that unit, as in *picosecond*. The symbol p is used for pico-; picosecond can thus be written ps. Pico- is part of a sequence

milli-, micro-, nano-, pico-, ...
of decreasing powers of 1000.

PICS /piks, pee-ÿ-see-**ess**/ (**platform for Internet content selection**)   A specification of the format conventions used by Internet rating systems. This was developed to enable parents, teachers, and administrators to limit the type of material accessed on the Internet, especially that available to children. It now also includes the protection of privacy.

PICT /pikt, pee-ÿ-see-**tee**/   A Macintosh file format for encoding graphical images, which can hold both object-oriented images and raster graphics images. Although now superseded by more sophisticated formats, it is still widely used.

picture processing   *See* image processing.

PID   *See* personal identification device.

Pilot (**programmed inquiry, learning or teaching**)   An early programming language that was used largely in CAL.

PIM (**personal information manager**)   Software that organizes personal information, such as addresses, diary notes, and appointments, in a useful manner.

PIN /pin/   *See* personal identification number.

ping /ping/ (**packet Internet groper**)   **1.** A basic Internet program that verifies that a particular Internet address exists and can accept requests. It sends a special signal message to the address that the host echoes back. Ping informs the user not only whether the host is available but also how long the signal took to return.
2. The process of sending a message to all the members of a mailing list requesting an acknowledgment (ACK). It is used in e-mail systems to test which users are current.

pin header   A device similar in form to a DIP but containing no circuitry. Instead each leg is extended vertically through the package allowing pins to be connected together in any configuration by soldering small pieces of wire across the relevant pins. A pin header is more flexible but clumsier than a DIL SWITCH.

pipe   *See* pipeline processing.

pipeline processing (pipelining)   1. (piping) A technique that allows two or more PROCESSES that would otherwise run sequentially to execute in parallel. Without pipeline processing the output of one process would be written to a temporary disk file that would form the input for the next process. Instead, the operating system provides an area of shared memory that emulates a file, called a *pipe*, to which the first process writes its output and which is read by the next process as it is filled. The processes can thus be executed in parallel and without the overhead of writing and reading from disk, which results in a lower overall execution time.
2. A technique used to increase the throughput of a CPU by beginning the execution of the next operation(s) before the execution of the current one is complete. For example, all MACHINE INSTRUCTIONS must be loaded into the CPU from main store before they can be executed, and so time can be saved by loading the next instruction while the current one is being executed. Further optimizations are possible because many operations specified by a single machine instruction in fact break down into several sub-operations (*see* microprogram). The technique does not always work, for example when the outcome of an instruction affects an OPERAND of the next one, or when a conditional JUMP may or may not transfer control to a different part of the program. However, it is successful frequently enough to bring significant performance gains. Formerly a feature only of SUPERCOMPUTERS, pipeline processing is now standard in modern microprocessors.

pipelining   *See* pipeline processing.

piping   *See* pipeline processing.

piracy   Illegal copying, distribution, or use, usually of software. Types of piracy include 'softlifting' (users sharing software), hard-disk loading (the installation of unauthorized copies of software on computers for sale), counterfeiting (making counterfeit copies), and on-line piracy.

pixel /**piks**-el/   In computer graphics, the fundamental graphical element of a VDU or a raster graphics image. In general it is an element in a large ARRAY that is holding pictorial information. It contains data representing the brightness, the color, or some other property of a small region of an image. The RESOLUTION of a VDU is measured in terms of how many pixels there are across and down the screen. The resolution of an image is measured in pixels per inch. The word is an abbreviated form of 'picture element'.

PKZIP /pee-kay-**zip**/   One of the most widely used shareware compression programs. PK stands for Philip Katz, the author of the program. PKZIP compresses one or more files into a single file, called a *zip file*, with the extension zip. The file can be decompressed by PKUNZIP. The zip file format is so widely used that transparent support for it is built into Windows XP and it is used as a standard means of distributing JAVA programs.

plaintext /**playn**-tekst/   Ordinary readable text that has not been encrypted.

plain text   Text with no formatting or other special information. A *plain text file* or *text file* is a file that contains such text. Plain text has a very simple structure, being a series of CHARACTERS, and can generally be read without special software. It is easily transferred between different operating systems. Many applications contain options to save data to or load it from plain text files. Data saved in this way will lose any special information the application associates with it; for example spreadsheets saved as plain text will contain the numbers or words in the cells but will lose the formulae that calculate the values of some cells. Despite these limitations such files can be useful in transferring data between otherwise incompatible applications.

plasma display (plasma panel display) *See* display.

platen /**plat**-ĕn/ The cylinder around which the paper is wrapped in most impact printers.

platform 1. The underlying hardware or software system on which application programs can run. Thus one speaks, for example, of a Windows platform as opposed to a Macintosh platform.
2. Any base technology on which other technologies are built. In a computer, which is a layered device, the layer at the bottom is often referred to as a platform. Once a platform has been defined developers can produce appropriate software.

platform for Internet content selection *See* PICS.

platform-independent Describing software that is independent of operating system and hardware.

PL/I A high-level PROGRAMMING LANGUAGE whose name is derived from Programming Language I (one). It was developed by IBM in the late 1960s to combine what were considered the best features of FORTRAN, ALGOL, and COBOL, and was intended to replace all three. PL/I is a very large comprehensive language, and although it was used fairly widely in IBM installations it was never taken up to any large extent by other manufacturers.

plotter An OUTPUT DEVICE that converts coded information from a computer into graphs or pictures drawn on paper or transparent film by means of one or more pens. There are many types of plotter, varying in the accuracy, quality, and speed of production of the graphs or pictures, the size produced, and the use (if any) of color.

Two basic designs of plotter are the *flatbed plotter* and the *drum plotter* (see illustration overleaf). In both cases the pen is mounted on a bar that spans the width of the paper. The pen can be moved to precise positions along the bar. In the flatbed plotter the bar can be moved on tracks running up and down the length of the sheet of paper or film. The pen can therefore be moved to any specified point on the surface of the paper or film to make a mark. It can either touch the surface as it moves so that it makes a line, or it can be lifted off the surface as it moves.

In the drum plotter the bar is fixed in position and it is the paper that is moved. The bar is parallel to the axis of a drum. The paper is wrapped around part of the surface of the drum; the holes down each side of the paper engage with pegs on the drum. Rotating movements of the drum cause the paper to move, and different positions down the length of the paper can be brought precisely under the pen. As in the flatbed plotter the pen can be moved along the bar to the required position across the width of the paper, and it can be lifted off the surface of the paper if necessary. The drum plotter is used when large numbers of drawings are required; these can be plotted in sequence down the paper and then split up. Several pens can be mounted on the bar of a flatbed or drum plotter to provide a selection of colors, which can be changed by computer control.

To control the movement of the pen of a plotter, the plotter is sent a stream of commands from the computer. These tell it where to move the pen to and whether the pen is to be up or down during the movement. Most plotters also have built-in *hardware characters*, i.e. the plotter will draw a character at the current pen position without being given the detailed movement required to construct it. Some plotters can also draw arcs, circles, and ellipses given only the start and end positions, radii, etc.

In recent years plotters have adopted ink-jet technology, and indeed large ink-jet printers have superseded plotters in many situations.

plug *See* connector.

plug and play The ability of an operating system to recognize new hardware that has been attached to computer and to configure the system automatically to use the hardware. If a suitable device driver cannot

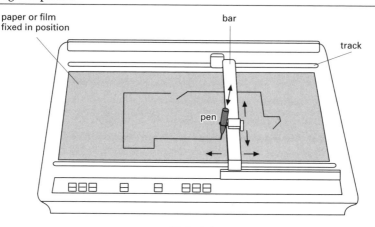

paper or film fixed in position

bar

track

pen

Flatbed plotter

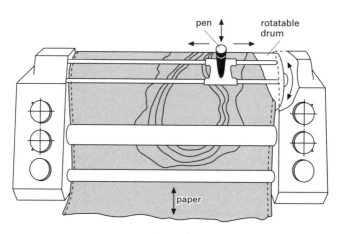

pen

rotatable drum

paper

Drum plotter

be found the user will be prompted to install one.

plug compatible   *See* compatibility.

plug-in   A small program that can be installed as part of other software to increase the functionality. It is often used to refer to such programs that can be used as part of a Web browser, allowing animation, video, and audio files embedded in HTML documents to be recognized.

PNG (portable network graphics)   A file format for compressed raster graphics files that may eventually replace the widely used GIF format. The PNG format was developed to be patent-free by an Internet committee and does not have the legal restrictions then associated with GIF. PNG uses lossless compression and a PNG file can be 10–30% more compressed than the equivalent GIF file.

podcasting   The packaging of audio material for download over the Internet and playback on a computer or MP3 PLAYER. This technique was first widely used with music tracks but has been extended to other types of audio content. An item made available in this way is called a *podcast*, the act of doing so is to *podcast*, and a person or organization that does so is a *podcaster*. Many radio broadcasters are making their programs available as podcasts and are expecting podcasting to become an important distribution method. However, podcasting is not confined to established media companies: anybody with appropriate hardware and software – both of which are cheap and widely available – can create an MP3 file, post it on the Web, and advertise its content through a WEB FEED. The term 'podcasting' was first used in 2005 and is a blend of 'iPod' and 'broadcasting' – a tribute to the IPOD's iconic status as the archetypal MP3 player, although neither an iPod nor any other MP3 player is essential to listen to podcasts. It is expected that podcasts will soon encompass video as well as audio content.

point **1.** In typography, a unit equal to just under 1/72 of an inch that is used to measure the height of a character and the distance between lines of text (leading). In computer DTP software, fonts, and page description languages, a point is usually regarded as equalling exactly 1/72 of an inch. **2.** A single pixel on a screen, which is identified by its row and column numbers.

point-and-click To select data or an object on a screen by using a mouse (or similar device) to move the cursor to the object (point) and then pressing a button (click) on the mouse to select the object. The term applies to any pointing device.

pointer (**link**) A character or group of characters that indicates the LOCATION of an item of data in memory.

pointing device Any device that identifies a point on a VDU and transmits its location to a computer. Examples are CURSOR KEYS, a MOUSE, JOYSTICK, TRACKER-BALL, TOUCH-SENSITIVE DEVICE, or DIGITIZING PAD.

point of presence (**POP**) A point in a WAN to which a user can connect with, for example, a local telephone call. For example, an ISP has a point of presence on the Internet.

point of sale (**POS**) The place where goods are paid for in a store, etc., often with a POINT-OF-SALE TERMINAL linked to a computerized transaction system.

point-of-sale terminal (**POS terminal**) A specialized cash register, credit-card recording system, or ticket dispenser that records the details relating to the sale of goods, generally using a scanner to read bar codes or tags, and feeds the information into a central computer. This system improves stock, cash, and credit control.

point-to-point protocol (**PPP**) A protocol used by machines that support serial interfaces, such as modems and ADSL transceivers. PPP is commonly used to provide and access dial-up services to Internet service providers. PPP can share a line with other users, and unlike SLIP it can handle asynchronous as well as synchronous communication and has error detection included.

policy In policy-based networking, a set of definitions showing how the network's resources are to be allocated among its users. Parameters for allocation, which can be static or dynamic, include items such as time of day, client priority, and resource availability. Policies are created by network managers and used by network-management software to make decisions. Policies can also be used to control how much freedom networked users are permitted in other areas, for example to what extent they may customize their DESKTOP.

polling A technique whereby a computer checks each of a number of input devices in rotation to see if any data is waiting to be read. Polling depends for its success on each device being checked at least as frequently as the data is arriving; in addition the time spent processing any data that has arrived must not be too long so that data on other devices is missed. Polling can be contrasted with the use of INTER-RUPTS, where it is the devices that inform the processor when they have data available. In some systems an interrupt will alert the processor to the fact that one of a number of devices has data available, and the processor then polls the devices to find out which one it is.

pop *See* stack.

POP /pop/ *See* point of presence.

POP3 /pop-**three**/ (**Post Office Protocol 3**) The version of the Post Office Protocol currently in use on TCP/IP networks. Most e-mail applications use this protocol, although some use *IMAP* (Internet message access protocol).

pop-up menu In a graphical user interface, a menu that appears on screen when a user performs a certain action and usually disappears when the user selects one of the menu items.

port 1. A point at which a connection can be made between an input/output device and the central processor of a computer system, allowing data to be passed. On personal computers ports usually have particular types of socket for connecting the mouse, keyboard, printer, etc.
2. A point through which data can enter or leave a computer NETWORK. It forms part of a NODE.
3. To move a program, usually written in a PORTABLE form, to another computer system.

portable Denoting programs that can be readily transferred from one computer to other computers. For maximum portability a program should be written in a HIGH-LEVEL LANGUAGE possessing an internationally agreed standard, and any extensions or deviations permitted by a particular COM-PILER should be avoided. Where it is absolutely necessary to make some reference to particular hardware or to include system-dependent features, then these should be concentrated in a few well-documented routines and not scattered throughout the

program. Portable programs are different from PLATFORM-INDEPENDENT programs in that the compiled versions of the latter will run on any supported platform unchanged, while a portable program will at least be recompiled for the new platform. *See also* configure.

portable computer A small computer that is designed for ease of transportation. LAPTOPS, NOTEBOOKS, and PALMTOPS are all types of portable computer.

portable document file (PDF) *See* Acrobat.

portable network graphics *See* PNG.

portal A gateway to the Internet provided by a Web site. There are general and specialized portals. General portals include Yahoo!, MSN, Netscape, Excite, and AOL.

portrait The arrangement of printed matter on a page so that the top of the page is on the shorter side. *Compare* landscape.

POS /poss, pee-oh-**ess**/ *See* point of sale.

positional notation *See* number system.

positive acknowledgment *See* acknowledgment.

POSIX /**poz**-iks/ (**portable operating system interface for UNIX**) A defined set of operating system interfaces, agreed by the IEEE, based on the UNIX operating system.

post To publish a message in an on-line forum or newsgroup.

POST /pohst/ (**power-on self-test**) The first task performed by the BIOS when a PC is switched on. It checks that all the basic hardware components, such as the keyboard and display, are running and that the CPU and memory are functioning properly. If an error is found a message will be displayed or if the monitor is faulty a series of beeps will be heard. After a successful POST, control is passed to the system's bootstrap.

Post Office Protocol 3   *See* POP3.

PostScript   A page description language introduced by Adobe in 1985. PostScript has a series of instructions for drawing graphics and placing text on a page. The FONTS used (known as *type 1 fonts*) are defined as outline fonts and can be embedded in the PostScript file. PostScript is an interpreted language; the interpreter, known as a *RIP* (raster image processor), is present in all PostScript printers or imagesetters. Most applications, especially word-processing or desktop-publishing applications, are able to generate PostScript output. *See also* Acrobat.

POTS (plain old telephone service)   The basic public switched telephone network with no additions. The main differences between POTS and non-POTS services, such as ISDN and ADSL, are bandwidth and speed.

power-fail recovery   *See* recovery.

Power Macintosh   A Macintosh computer based on the PowerPC chip. The 6100/60, 7100/66, and 8100/70 were available in 1994. The latest generation, the G5, has been released in various versions since 2003. *See also* iMac.

power on   To switch on a piece of computing equipment.

power-on self-test   *See* POST.

PowerPC   A RISC-based microprocessor open architecture developed jointly by IBM, Apple, and Motorola. The name is derived from Performance Optimization with Enhanced RISC. PowerPC offers an alternative to Intel architectures and is used primarily in some IBM machines and in Apple Macintosh computers.

PowerPoint   (Microsoft PowerPoint) *Trademark* An application used to create and run visual presentations. Developed by Forethought, Inc., it was first released for the Macintosh in 1987, then purchased by Microsoft and released for Windows in 1988. It has been upgraded several times and the current version is PowerPoint 2003. PowerPoint is very widely used for business and other presentations and forms part of Microsoft OFFICE.

PPP   *See* point-to-point.

PRAM /pee-ram/ **(parameter RAM)**   A battery-operated form of RAM used in some Macintosh computers in which information such as date, time, and computer configuration are stored. Occasionally PRAM gets corrupted and needs to be restored.

precedence   *See* order of precedence.

precision   The degree of exactness to which a numerical value is stated, usually based on the number of digits to which the number is represented. The value of some quantity given to 6 digits thus has 6-digit precision. For example, π to 6-digit precision and to 8-digit precision is respectively

$$3.141\ 59 \text{ and } 3.141\ 5927$$

In a computer each item of information must fit into a WORD, which contains a fixed number of bits. This sets a limit on the precision to which a number can be represented.

Many programming languages offer *extended precision* (for FLOATING-POINT NUMBERS). More powerful computers will perform extended-precision floating-point arithmetic in hardware. In the simplest case two words can be operated as a single *double-length word*; this doubles the number of bits available to represent a number, and leads to what is known as *double precision*. (The use of only one word is then called *single precision*.) The increased number of bits is normally added to the mantissa section of the word, leaving the exponent section unchanged in length.

preemptive multitasking   *See* multitasking.

presentation graphics The use of such techniques as bar or line charts and pie graphs to present information. Presentation-graphics applications usually produce high-quality output used principally for presentations on projectors for lectures, meetings, etc.

presentation layer The sixth of the seven layers in the OSI communication reference model, which ensures that the communications passing through are in the appropriate form for the recipient.

preventative maintenance MAINTENANCE that is performed on a regular basis and is intended to prevent failures or to detect failures that are just about to happen. When the performance of, say, an electric circuit is suspect, the circuit can be checked by deliberately trying to induce a failure. *Compare* corrective maintenance.

preview To display a document in its printed form on the screen before sending it to the printer.

primary color One of a set of three or four colors that can be mixed to produced many other colors. There are two sets of primary colors: the RGB SYSTEM - red, green, and blue - used in such video equipment as televisions and computer monitors; and the CMYK SYSTEM - cyan, magenta, yellow, and black - used for color printing.

primary store *See* main store.

print To output information from a computer to a printer. Application programs can also output printed information to a file for printing at a later date, rather than directly to the printer. This is known as *printing to file.*

PRINT *See* Basic.

printed circuit board (PCB; pcb) A thin rigid board of insulating material, usually fiberglass, with electronic components and interconnections on one or both sides. A specific pattern of metal strips is chemically formed on the board in accordance with the circuit design; the metal strips are conducting tracks. The components, usually INTEGRATED CIRCUITS, are then soldered into position between the appropriate conducting tracks to complete the circuit. Double-sided PCBs are commonly produced; these have components and some of the conducting tracks on one side of the board with additional tracks on the other, interconnections being made through the board.

A PCB connects via an appropriate socket to the internal wiring of, say, a computer system. Smaller modular PCBs may be connected to a PCB to augment its function. *See also* expansion card; edge connector; motherboard.

printer An OUTPUT DEVICE that converts the coded information from a computer into a readable form printed on paper. There are many types of printer, varying in the method and speed of printing and the quality of the print.

Some printers print a single character at a time and are thus often referred to as CHARACTER PRINTERS; the DOT MATRIX PRINTER is an example. There are also LINE PRINTERS, which produce a complete line of characters at a time. Page printers, such as the LASER PRINTER, produce a complete page at a time.

Printers can produce their printed characters either by mechanical impact or by nonimpact; in some devices different colors can be printed. With *impact printers* a character is produced by a hammer action causing an ink ribbon to press against the paper. In some cases the printed characters have a solid form, being created by an embossed character struck against the ribbon; the band printer and barrel printer are examples of this type of impact printer. Alternatively, characters may be built up as a pattern of closely spaced (or possibly overlapping) dots, each dot produced by the impact of a thin rod. In *nonimpact printers* characters are produced without the use of mechanical impact; examples include INKJET, THERMAL, and laser printers. The characters are usually composed of fine dots.

Printers – either impact or nonimpact – that produce characters as a pattern of dots

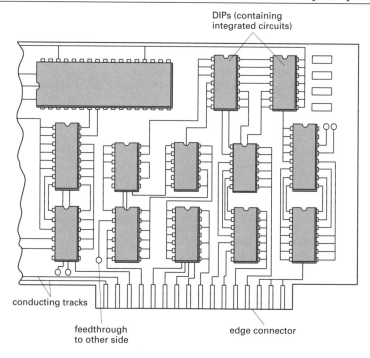

DIPs (containing
integrated circuits)

conducting tracks

feedthrough
to other side

edge connector

Part of a printed circuit board

are known as *matrix printers*. Matrix printers can produce a large selection of shapes and styles of letters and digits, and may also print Arabic characters or the ideograms of oriental languages. In addition to characters it is possible to produce diagrams, graphs, etc.

printer control language   *See* PCL.

printer format   *See* format.

printer port   A port to which a printer can be connected. On PCs this is usually a parallel port and is given the logical name LPT by the operating system. Serial ports can also be used to connect printers (logical device name COM), but they often require configuration. Modern printers increasingly connect to a computer via a USB port. On Macintosh computers printer ports were formerly serial and are now USB.

printout   The output of a PRINTER, either the printed matter itself or the paper sheets or stack of CONTINUOUS STATIONERY on which the matter is printed.

print queue   Part of a computer's memory set aside as a buffer, to hold documents waiting to be printed. When the printer is ready the next document will be taken from the queue.

print screen key   A key on PC and compatible keyboards that, when depressed, will place a raster graphics image of the screen on the Windows clipboard. *See also* screen dump.

print server   On a network, a SERVER that makes one or more printers connected to it available to network users. Print servers are becoming less common as printers increasingly act as their own servers and connect directly to the network.

print spooler   Software that takes documents to be printed and stores them temporarily in memory or on disk until the printer is ready for them.

privacy The protection of stored data against unauthorized reading. When privacy is required for data stored in a computer it implies that the data is confidential and that access to it should be limited. The data may be 'owned' in some way by a particular person or organization, for example it may be data derived from commercial research and development, or it may be data about a particular person or organization. Personal data is protected by legislation in many countries (*see* data protection).

If a system has poor SECURITY then the privacy of data is at risk. If the INTEGRITY of the data is to be maintained in some way by the system, then the privacy is reduced because of the number of copies of the information that safe operation of a computer system demands.

Privacy has become a particular issue regarding the Internet. As the sophistication of computers' interaction with the Internet has grown, it has become increasingly possible for personal information stored on a computer or information about the configuration or use of the computer to be acquired by Internet sites without the user's knowledge or permission.

privacy-enhanced mail (PEM) E-mail that provides confidentiality and authentication using various techniques.

private key One of two keys used in PUBLIC-KEY ENCRYPTION. It is usually a very large integer that is used to decrypt messages encrypted with the appropriate PUBLIC KEY and for encrypting a user's DIGITAL SIGNATURE. It is vital that private keys be kept secret and secure.

problem-orientated language *See* high-level language.

procedural language 1. A PROGRAMMING LANGUAGE where the programmer must specify in detail the steps (procedure) to be used in solving the problem. All LOW-LEVEL and most HIGH-LEVEL LANGUAGES are procedural languages. *Compare* nonprocedural language.

2. A programming language where the source code is mainly a set of named PROCEDURES, FUNCTIONS, and SUBROUTINES.

Fortran, Algol, Pascal, and C are examples of procedural languages in both senses.

procedure A section of a program that carries out a well-defined operation on data. The same procedure can be used at different places in a program. Rather than repeating the same set of instructions at each place, the procedure appears only once and is brought into effect by CALLS: control is transferred to the procedure and on its completion control reverts to the program statement following the call. The actual data values to be used in the operation (or their addresses in memory) are specified at the time of call; these values or addresses are known as PARAMETERS. Different parameters can be provided for each call. In some languages a distinction is made between a procedure and a FUNCTION, in that a procedure does not return a calculated value whereas a function does. Some languages, such as Fortran, use the term SUBROUTINE instead of procedure.

In a large computer system with many users, a collection of procedures can be held on backing store and made available for general use. *See also* program library; subroutine library.

process 1. In general, to perform a sequence of operations on something in order to produce a particular result. In computing, the operations are arithmetic calculations and LOGIC OPERATIONS that are performed on data in accordance with instructions in a computer PROGRAM. The data is stored within the computer and is processed by the PROCESSOR (or processors) in the computer.

2. (task) A sequence of operations or events defined by the result it produces or by its purpose. Simple PROGRAMS consist of one process, which performs the necessary sequence of steps to accomplish the program's purpose. More complex programs may divide their work into two or more

distinct processes that run simultaneously and independently with their own resources (memory, files, etc.), communicating only to coordinate their activities. *See also* multitasking.

process control The computer-based monitoring and control of a manufacturing or industrial process. Sensing instruments measure one or more variables. The input from these instruments is compared with the optimum values, and control devices are adjusted to maintain the process in an efficient and safe state. Any serious problems will cause automatic safety features to operate and human supervisors will be notified.

In industries concerned with fluids and bulk solids, such as chemical manufacture, petroleum refining, and food processing, the variables to be controlled include temperature, pressure, density, consistency, and composition. If sheeted or webbed materials are involved, as in the paper, plastics, and textile industries, the major variables are temperature, thickness, width, length, and weight per area. When discrete items are handled or manufactured, as in the car and aircraft industries, then position, dimensions, and other properties are controlled.

processing The performance of ARITH-METIC and LOGIC OPERATIONS by the PROCESSOR (or processors) in a computer on data stored within the system. The operations are performed in accordance with the instructions in a computer PROGRAM.

processing unit *See* processor.

processor A piece of HARDWARE, or a combination of hardware and FIRMWARE, whose function is to interpret and execute instructions. It may be the principal operating part of a computer, in which case it is also known as the CENTRAL PROCESSOR. In more powerful computers there may be several processors, acting independently, each one performing some of the processing tasks or possibly having a specialized function. The processor or set of processors in a computer is often called the *pro-*

cessing unit. A processor may form part of a unit, as with an intelligent TERMINAL. The words processor and computer are often used as synonyms. *See also* front-end processor; microprocessor.

program 1. A set of STATEMENTS that can be submitted as a unit to a computer system, and used to direct the way in which that system behaves. Since the program will be followed slavishly by the computer system, it must indicate precisely what is to be done, and is thus expressed in a formal notation. The PROGRAMMING LANGUAGES used for this purpose can be picked from a selection of HIGH-LEVEL LANGUAGES or LOW-LEVEL LANGUAGES (usually ASSEMBLY LANGUAGE). Before it can be executed by a computer, however, the program must be converted into the appropriate MACHINE CODE.

The major part of any program usually consists of one or more ALGORITHMS, each setting out the steps by which some calculation or other task can be performed. With all but the simplest programs, there are various stages in producing a program and maintaining it in the desired form. The first stage is to understand exactly the whole task under consideration so that the *requirements* of the program can be clearly stated. During the process of *program design*, a description is produced of the program itself, the major components of the program, the way in which the components are interrelated, and the main algorithms employed by the components.

The design acts as the basis from which a working version of the program is created. This *implementation* stage involves writing the program in the chosen language(s) as well as TESTING and DEBUGGING it. If at any stage the design proves to be inadequate or wrong, then a return to the previous stage is made. Once testing and debugging have been satisfactorily completed the program can be released for *operational usage*, together with its DOCUMENTATION. During its working lifetime it may have to be modified or upgraded, and its documentation suitably amended. This forms part of *program maintenance*.

See also stored program.

2. To design, write, test, and debug programs.

program counter *See* instruction address register.

program crash *See* crash.

program design *See* program.

program file A FILE containing one or more programs or program fragments, which may be written in source code or object code. *Compare* data file.

program flowchart *See* flowchart.

program library An organized collection of computer programs and MODULES held on BACKING STORE. It may be available for general use by all users of a particular computer system, in which case it is called the *system library*. The individual programs and modules are known as *library programs*. A typical library might contain COMPILERS, UTILITY PROGRAMS, APPLICATION PACKAGES, and PROCEDURES, FUNCTIONS, or SUBROUTINES. Usually it is only necessary to make a reference to a particular library program, i.e. indicate its NAME, to cause it to be executed or automatically incorporated in a user's program by the LINK EDITOR.

program listing A printout of a computer program.

programmable ROM *See* PROM.

program maintenance *See* program.

programming The activities involved in producing a computer program. In the broadest sense of the word, this includes producing an accurate description of the requirements that some envisaged program is expected to meet, and all stages of program design and implementation (*see* program). In a much narrower sense, the activities involve simply the writing and testing of a program from some given design. This narrower usage is most common in commercial programming, where a distinction is often drawn between systems analysts and programmers. *See* systems analysis.

programming error *See* error.

programming language Any of a wide variety of notations designed for the precise description of computer PROGRAMS. With any programming language, the notation consists of a set of letters, digits, punctuation marks, and other CHARACTERS that can be assembled into various combinations, together with a strictly defined set of rules describing exactly which combinations are permitted. This set of rules is the SYNTAX of the language. The meaning of text constructed according to the syntax is also strictly defined – by the SEMANTICS of the language. Programming languages are thus artificial languages in that there is no freedom of expression characteristic of a natural language like English. *See also* high-level language; low-level language.

program module *See* module; modular programming.

program unit (subprogram) A part of a program that performs a specific task and so in some sense is self-contained. A PROCEDURE, FUNCTION, SUBROUTINE, and program MODULE are examples of program units.

Prolog /**proh**-log/ A PROGRAMMING LANGUAGE whose name is derived from **pro**gramming in **log**ic. It was developed in Europe in about 1970 and is based on symbolic logic. It is used in the field of ARTIFICIAL INTELLIGENCE. It is a FIFTH GENERATION LANGUAGE and its structure is not at all like that of more conventional languages such as C++ or Basic. It consists mainly of statements about relationships between objects, or implications derived from the statements, rather than instructions about how the objects are to be manipulated.

PROM /prom/ (**programmable read-only memory**) A type of SEMICONDUCTOR MEMORY that is fabricated in a similar way to ROM. The contents required, however, are added after rather than during manufacture and cannot be altered from that time. These contents are fixed electronically by a device known as a *PROM programmer*. It is only outside a computer that a PROM can be programmed. Within a computer the contents cannot be changed; they can only be read. *See also* EPROM.

PROM programmer *See* PROM; EPROM.

prompt A short message displayed on a VDU by a COMMAND LINE INTERFACE to indicate that additional data is required from the user to allow processing to continue. The prompt frequently has some mnemonic content to describe the type of data expected. Ideally a prompt should be explicit and informative for those unused to the system, but capable of being shortened to increase the speed of use for an experienced user. For example, a novice prompt may take the form

Enter the name of a file,
1 to 8 characters starting
with a letter:

The equivalent expert prompt could then be

Filename:

This state of affairs is seldom realized, and many operating systems in particular prompt with a single character such as > or $.

property A data item that forms part of a CLASS or software component. Assigning different property values to different instances of the class or component can alter its behavior to suit the precise circumstances. Some properties are transient, holding temporary values generated during the execution of a program. A games program, for example, might contain a class called 'Auto' that models the characteristics of an automobile, with such properties as 'current speed', 'fuel remaining', etc. These would change in response to user input and elapsed game time. Other properties might define more fundamental characteristics of the particular type of automobile being modeled, such as maximum speed and maximum rate of accelera-

tion. These would typically be set at the beginning of the program, when the user chooses the desired type of automobile, and remain constant thereafter. Software components usually have properties whose values are set when the program is designed. A clock CONTROL, for example, might have properties determining whether it displays the time in 24-hour or 12-hour format, in analog or digital format, etc.

protection *See* data protection; file protection; storage protection.

protocol /proh-tŏ-kôl/ An agreement that governs the procedures used to exchange information between entities in a computer NETWORK. In general, a protocol will govern the way in which information is encoded (*see* code), the generation of checking information (*see* check), and the FLOW CONTROL of information, as well as actions to be taken in the event of ERRORS.

proxy server An intermediate server between a LAN and another network, especially the Internet, that provides security (e.g. a FIREWALL), administrative control, and caching facilities. Performance can be improved with a proxy server as its cache can serve all users and supply frequently requested data, such as popular Web pages.

pseudocode /sew-doh-kohd/ A programlike but informal notation consisting of text in natural language (e.g. English) and used to describe the functioning of a procedure or a program. The flow of control is usually expressed in programming terms, e.g.

if ... then ... else ...
repeat ... until ...

while the actions are elaborated in prose. Pseudocode is used mainly as a design aid and is an alternative to FLOWCHARTS.

pseudorandom numbers /sew-doh-**ran**-dŏm/ *See* random numbers.

PSN (packet switching network) *See* packet switching.

PSTN (public switched telephone network) The worldwide public telephone system. The PSTN provides much of the Internet's long-distance infrastructure. ISPs pay long-distance providers for access and share circuits among many users by packet switching. Thus Internet users only pay for usage to their ISPs and for local telephone calls.

public-domain Describing any work that is not copyrighted. Such work can be copied freely by anyone and used for any purpose. Much of the information on the Internet is of this form but note that copyrighted information on the Internet is not in the public domain.

public-domain software Uncopyrighted software available for anyone to use. *See also* freeware; shareware.

public key One of two keys used in PUBLIC-KEY ENCRYPTION. A value is provided by some designated authority (the user) and released to the public as a key that can be used to encrypt messages to be sent to the user or for decrypting the user's DIGITAL SIGNATURE. Public keys are usually very large integers.

public-key encryption The use of different PUBLIC and PRIVATE KEYS for encryption, known as asymmetric cryptography. A public key encrypts data and a corresponding secret private key decrypts data. The process is reversed for digital signatures.

public switched telephone network *See* PSTN.

pull-down menu A menu of commands or pointers to other menus that is pulled down from the menu bar and remains open until either another menu is pulled down or it is closed by the user.

punched card An obsolete data MEDIUM, consisting of a rectangular paper card on which data could be encoded in the form of holes punched in columns. Since a hole could either be present or absent, it could be used to represent a binary 1 or a binary 0. A pattern of holes and spaces could therefore represent a sequence of 1s and 0s that in turn represented a character (*see* binary representation). Punched cards were employed extensively for the input, output, and storage of data in early computers. They were in fact used for data processing before computers were available.

Each letter, number, or other character was represented on a card by holes produced mechanically – by a *card punch* – in one or more rows within a column. When used for the input of data, the card punch was operated by means of a keyboard; it was then known as a *keypunch*. The presence of holes in a column, and their positions, was sensed (photoelectrically) by another machine – a *card reader*. The reader converted the punched data into a binary coding of 1s and 0s that could be fed into a computer. Binary data from a computer could be punched on to cards – again by a card punch – as output.

purge To clean up, usually by removing old or redundant data or files.

push *See* stack.

pushdown list *See* stack.

pushup list *See* queue.

Q

QBE (query by example) A simple database querying technique in which the criteria are set in a blank database record. This method is easier than using a query language.

quadbit /**kwod**-bit/ *See* nibble.

quantum computing /**kwon**-tŭm/ Research into the use of quantum mechanical effects as the basis of computer processors.

query by example *See* QBE.

query language A specialized language that is used for obtaining information from a database.

query processing The retrieval of one or more values from a FILE or DATABASE, leaving the contents unchanged.

queue (pushup list) A LIST that is constructed and maintained in computer memory so that all insertions are made at one end of the list and all removals are made at the other end. The next item to be removed is thus the earliest to have been inserted, i.e. the queue works on a first in first out (FIFO) basis. In contrast, a STACK works on a last in first out (LIFO) basis. Queues may

be used to store requests for access to peripherals, such as disks or printers.

QuickDraw *Trademark* A graphics display system for Apple Macintosh computers that enables programs to create and display graphical objects. It has been superseded but is still widely used.

QuickTime *Trademark* A standard developed by Apple Macintosh for displaying multimedia on computers. QuickTime is built into the Macintosh operating system. PCs can also run files in QuickTime format if they have the appropriate software. QuickTime files usually have the file extension mov.

quit To exit from a program or system in a clean and orderly way.

QWERTY /**kwer**-tee/ Describing the standard form of a Latin-alphabet-based keyboard, laid out in a typewriter style, in which the first six keys on the upper left of the alphabetic keys are QWERTY. It originated in the 1880s to reduce the jamming of type bars in typewriters by fast typists. Various alternatives have been propounded but QWERTY is so entrenched that these have had little impact.

R

radio button One of a set of small icons used to select an option in a graphical user interface. Typically radio buttons are small circles. Only one button in any set can be selected at any one time.

radix /**ray**-diks/ *See* base.

radix point A symbol, a period (full point) in the USA and UK but often a comma elsewhere, used to separate the integral part of a number from the fractional part. The radix point in decimal notation is called the *decimal point*.

RAID /rayd/ (**redundant array of independent disks**) A method of improving disk fault tolerance and performance. Several HARD DISKS are combined in a special unit controlled by a *RAID controller*, which presents them to the rest of the system as one large disk and stores data by STRIPING. However, not all the available space is used to store genuine data: sectors on at least one disk always store data calculated from the contents of the equivalent sectors on the other disks, for example PARITY information. This data is *redundant* because it is not necessary to store it since it can be calculated; however, the act of storing it means that, should any of the other disks fail, this redundant data can be used to calculate the contents of the missing disk. Thus no data is lost when a disk fails and the system suffers no DOWNTIME. A new disk can be swapped in and its contents generated automatically. There are various types of RAID, called *levels*. Levels 2 to 6 use variants of the system just described, typically with from 3 to about 14 disks. Level 1 is disk mirroring (*see* mirror) and can be used with only 2 disks. Level 0 uses all available space for genuine data and has

no redundancy at all, but is very fast; it is sometimes considered not to be a proper form of RAID storage. RAID disk systems are becoming standard on all but the smallest FILE SERVERS.

RAM /ram/ (**random-access memory**) A type of SEMICONDUCTOR MEMORY that can be read from and written to by the user, i.e. it is READ/WRITE MEMORY. The basic storage elements are microscopic electronic devices fabricated as an INTEGRATED CIRCUIT. These elements, often referred to as *cells*, are arranged in a two-dimensional array, i.e. in rows and columns. A single cell can store one bit of information – either a binary 1 or a binary 0. Each cell can be identified uniquely by its row and column. It can therefore be accessed directly (and extremely rapidly), i.e. there is RANDOM ACCESS to any cell. Since it is read/write memory, data can be both read from and written to the cells in the array.

It should be noted that all kinds of semiconductor memory provide random access, including ROM (read-only memory). The term RAM is used for random-access memory that can be written to. Because a computer's MAIN STORE consists of RAM, the term RAM is often used as a synonym for main store.

See also dynamic RAM; static RAM.

RAM cache Cache memory that is used for the temporary storage of data retrieved from RAM to speed up access.

RAM disk An almost obsolete technique in which a portion of main store is set aside to be used in place of a disk to provide faster processing. This part of memory is treated as a logical drive and data is stored as though it were a real phys-

RAID CONFIGURATIONS

RAID Mode	Implementation	Security/Performance
RAID 0	Straight data striping	No data security but gives maximum performance.
RAID 1	Duplication	This gives the best security but no performance gain. It is the most expensive as 2 times the number of disks is needed.
RAID 2	Not implemented yet. Requires extra information to be added to data disks.	
RAID 3	Straight parity	Gives security against a single hard drive failure and is reasonably economical.
RAID 4	Although a dedicated disk is used for the parity information data disks are accessed independently of it, so performance is not affected.	Not used as much as RAID 3.
RAID 5	Parity based data security where the parity information is stored on each of the data disks.	Gives security against a single hard drive failure with the minimum of performance loss.
RAID 6	Parity based with 2 disks holding parity information.	Gives security agianst failure of both the 2 main disks. RAID 1 gives the same security with more speed at the same reliability level.
RAID 10 and RAID 1+0	RAID 0 striping together with RAID 1 duplication together.	Gives both the performance of striping and the security of duplication. The simplest implementation is with 4 hard drives grouped into identical pairs.
RAID 30 and RAID 3+0	RAID 3 plus striping. Parity is calculated before striping.	Data is recoverable even if one of the RAID 0 disks fails.
RAID 50 and RAID 5+0	RAID 5 parity plus striping of all the data disks. Parity is calculated before striping.	Data is recoverable even if one of the RAID 0 disks fails.

ical drive. When the computer is switched off all the data is lost. This system has been superseded by the use of cache memory.

random access   A method of retrieval or storage of data that does not require any other stored data to be read first. The storage locations can be accessed (read or written to) in any order. There is random access with MAGNETIC DISK storage. There is also random access with main storage, i.e. with RAM and ROM, but with a much shorter ACCESS TIME than with disk storage. *See also* file. *Compare* serial access.

random-access file *See* file.

random-access memory *See* RAM.

random numbers Numbers that are drawn from a set of permissible numbers and that have no detectable pattern or bias. Any number in the set should therefore have an equal probability of being selected. Random numbers are required, for example, when selecting the winners of a lottery. In a computer true random numbers are difficult to obtain. Instead programs are designed to generate what are known as *pseudorandom numbers*. In principle the numbers produced depend on their predecessors and so are not truly random; in practice the pseudorandom numbers generated by a particular program are sufficiently random for the purpose intended.

raster graphics /rass-ter/ A method of producing pictorial images in which the desired shape is built up line by line. Each line is composed of closely spaced elements that can be any one of a number of colors or shades. This is the technique used in most VDUs, in TV monitors, and also in some plotters. The term is also used of a type of COMPUTER GRAPHICS image that uses these principles. *Compare* vector graphics.

raster image processor *See* RIP.

raw data Data in the form in which it reaches a computer system from the outside world. It is data that has not been vetted for correctness, nor sorted into any particular order, nor processed in any other way. *See also* data cleaning.

ray tracing A technique used mainly in three-dimensional graphics for showing objects with realistic shadows cast by an imaginary light source. In ray-tracing applications primitive objects (e.g. cubes, cylinders, spheres, etc.) can be placed in position, sized, distorted, rendered (*see* rendering), and illuminated and viewed from different directions.

RDF Site Summary *See* RSS.

read /reed/ **1.** To sense and retrieve data from a storage medium such as MAGNETIC DISK or SEMICONDUCTOR MEMORY. This data can then be transferred to another storage medium or to an output medium. *See also* readout.
2. To sense and interpret data on an input medium such as a document or a BAR CODE, and to convert the data into an electrical signal that can be fed into a computer.

READ /reed/ *See* Basic.

reader An INPUT DEVICE, such as a DOCUMENT READER or bar-code reader, that is able to READ data, i.e. sense and interpret data, on an input medium.

read head /reed hed/ *See* tape unit.

read instruction A MACHINE INSTRUCTION that causes an item of data to be READ from a specified location in a memory device and copied into a BUFFER store or into a REGISTER.

read-only memory *See* ROM.

readout **1.** Information retrieved from MAIN STORE after processing and displayed on a screen or copied into backing store.
2. The process of reading data from a memory device. In most types of memory this does not alter the actual representation of the data; the readout is then described as *nondestructive*.

read/write head *See* head. *See also* disk drive.

read/write memory A type of memory that in normal operation allows the user to read from or write to individual storage locations within the memory, i.e. to retrieve and store information. It may involve SERIAL ACCESS or RANDOM ACCESS. RAM is read/write memory.

real address The absolute machine address that specifies a physical memory location.

RealAudio /ree-ăl-**aw**-dee-oh/ *Trademark* A continuous or streaming sound technology from RealNetworks. A RealAudio player or client program is needed to use this technology, which may be included in a Web browser or which can be downloaded from the Internet. To deliver RealAudio from a Web site either the user or the user's ISP needs to have a RealAudio server. Real audio files have the extension ra or ram.

Really Simple Syndication *See* RSS.

real number (**real**) In computing, any number with a fractional part, such as 0.75, 57.3654, or 3.141 59... (i.e. π). As with INTEGERS, reals can be unsigned or can carry a plus or minus sign. Real numbers are represented in a computer in FLOATING-POINT NOTATION or sometimes in FIXED-POINT NOTATION; they are thus represented and manipulated differently from integers. Only a subset of the real numbers can be represented in a computer. If n bits are reserved to store each real number then only 2^n different real numbers can be represented. (In practice the number will be smaller than this since not all combinations are valid floating-point numbers.)

real-time clock A battery-powered clock included on a computer motherboard.

real-time processing A method of using a computer where the time at which the output is produced is of importance. The computer must be able to respond to some external event (for example, some movement in the physical world) and be able to perform calculations and control functions within a specified and usually very brief time limit. This response time, i.e. the lag between input and output time, can range from a few seconds to less than a microsecond.

real-time system Any computer-based system, such as air-traffic control, PROCESS CONTROL, or defence systems, that uses REAL-TIME PROCESSING to implement its core functionality. Often such systems use special hardware and software to ensure response times within the specific tolerances.

reboot To reuse a BOOTSTRAP program in a computer. An instruction to reboot the system can be given to the system when, say, the OPERATING SYSTEM is not running properly or has been corrupted in some way. Rebooting loads a new copy of the operating system from backing store and allows the system a fresh start.

reconfiguration /ree-kŏn-fig-ŭ-**ray**-shŏn, -fig-yŭ-/ A change or redefinition of the configuration of a computer system, i.e. of the pieces of hardware making up the system and in some cases of the way in which they are interconnected. A reconfiguration may be necessary, for example, to bypass a defective unit or to provide a different overall function. The software may need to be reconfigured to reflect a change in the hardware (*see* configure).

record A collection of related items of data, treated as a unit. It consists of a number of parts, called *fields*, each field holding a particular category of data. In a person's hospital record, for example, there could be a field for the person's name, another for date of birth, and another for, say, the clinic attended; the *key field* holds a unique identifier, for example a number allocated to the patient for ease of identification, and is used to find the record in a group of hospital records or to sort the group into patient order (*see* sorting). Normally one particular field in a record is used as the key field. It is also possible for there to be several key fields in a record.

A field may contain either a fixed or a variable number of CHARACTERS, and hence be of a fixed or variable length. A record may thus also be of fixed length or of variable length. Fixed length fields and records are easier to implement (e.g. a given field will occupy the same part of every record), but can waste space if used to hold data items of intrinsically variable length, such as names. Variable length records and fields must be delimited by using specified characters to mark their beginnings and/or ends or a separate data structure must be maintained that records their starting positions or lengths.

This collection of related data items is considered in terms of its contents or function rather than on how it is stored and moved in a computer. It is often referred to as a *logical record*. In contrast, a *physical record* is the unit in which data is transferred to and from peripheral devices; it is also called a BLOCK. One logical record may be split up and held in different physical records. Several logical records or parts of logical records may be located in one physical record.

The record is one of the fundamental ways of collecting and organizing data. DATA FILES held in backing store are often treated as sequences of records. Many programming languages permit operations on an entire record as well as on its individual components. *See also* database.

record length The storage space needed to contain a record, usually stated in bytes.

recoverable error An error that is not fatal, i.e. one from which the program can recover and continue running.

recovery The procedure whereby normal operation is restored after the occurrence of a FAULT. *Power-fail recovery*, for example, deals with a loss of the incoming power supply to a computer system. The system is equipped to detect any long-term deviation of the power-supply voltage from acceptable limits. When such a deviation is detected, and before it can cause any problems, a signal (INTERRUPT) is sent to the processor: the sequence of instructions being carried out is suspended, relevant information is dumped in NONVOLATILE MEMORY, and processor activity is halted. When the voltage is restored the program or programs that had been running can be RESTARTED.

recursive routine A PROCEDURE, FUNCTION, or SUBROUTINE that CALLS itself, resulting in a continued repetition of the operation that the procedure, function, or subroutine is designed to perform. The recursive call must be CONDITIONAL, other-

wise there would be an endless series of calls. Recursion is not allowed in every programming language. Improper use can cause problems during program execution, being difficult to detect during compilation.

reduced instruction set computing *See* RISC.

redundancy The provision of additional components in a system, over and above the minimum set of components required to perform the function of the system. This is done to increase the reliability of the system and, in the event of errors occurring, to increase its ability to recover from such situations. For example, two or three identical pieces of hardware may be used in a computer so that if a fault occurs in one piece that item can be replaced. Again, additional bits are attached to data when it is transferred from one location to another, enabling a CHECK to be performed – a *redundancy check* – and any errors occurring in the data during transfer to be detected.

reentrant program /ree-en-trănt/ A program that uses code that can be shared by other programs at the same time. Operating systems often use reentrant code so that only one copy is needed in memory to service many live applications.

reference file A FILE that contains reference material and thus changes infrequently.

reformat To FORMAT a disk that contains data. Reformatting will totally remove any data on the disk.

refresh 1. To recharge a dynamic memory bank so that information is not lost. This is an automatic function of the memory board. *See also* dynamic RAM; refresh cycle.
2. To retrace a video screen after a certain interval. *See also* refreshed display; refresh rate.

refresh cycle The repeated cycle of electrical pulses used to renew stored electrical charges in DYNAMIC RAM, without which data would be lost.

refreshed display A display in which the image lasts only a very short time and therefore has to be repeatedly redrawn electronically. An ordinary TV screen is a refreshed display. If the same image is redrawn on a refreshed display then the display appears continuous, without flickering. Alternatively different information can be fed to the screen to give the appearance of motion in the image. A refreshed display can be used for instance to produce an animated cartoon.

refresh rate The number of times a display's image is redisplayed per second. This is expressed in hertz. The refresh rate for each screen depends on the video card used and can be changed in the display settings. However, if an incompatible rate is chosen the display will go blank or the image will be distorted. The older and slower refresh rate of 60 Hz caused image flickering and consequently operator tiredness and headaches. Newer rates of 75 to 85 Hz are commonly used.

register A storage device in the processing unit of a computer in which data has to be placed before it can be operated upon by instructions. It is therefore a temporary location, and has a CAPACITY of only a WORD, a BYTE, or a BIT. The registers within a processing unit are quite separate from the storage LOCATIONS in main store and backing store. Registers are either general-purpose devices or are intended for a special purpose. Special-purpose registers include the INSTRUCTION ADDRESS REGISTER and INSTRUCTION REGISTER in the CONTROL UNIT. In larger computer systems there may be several hundred registers.

Information must be stored in and retrieved from registers extremely rapidly. They are therefore high-speed semiconductor devices (usually a group of FLIP-FLOPS, each flip-flop storing one bit).

register transfer An operation whereby information is transferred between two particular REGISTERS or between a specified storage LOCATION in main store and a particular register. For example, loading the contents of address 10 in main store into the ACCUMULATOR would involve
  (1) loading address 10 into the MEMORY ADDRESS REGISTER,
  (2) copying contents at address 10 to accumulator.
This can be written in symbolic form:
  (1) $10 \Rightarrow MAR$
  (2) $MEM \Rightarrow ACC$
where the arrow symbol means 'is transferred to'.

registry A database used by Windows systems to hold hardware and software configuration information. When a new piece of a program is installed it usually writes entries in the registry. When they are no longer required they should be uninstalled in a proper manner. In this way the registry entries will also be removed. The registry can be edited by using the REGEDIT.EXE program provided with Windows. However care must be taken when doing this as the system can be disabled if the registry is altered incorrectly.

regression testing The testing of a complete program after modification of some part rather than only testing the modified routines, to ensure that the program still works in a proper manner.

regular expression A compact method of specifying more than one character STRING that is often used in searching. For example the regular expression (or *pattern*) 'abc[def]g' represents the three strings 'abcdg', 'abceg', 'abcfg': enclosing the characters 'def' in square brackets means that they are alternative values for the fourth character in the string. It is often more useful to perform one search using a regular expression defining a set of strings, which produces one set of results, than to search for each string separately. There is no standard syntax for regular expressions but some features, such as the use of square brackets described above, are universally recognized.

relational database /ri-**lay**-shŏ-năl/ A database management system that is a collection of data items organized as a series of tables. The relationships between the tables are formally defined together with any constraints on the data, and data can be accessed or reassembled in many different ways so that special views of selected data can be obtained to suit a user's needs. A query language, for example SQL, is used for interactive investigations and also for gathering information for reports. Such databases can be relatively easy to create and use, but can be highly complex and powerful.

relational operator (relational) A symbol representing a comparison operation.

RELATIONAL OPERATORS	
Comparison	Operator
less than	<
less than or equal to	<=
equal to	=
greater than or equal to	>=
greater than	>
not equal to	<>, #, !=, ¬=

These operations take two OPERANDS and provide a logical value *true* or *false* as their result. The operands may be of any type so long as like is compared with like. The common comparison operations are shown in the table, together with the relational operators normally used in computing; 'not equal' has many denotations. An example of how relationals are used, in Pascal, is as follows:

    if (b<a) then
    c := b
    else
    c := a;

See also order of precedence.

relative address An ADDRESS expressed as a difference with respect to some numerical value used as a reference. This reference is known as a *base address*. Addition of a base address to a relative address gives an ABSOLUTE ADDRESS. The base address may, for example, be the address specified in the INSTRUCTION ADDRESS REGISTER, or it may be stored in a special-purpose register, the *base address register*. See *also* relative addressing.

relative addressing An ADDRESSING MODE in which a RELATIVE ADDRESS is used in a MACHINE INSTRUCTION in order to identify the storage location to be accessed. This mode is often used to access the contents of complex data structures, with the base address register containing the address of the whole structure and the relative address being the offset of the desired item within it. It is also used to control the flow of the program code: an instruction to skip the next instruction would then merely cause an extra 1 to be added to the INSTRUCTION ADDRESS REGISTER (IAR) rather than replacing its contents with an explicit new address. A backward jump could be made in a similar way, by subtracting the relevant amount from the IAR.

reliability A measure of the ability of a system – hardware or software – to provide its specified services for a given period or when demanded.

reload To load a program into memory again for running after some system interruption or crash has stopped the program.

relocatable code The code making up a program or part of a program that can be loaded anywhere in main store. No ABSOLUTE ADDRESSES are used, except perhaps those of known operating system or I/O functions. All JUMPS are made relative to the current instruction, the start of the code, or the contents of a REGISTER, examples being
jump to the next instruction but one,
jump to the second instruction of the program,
call the procedure whose address is in register 5.

REM /rem/ *See* Basic.

remainder check *See* modulo-*n* check.

remote Describing a workstation, terminal, etc., that is situated some distance from a main computer or network but is connected by a link.

remote access The ability to obtain access to a computer or network from a remote location.

remote procedure call (RPC) A protocol using the client/server model that can be used by a program to request a service, usually a PROCEDURE or FUNCTION call, from another program located on another computer in a network.

rendering The process of converting a wireframe three-dimensional image into a solid image. Three-dimensional drawing programs have a wide range of options for 'covering' wireframe structures with different textures and colors. *See also* ray tracing.

repaginate /ree-**paj**-ă-nayt/ To rearrange the pages of a desktop or word-processing document after editing, to take account of inserted and deleted data. This is usually an automatic procedure in the application, which the user can override if necessary.

repeater A device that receives an electromagnetically or optically transmitted signal, amplifies it, and then retransmits it along the next leg of the medium. Signals can be sent over long distances by the use of repeaters. They are used to interconnect segments in LANs and to extend WAN transmission.

repeat until loop *See* loop.

replace To substitute one set of data for another: often used in word processing in search and replace operations.

replication The process of creating and managing duplicate versions of a database. Changes made in one copy are reflected in all other copies. These changes are synchronized so that many users can work on their own local copy of the database in geographically dispersed situations, but ap-

pear to be working on a single centralized database. Lotus Notes was one of the first systems to make replication a central part of its design.

report generator *See* generator.

request for comments (RFC) A formal Internet document or standard that has evolved from drafts and reviews by interested parties. Some RFCs are for information only.

requirements description *See* program.

rerun **1.** To RUN a program again from the start, usually following a malfunction of the computer. *See also* restart. **2.** A repeat run.

reserved character A keyboard character, for example an asterisk (*), forward slash (/), or backward slash (\), that has a special meaning to a program and cannot be used for other purposes (such as naming files or keyboard shortcuts).

reserved word A word that has a specific role in a programming language and cannot therefore be used as an IDENTIFIER. For example, in many languages the words *if*, *then*, and *else* are used to separate the different parts of conditional statements. These are thus reserved words.

reset (restore) **1.** To cause a device to take up an earlier value. For example, when a program is about to be RESTARTED on a processor, the contents of the working registers of the processor must be reset to the values they last held when the program was previously running. **2.** To set the whole computer back to its initial condition when it was switched on. This process is used in smaller computers to clear faults that have corrupted the OPERATING SYSTEM and made further processing impossible.

resident Describing a program that is present in memory.

resolution /rez-ŏ-**loo**-shŏn/ In general,

the amount of information or detail that can be produced by a device such as a VDU or plotter or can be revealed in an image. The device or image can be described qualitatively as having a *high resolution* (*hi res*) or a *low resolution*, etc. The resolution can also be given in numerical terms. The resolution of a display screen, for example, can be expressed as the number of PIXELS (picture elements) that are available in the horizontal and vertical directions of the screen, e.g. 1024 by 768 pixels. The resolution of a graphics image can be given in dots (i.e. pixels) per inch, e.g. 300 dpi. The resolution of a plotter can be given as the smallest possible pen movement. The resolution of a printer output can be in dots per inch, e.g. 2400 dpi.

resource Any of the facilities that are available in a computer system and that are required by a processing activity. The resources in any computer system include the following devices: one or more PROCESSORS, some form of memory – MAIN STORE and BACKING STORE – in which to store the instructions for the processor(s) and the data awaiting manipulation, and INPUT/OUTPUT DEVICES through which instructions and data are fed to the computer and information is fed back to the outside world. Program instructions and data are held on backing store as FILES. Files are also resources.

Any facility required from the OPERATING SYSTEM of a computer is regarded as a resource. These include programs for scheduling and supervising the running of users' programs on the computer.

resource allocation (**resource sharing**)
1. The sharing of the RESOURCES of a computer system between various processing tasks. A resource, such as the processor, that can only be used by one task at a time, is shared by allowing each task in turn a set time period. In the case of storage devices, for example magnetic disks, requests from tasks are queued and serviced one at a time. Resource allocation is under the control of the OPERATING SYSTEM of the computer.

2. The sharing of RESOURCES among the users of a computer system to prevent or reduce the unfair use of the system by some users at the expense of others. It can be done, for example, by allocating units of processor time to each user, depending on their status, work, etc., and by allocating fewer units at peak time rather than, say, at night.

resource information file format *See* RIFF.

resource sharing *See* resource allocation.

response time Usually, the time that elapses from some action by a computer user to the receipt of some response or feedback from the computer. With a personal computer, with one user, the response time depends entirely on the size of the task to be performed; there is a fast response for simple tasks like displaying the contents of a file on the screen and a slow response for, say, compiling a large program. When there are a large number of terminals linked to a mainframe or minicomputer the response time increases as the number of users increases; a lightly loaded mainframe responds extremely rapidly. In a network, the response time depends on how heavily the server responsible for servicing the requested function is loaded, and in extreme cases the loading of the network itself. *See also* interactive.

restart The resumption of the execution of a program after a temporary halt, using data dumped at a particular point in the program (*see* dump). The program is said to be *restarted*. Much of the data from the previous RUN is therefore preserved, and a restart is often described as a *warm start*. In contrast, there is a *cold start* when a program has to be rerun from the start. Some programming languages allow the programmer to specify the points at which a program can be restarted.

restore *See* reset.

retrieve To locate and return data to the user or program requesting it.

return 1. In programming, the transfer of control from a called routine or program back to the calling routine or program. 2. *See* return key.

RETURN *See* Basic.

return instruction *See* exit.

return key (**enter key**) A key on a KEYBOARD that sends a carriage return character to the computer. It is often used to signal that the current line of typing is complete and may be processed. In GUI interfaces it often initiates the default action of the currently selected object, e.g. a program is executed, a word-processing document is opened in the appropriate word processor, etc.

reverse Polish (**RP**) A notation used in a computer to evaluate arithmetic expressions. In RP the ARITHMETIC OPERATOR (e.g. $+, -, *, /$) is placed after its operands rather than between them as in conventional notation. The addition of two numbers, a and b, would thus be expressed as $ab+$.

Expressions in RP are scanned from left to right and the values of the operands stored until an operator is encountered. The operator is then applied to the last two values and replaces them by a single result, which may be used as a value for the next operator, and so on. Hence
$$ab-cd+*$$
would be evaluated as
$$(a - b) * (c + d)$$
i.e. b would first be subtracted from a; c and d would then be added together and the sum would finally be multiplied by $(a - b)$.

The translation of conventional notation to RP is achieved by an algorithm. A programmer would not have to write expressions in RP but a COMPILER producing machine code from a high-level language might well generate RP for expression evaluation.

reverse video (**inverse video**) A display attribute of most VDUS, used to draw attention to a character. The character is displayed on the screen in the opposite contrast to surrounding information. For instance, with a screen that normally has dark characters on a bright background, a character in reverse video will be a bright character within a dark character-sized rectangle.

RFC *See* request for comments.

RGB system A method of defining colors by the proportions of R (red), G (green), and B (blue) present. *See also* CMYK system.

ribbon cable An electric cable in which a number of similar wires lie side by side in a flat plastic strip and are electrically insulated from each other. A ribbon cable provides a continuous path along which electrical signals can be conveyed from one point in a system to another. It can be used for PARALLEL TRANSMISSION of data.

Rich Site Summary *See* RSS.

rich text format *See* RTF.

RIFF /riff/ (**resource information file format**) A tagged file-format specification used to cover different types of multimedia files. Tags are used to identify data elements in a file by length and by type. The specification can be extended to cover new kinds of data elements when necessary while still supporting older formats.

right-justified *See* justify.

ring network *See* network.

RIP (**raster image processor**) Hardware or a combination of hardware and software that converts vector graphics images into raster graphics images. *See also* PostScript.

RISC /risk/ (**reduced instruction set computer**) A design philosophy for CPUs. In contrast to CISC, RISC processors have a smaller and simpler INSTRUCTION SET; they therefore operate more quickly, because it takes less time to decode and execute each instruction, and can be manufactured more cheaply. Also, research has shown that many of the missing CISC instructions were rarely used in practice, and so were needless complications. The MACHINE CODE of programs for RISC processors tends to be longer and more complex, because operations that are intrinsic to CISC processors must be performed as a series of smaller operations. Elements of the RISC philosophy have been incorporated into modern CISC chips, which often work as RISC devices internally through MICROPROGRAMS. PowerPC chips are the most widely used RISC-based chips in computers.

RMON /**ar**-mon/ (**remote monitoring**) A network management protocol that allows a single workstation to gather network information (e.g. usage statistics). The routers, hubs and switches, and other devices must support RMON for this to work.

robot /**roh**-bot/ A programmable device that has been engineered to perform many simple repetitive tasks formerly done by human labor. It consists of one or more articulated mechanical manipulators, typically an arm and hand, that are linked to a computer and have considerable freedom of movement. In more advanced types there is also some form of sensory mechanism, such as a TV camera acting as an artificial eye. Most robots in current industrial use follow a fixed sequence of instructions but can be reprogrammed to carry out another task. They have little if any sensory capability. They are designed to be flexible and transportable and can be used, for example, for loading and unloading machines, positioning and transferring parts, and welding pieces of metal or plastic.

robustness A measure of the ability of a system – hardware or software – to recover from conditions caused by ERRORS. The errors may be external to the system or may occur within the system itself. The system can be described as robust or possibly

highly robust if it recovers and continues to operate despite numerous error conditions.

rogue value *See* terminator.

roll-out roll-in *See* swapping.

ROM /rom/ (**read-only memory**) A type of SEMICONDUCTOR MEMORY that is fabricated in a similar way to RAM but whose contents are fixed during manufacture and cannot therefore be modified. The contents can only be read. As with RAM, ROM is composed of an array of cells, each of which can be identified uniquely by its position in the array. There is thus direct and extremely rapid access to any cell in ROM, i.e. there is RANDOM ACCESS to any of the storage locations. Since the contents cannot be altered, ROM is NONVOLATILE MEMORY. *See also* PROM; EPROM.

ROM cartridge (**ROM pack**) A module containing software that is permanently stored as ROM. The module can easily be plugged into and later removed from a computer, games console, etc. without the integrated circuitry being handled.

root directory The main directory in a HIERARCHICAL FILE SYSTEM. It is the directory that includes all other directories.

ROT-13 /rot th'er-**teen**/ The encryption of a message by exchanging each of the letters in the first half of the alphabet with the corresponding letters in the second half. Thus A becomes N, B becomes O and so on. Other characters are not changed. ROT-13 is sometimes used to conceal sensitive parts of on-line documents, for example offensive words, the ending of a movie or book, etc.

rotate *See* shift.

rounding The process applied to a number whereby its length is reduced to a specified number of significant digits, the nearest approximation to its true value being taken. For example, rounding the number 3.257 69 to 5, to 4, and to 3 digits yields 3.2577, 3.258, and 3.26. The REGIS-

TERS in a computer can hold only a fixed number of binary digits (bits). When the result of an arithmetic operation requires more than this fixed number of bits, either rounding or TRUNCATION will take place. Since the rounded value is an approximation to the true value of the number, a *rounding error* is introduced; on average this is less than the truncation error. Repeated rounding can cause a build-up of such errors and the programmer should try to minimize this problem.

router /**rowt**-er, **root**-er/   A hardware device that routes data from a LAN to a telephone line or another connection that leads to another network. It also handles errors and gathers network statistics. A router is connected to at least two networks and decides which way to send each information packet according to its 'understanding' of the current state of those networks. Routers will only allow authorized machines to send data through a LAN so that private information is kept secure.

routine /roo-**teen**/   A subroutine (in assembly language) or procedure or function (in high-level language); the term is usually used in combinations, as in *input routine, recursive routine, diagnostic routine.*

RP   *See* reverse Polish.

RPC   *See* remote procedure call.

RS-422, 423, 449 The standards adopted for serial communications over distances greater than 50 feet. RS-449 incorporates RS-422 and RS-423.

RSA (Rivest-Shamir-Adleman) The most commonly used Internet encryption and authentication system. It is a PUBLIC-KEY ENCRYPTION system developed in 1977 by Ronald Rivest, Adi Shamir, and Leonard Adleman.

RS232C interface   An INTERFACE developed by the Electronic Industries Association (EIA) in the USA. It can be used for making the connection between two computers, between a computer and a MODEM,

or between a computer and a printer or some other peripheral device. Although superseded by the USB interface, RS232C ports are still supplied as standard on most PCs.

RSS   (**Really Simple Syndication; RDF site summary; Rich Site Summary**) Several widely-used formats for WEB FEEDS. They use XML to provide a standardized concise summary of an item on a Web site and a link to the full item. RSS originated in 1999 and has since split into two incompatible series of formats; however, most RSS-aware software can understand both of them.

RTF (rich text format)   A file format for saving word-processor files in ASCII form with additional information that retains the formatting, font information, color, etc., of the original document. RTF files are often used for transferring files between different word-processing applications. They have the file extension rtf.

RTS (request to send)   A hardware signal that is often used in serial communications and sent over an RS-232 connection requesting permission to transmit to a device.

rubber banding   In computer graphics, the action of grabbing a point on an anchored line of a graphics object and pulling it to a new position to change the shape of the object.

rule-based system   *See* expert system.

run 1. To execute a program, i.e. to carry out the program instructions. A program or part of a program that is currently being executed by the processor (or processors) is said to be *running* or *active*. Only one program or program section can be running at any one instant on a processor.
2. The act of executing a program, a set of programs, or some part of a program.

running head   A portion of text that appears in the top margin of each page of a document. *See also* header.

run time The period of time during which a program is actually being executed, or the time taken for execution. It does not include the time during which the program is loaded, or is compiled or assembled.

run-time error (**execution error**) An ERROR detected at RUN TIME. The error may be *fatal*, in which case execution is terminated; a fatal error would result, for instance, from an attempt to divide by zero or read past the end of a file. Alternatively the error may be *nonfatal*, in which case a warning is generated and execution continues; a nonfatal error would result, say, from an attempt to calculate a real number too small to be stored – it would be replaced by zero.

run-time system A collection of PROCEDURES that support programs derived from a high-level language at RUN-TIME, providing services such as STORAGE ALLOCATION, INPUT/OUTPUT, DEBUGGING, etc., but are not actually a part of the program code as stored on disk. This avoids storing large numbers of copies of commonly used procedures, and allows new versions of them to be installed without the need to recompile all the programs that might use them. In computer systems where a number of high-level languages are supplied by a common manufacturer, it is often possible for a single run-time system to service programs written in any of these languages.

S

sampling A process by which the value of an analog (i.e. continuous) signal is 'examined' at distinct intervals of time. The sampled values can, for example, be converted into a digital form using an A/D CONVERTER for subsequent processing by a computer.

SAN *See* storage area network.

SATA (S-ATA) *See* Serial ATA.

satellite computer A computer that is attached to another computer by a communications link and is controlled by the main computer.

save To write data to a storage medium.

SAX (Simple API for XML) A de facto standard for software that processes an XML file in discreet units, contrasted with software that constructs a representation of the whole file in the computer's memory before beginning processing. SAX-based programs are generally faster and use less memory but cannot perform some complex types of processing. SAX was first released in 1998 and the current version is version 2 (*SAX2*).

scalable **1.** Denoting something that works as well at large sizes as it does at small sizes. A piece of software, for example, will generally run very efficiently if it only has a small number of simultaneous users. Software that can handle an unpredictably large number of simultaneous users with a minimal decline in performance is said to be scalable, to *scale well*, or simply to *scale*. Any software that either has a fixed, and relatively small, maximum number of users or whose performance declines unacceptably as users increase is said to be not scalable or to *scale badly*. It is important that such types of software as OPERATING SYSTEMS, DATABASE MANAGEMENT SYSTEMS, and WEB SERVERS scale well. The term is also applied to other aspects of computing, for example hardware.
2. Denoting something that is designed to work equally well at more than one size. For example, OUTLINE FONTS and vector graphics are scaleable, whereas BITMAPPED FONTS and raster graphics are not.

scalable font Any font that can be scaled to produce characters of different sizes without any loss in quality. *See also* outline font.

Scalable Link Interface *See* SLI.

scale *See* scalable.

scanner A device for capturing diagrams and photographs and converting them into electrical signals, so that the image can be stored in digital form. Scanners are also used as optical character readers. For very high-quality work, as in the reproduction of photographs for book and magazine printing, *drum scanners* are used. In these the image to be scanned is wrapped around a cylindrical drum. FLATBED SCANNERS are suitable for many applications, and HAND-HELD SCANNERS are also in use.

scheduling The method by which the use of a PROCESSOR is controlled, i.e. by which time on the processor is allocated. Scheduling is necessary in a system where the processor has to be shared between a number of programs; this occurs for example in a MULTITASKING system. The choice of which processing activity should be granted access to the processor, and can safely be granted access, is determined by the *scheduling algorithm*. This is designed to ensure that the computer is used as efficiently as possible.

schema 1. In XML, a file that defines the structure of a document or a set of similar documents. It details such matters as which ELEMENTS are allowed in the document, in what order they must appear, how many of each type can be used together, what type of data they can contain (STRING, INTEGER, FLOATING-POINT NUMBER, etc.). Documents that are correctly structured are said to *conform* to their schema, and schemas can be used to create software that will correctly process any conforming document. Schemas thus perform the same role as DTDS but in a more sophisticated and flexible way, and they use normal XML syntax rather than the special syntax (inherited from SGML) used by DTDs. However, there are currently several varieties of schema whereas the syntax for DTDs is standardized.
2. Any abstract model that describes the structure of data, operation of hardware or software, etc.

Schottky TTL /**shot**-kee/ *See* TTL.

scientific notation A method of representing real (floating-point) numbers that is much simpler for very large or very small numbers. Instead of writing the full number, scientific notation shows values as a number between 1 and 10, multiplied by 10 to some power where the 10 is usually replaced by E. For example, 1.234E6 is equivalent to 1 234 000 and –1.234E–6 is equivalent to –0.000 001 234.

scope The range of program code over which a VARIABLE or some other element of a program is defined. The element can be described as *local*, *nonlocal*, or *global*. A local element is accessible only in a restricted part of the program, typically in a PROCEDURE or FUNCTION. By contrast, nonlocal elements are accessible in a wider scope and global elements are accessible from all parts of the program. The scope of

a variable or other element is normally indicated by the positioning of or a qualification within its DECLARATION.

scrapbook A Macintosh file used to hold text and graphics images for later use. *See also* clipboard.

scratchpad A temporary storage area in a computer.

screen The front surface of a VDU or monitor on which computer text or graphics can be displayed. *See* display.

screen dump A way of transferring the entire graphical or textual contents of a VDU screen to a printer or to a raster graphics file. *See also* dump; print screen key.

screen editor *See* text editor.

screen saver A utility program that displays a moving image on the screen after a set time interval if no keyboard or mouse actions have been performed. On older monitors this stopped the possibility of a permanent static image being burned onto the screen. Modern monitors do not suffer from this problem.

screen turtle *See* turtle graphics.

script *See* scripting language.

scripting language A language that allows the user to execute a sequence of operations automatically that would normally be invoked individually, by specifying them in a text file called a *script*. For example, COMMAND LANGUAGES are scripting languages by which users can automate a series of commands to the operating system. Applications, such as word processors, spreadsheets, etc., often have their own scripting languages. As well as specifying operations, scripting languages usually contain features of programming languages, such as CONTROL STRUCTURES, VARIABLES, and ARITHMETIC and LOGIC OPERATIONS, that allow sophisticated programs to be written with them.

scroll To move the information displayed on a screen or in a window in a vertical or horizontal direction. As information disappears at one edge of the window, new information can become visible at the other edge or alternatively space is provided for the entry of new data.

scroll bar A horizontal or vertical bar shown at the bottom or side of a display area, such as a window, containing controls that can be used to SCROLL the text or image in the area.

SCSI /skuz-ee/ (small computer systems interface) A standard high-speed parallel interface used to attach such devices as hard disks, scanners, and CD writers to a PC. Up to 15 devices can be connected to a SCSI interface. There are several different SCSIs available giving different performance levels; the fastest supports data-transport speeds of 320 Mbps. There are now versions of SCSI that operate over serial links and TCP/IP networks.

SDLC (synchronous data link control) The communications protocol used by a network conforming to the IBM systems network architecture (SNA).

SDRAM /ess-dram/ (synchronous dynamic RAM) A type of DYNAMIC RAM that operates synchronously, i.e. its activities are coordinated with those of the rest of the system by the CLOCK. This synchronization enables it to use features that make it much faster than (asynchronous) DRAM. *DDR SDRAM* (Double Data Rate SDRAM) and its successor, DDR2 SDRAM, are faster still, the latter typically running at over 400 MHz.

SDSL (symmetric digital subscriber line) A variant of DSL technology that transfers data at high speed over an ordinary copper telephone wire. Unlike ADSL it uses the same speed in both directions – up to 2.3 Mbps, which is between the two speeds employed by ADSL – and cannot share the wire with a telephone. It is suitable for businesses and other organizations requiring high-speed data links that do not rely on data being predominantly transferred in one direction. *See also* VDSL.

search engine An ENGINE that searches for specified data in a file or database. The term is used particularly of Internet search engines. Large Web sites usually have their own search engine, which allows the user to search the whole site for documents containing certain key words. More generally, a number of search engines exist that allow the user to find Web pages on any topic and have become important Web sites in their own right. Examples of these are Google, Yahoo!, AltaVista, Lycos, and Excite. They actually search the World Wide Web (and also FTP archives, Gopher lists, etc.) using a CRAWLER (spider) program to accumulate a very large index of Web pages, etc., by key word. When a user uses such a search engine, he or she is not searching the Web but rather the index that the provider has generated.

searching The process of locating information held in a FILE. In the case of data files or database TABLES this is done by reference to a particular field of each RECORD in the file or table. This field holds a *key* by which the record can be identified. The goal of the search is to discover all records (if any) with a given key value. Many different ALGORITHMS can be used for searching.

The most simple search is a *sequential search*, in which the keys in the file or table are searched sequentially from beginning to end and records with matching keys extracted. (If each record has a unique key, the search can stop at the first match.) In a *binary search* the records in the file or table must be arranged with their keys in ascending order. The middle key is examined and, depending on whether this is less than or greater than the sought-for key, the top or bottom part of the file or table is examined. Again the middle key is checked against the desired key, and a smaller section – top or bottom – is indicated and re-examined. The process continues until the desired records are found or their absence discovered. Other searching algorithms use HASHING or various types of INDEX.

When a record is to be inserted in a table or file, with its own unique key, then a search must first be made to ensure that no existing record has the new key.

For unstructured data, e.g. text, the options are more limited. Either a *full-text index* can be used, which indexes the position of every word in the file, or else the file must be searched sequentially from the beginning.

secondary store *See* backing store.

second generation computers Computers designed in the mid- and late 1950s and early 1960s, i.e. after the FIRST GENERATION, and characterized in particular by the increasing use of discrete TRANSISTORS in addition to valves in their electronic circuitry. A wider range of INPUT/OUTPUT equipment became available in the period, together with higher performance MAGNETIC TAPE and the first forms of on-line storage – MAGNETIC DRUMS and early MAGNETIC DISKS. Magnetic CORE STORE was introduced for main store. Initial efforts were made at HIGH-LEVEL programming languages, the earliest versions of FORTRAN, for example, being issued in 1956 and 1958.

During the 1950s several companies, including IBM, had become active in developing computer systems and peripherals. In the late 1950s IBM designed and built a monster computer, called *Stretch*, which had very advanced hardware. It became available commercially in 1961 as the IBM 7030. The IBM 1401 (1959) and 1410 (1960) included early on-line disk storage.

Another important machine designed in this period was *Atlas*, which incorporated many features now considered standard. Its design was begun in 1956 by T. Kilburn at the University of Manchester and it became commercially available, from Ferranti Ltd., in 1963.

See also third, fourth, fifth generation computers.

sector A subdivision of a track on a MAGNETIC DISK. The computer records data on a disk one sector at a time. The sector is thus the smallest addressable portion of the

track: each sector has a unique ADDRESS that contains the track location and sector number. The disk is divided into sectors prior to any recording of data; the disk may be *hard-sectored*, in which case the sector positions are set at the time of manufacture, or *soft-sectored*, when sector lengths and positions are set by control information written on the disk by computer program. Once divided into sectors, the address plus other control information can be added to each one. This initial preparation of a disk is known as *formatting*.

security Prevention of or protection against (a) access to information by unauthorized recipients or (b) intentional but unauthorized alteration or destruction of that information. The measures taken to provide security in a computer system include the use of PASSWORDS when logging on and the classifying of information in order to restrict the users who can access it. The security provided may guard against accidental access to sensitive information as well as deliberate access attempts. *See also* firewall; integrity; privacy.

seek time *See* access time.

select To choose certain data, which becomes the target for formatting or content changes until it is deselected. Selection is often accomplished using a mouse; for example, in word-processing applications dragging the mouse across text selects that text. Selected data is usually HIGHLIGHTED.

self-extracting file An executable file that contains one or more compressed files. When the file is run, the compressed files are uncompressed and stored on the user's hard drive. Self-extracting files are a common way of downloading programs from the Internet.

self-test A set of diagnostic tests performed by a device on itself. For example, most printers have a self-test option available, which will print a test page.

semantics /sĕ-**man**-tiks/ Part of the description of a PROGRAMMING LANGUAGE, concerned with specifying the meaning of various constructs – STATEMENTS, CONTROL STRUCTURES, etc. These constructs must conform to the SYNTAX, or grammar, of the language; if they do not then they are meaningless. Most programming languages have their semantics defined by descriptions in English or some other natural language. Because natural language is imprecise and hence ambiguous, different interpretations of the semantics are possible in the early stages of development of a language.

semiconductor memory /sem-ee-kŏn-duk-ter/ (**solid-state memory**) Any of various types of compact memory composed of one or more INTEGRATED CIRCUITS fabricated in semiconductor material, usually silicon. The integrated circuit is composed of an array of microscopic electronic devices, each of which can store one bit of data – either a binary 1 or a binary 0. There may be millions of these storage locations within a single integrated circuit measuring only a few square millimeters in area. Data can be accessed extremely rapidly from these locations.

The categories of semiconductor memory include RAM (which is read/write memory), ROM (read-only memory), PROM (programmable ROM), and EPROM (erasable PROM). There is RANDOM ACCESS to all these types of semiconductor memory. Semiconductor RAM has been used to build the MAIN STORE in computers since the early 1970s.

sentinel *See* flag.

separator A symbol that separates STATEMENTS in a program. The symbol used depends on the programming language. A semicolon is the separator in several languages, including Pascal:
  begin read (x); y := x∗x; write (x,y) **end**
Note that the final statement needs no semicolon. *Compare* terminator.

sequential access /si-**kwen**-shăl/ *See* serial access; file.

sequential file *See* file.

sequential flow *See* control structure.

serial access (**sequential access**) A method of retrieval or storage of data that must be used for example with magnetic tape. Blocks of data are read from the storage medium in the actual order in which they occur until the required item or storage location is found. The ACCESS TIME thus depends on the location of the item. *Compare* random access.

Serial ATA (**SATA**; **S-ATA**) A development of and the successor to the ATA disk INTERFACE standard. Developed in the early 2000s and a formal specification since 2003, it offers greater data-transfer speeds by using a serial rather than a parallel BUS and is now used in most new PCs.

serial port An interface used for SERIAL TRANSMISSION. When used of microcomputers, the term denotes an RS232C INTERFACE, not a USB interface.

serial printer A printer that connects to a SERIAL PORT on a computer. Formerly standard on Macintoshes and sometimes used on PCs, they are being superseded by printers that connect to a USB port.

serial transmission A method of sending data between devices, typically a computer and its peripherals, in which the individual BITS making up a unit of data (e.g. a character) are transmitted one after another along the same path. For example, if two devices wish to communicate an 8-bit unit of data, the sending device will transmit the 8 bits in sequence (perhaps with a *start bit* preceding them and a *stop bit* following to aid synchronization). The receiving device can then reassemble the stream of bits into the original 8-bit unit. *See also* parallel transmission; handshake.

server A program that provides services to other programs, called *clients*. Usually a server and its clients run on different computers attached to a network. There are many types of server. For example, a *file server* – the original type of server – allows clients to access its computer's hard disks and read or store files; a *database server* runs a DBMS and handles requests from clients to query the database, update its contents, etc.; a *mail server* implements an e-mail system for the network, usually with an external connection to the Internet; and so on. The computer running a server program is often *dedicated*, i.e. it does nothing else. It is called the *host*, the *server computer*, or, simply, the *server*.

server-based application An application program that is installed on a server and is shared over a network.

server-side scripting A technology that allows the content of WEB PAGES to be varied each time they are accessed. It operates on the Web server rather than the user's machine (*compare* dynamic HTML). A normal (or *static*) Web page is written in HTML and must be edited to change its content or appearance. Server-side scripting embeds fragments of program written in a SCRIPTING LANGUAGE in the Web page; when processed by the Web server, this program produces HTML that varies in response to choices made by the user or to changing external data. Such pages are known as *dynamic Web pages* and are widely used for any purpose that requires either up-to-date information (e.g. current

stock market prices) and/or user selection of content (e.g. online shopping). *See also* ASP; PHP.

session A recording made on a CD-R or CD-RW at a particular time. The session has lead-in and lead-out information on the CD. Multisession recordings can be made on a single CD at different times, one session following on from another.

Session Initiation Protocol *See* SIP.

set-top box *See* Web TV.

seven-segment display A simple display in which the numbers 0–9 and some letters can be formed from a group of seven segments (see diagram). The segments can be individually darkened or illuminated to form the different characters. The figure eight for instance uses all seven segments; the figure one uses only two segments. *See also* LCD; LED display.

SGML (**standard generalized markup language**) A METALANGUAGE used to define MARKUP LANGUAGES adopted by the International Organization for Standardization (ISO) in 1986 that allows documents to be independent of any particular application. The text in such a document has 'tags' defining the type of information. For ex-

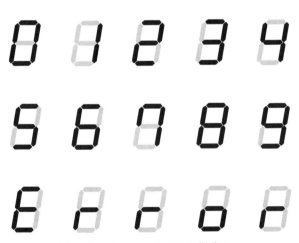

Characters in a seven-segment display

ample, in this book the basic text is tagged as follows

```
<en><hw>SGML</hw>
<vg>(<var>standard generalized
markup language</var>)</vg>
<def>A form of…</def></en>
```

Here <en> and </en> enclose the whole entry; <hw> and </hw> enclose the headword; and <def> and </def> enclose the definition. Also <vg> and </vg> denote a group of variant's of the headword. In this case there is only one variant, which is tagged using <var> and </var>. Note that the tagging is for information content rather than typeface, although often tags such as <b> and <i> are used for bold and italic type. Another feature of SGML is the use of ASCII characters only. Nonstandard characters are represented by strings, as in é for acute e (é), á for acute a (á), ° for degree sign (°), and so on. The rules for SGML are quite complex. A particular feature, which makes the data portable, is the use of a *document type definition (DTD)*, defining the way the tags are used in a document or class of documents. For example, HTML is defined in SGML with a DTD detailing the structure of Web pages. *See also* XML.

SGN *See* Basic.

share *See* shared folder.

shared folder (share) The term used for a NETWORK DIRECTORY on Windows- and Macintosh-based networks.

shareware /shair-wair/ Software that is available for use for a trial period, after which a decision is taken not to use it or it must be registered and paid for. Shareware programs are less expensive than commercial software but may be more error-prone and not supported.

sheet-fed scanner A SCANNER with a single-sheet feeding mechanism that can automatically scan multipage documents.

shell The part of an operating system

that provides a COMMAND-LINE INTERFACE. Some operating systems, in particular UNIX, have a choice of shells.

shift 1. A concerted movement of some or all of the bits held in a storage location in a specified direction, i.e. to the left or to the right. Shifts are performed in SHIFT REGISTERS. A left (or right) shift of, say, 3 bits means that 3 bits are removed from the left (or right) end of the register. The bits are said to be *shifted* left, right, out, etc.

There are three kinds of shift. In a *logical shift* the bits shifted from the end of the register are lost, and 0s are shifted in at the opposite end. In a *circular shift* (or *rotate*) the bits shifted from the end are reinserted at the opposite end, in the order in which they were shifted out. In an *arithmetic shift* the bit pattern is regarded as representing a binary integer, and only the bits representing the magnitude of the number are moved; this preserves the positive or negative sign (*see* complement). If the bits shifted out are all zero, a left shift of n bits is equivalent to multiplying by 2^n and a right shift of n bits is equivalent to integer division by 2^n.

The effects of different kinds of 3-bit shift on the same 8-bit word are shown in the table; the left arithmetic shift is equivalent to multiplication by 2^3 (i.e. 8), the right arithmetic shift to division by 2^3.
2. On a keyboard, a movement with a SHIFT KEY.

shift instruction A MACHINE INSTRUCTION specifying that the contents of a SHIFT REGISTER are to be shifted either to the left or to the right by a specified number of po-

original word	00010111
logical left	10111000
logical right	00000010
circular left	10111000
circular right	11100010
arithmetic left	10111000
arithmetic right	00000010

Effects of different 3-bit shifts on an 8-bit word

sitions. The bits that are introduced as replacements at the opposite end of the register depend on the kind of SHIFT. In the case of a circular shift, the bits shifted off one end of the register are reintroduced at the other end. With a logical shift and an arithmetic shift, zeros are introduced and the bits shifted out are often discarded.

shift key A keyboard key that when depressed at the same time as another key either gives an upper-case letter or the character marked on the upper part of the second key. The shift key is also used in combination with other keys to perform special functions.

shift register A REGISTER in which SHIFTS are performed.

Shockwave /shok-wayv/ A technology developed by Macromedia that enables Web pages to include multimedia objects. In order to see a Shockwave object with a Web browser a plug-in is usually needed, which is freely available on the Internet.

shortcut key *See* accelerator key.

short filename *See* long filename.

signal A means of conveying data through, say, an electric circuit, a computer, or a communications system. A signal is usually a sequence of values of an electric voltage recorded against time. The value – the *amplitude* – varies either continuously or discretely (i.e. in steps). It is the amplitude that represents the data: the data can therefore be represented in discrete form (e.g. as integers) or in continuous form (as 'real' numbers). The amplitude can be recorded or displayed over a continuous period of time or it can be *sampled*, i.e. examined at discrete intervals. Discrete representations of both amplitude (i.e. data) and time are required by DIGITAL COMPUTERS. ANALOG COMPUTERS employ continuous representations. *See also* digital signal; digitized signal; analog signal.

signature **1.** A set of bytes that can be used to identify a virus.
2. A set of characters used for identification.

sign bit A single bit used to represent the sign of a number; normally 0 represents + (plus) and 1 represents – (minus). *See* complement; sign-magnitude notation.

significant digits (significant figures) The digits in a number that make a contribution to the value of that number. The number 0.1234 has four significant digits: the zero does not contribute to its value. The number of significant digits can be reduced by TRUNCATION or ROUNDING.

sign-magnitude notation A notation that can be used to represent positive and negative integers in a computer. Normally the leftmost bit in a WORD is used to denote the sign (0 for +, 1 for –) and the remaining bits are used to represent the magnitude of the integer; the leftmost bit is called the *sign bit*. It is more usual, however, for a computer to use two's complement notation to represent signed integers. *See* complement.

sign off *See* log off.

sign on *See* log on.

Silicon Valley A nickname for an area of west California, south of San Francisco, that contains an unusually high proportion of computer companies.

SIMM /sim/ (single in-line memory module) A small slim circuit board containing surface-mounted memory chips, used as an extension to main store. SIMMS have now largely been superseded by DIMMS.

simple mail transfer protocol *See* SMTP.

simple network management protocol *See* SNMP.

simplex transmission /sim-pleks/ One-way transmission of data between two endpoints. There is no possibility of data traveling in the opposite direction. Simplex transmission could be used to drive an output device such as a printer when it is known that the output device is faster than the transmission line and can therefore always keep up. *Compare* half-duplex transmission; duplex transmission.

simulation /sim-yŭ-**lay**-shŏn/ The imitation of the behavior of a system, or of some aspect of that behavior, by another system. The imitation may be in the form of a model, possibly quite abstract, that can be manipulated by a computer. Again, the imitation may be achieved by a device under the control of a computer and often employing components from the system being simulated – such as the controls in the flight deck of an aircraft. Any computer device or program that performs a simulation is called a *simulator*.

Simulation is widely used as a design aid for both large and small computer systems and for computer networks. It can also be used for more specific purposes, for example in weather forecasting to predict likely developments in the weather pattern. Simulators are extensively used in the training of airline pilots, military commanders, etc. Simulation is an important application of digital computers and is also the major application of ANALOG COMPUTERS. *Compare* emulation.

single in-line memory module *See* SIMM.

single precision *See* precision.

SIP (Session Initiation Protocol) A protocol, defined in RFC 2543 (1999) and RFC 3261 (2002), for initiating and controlling an exchange of data over the Internet. It is particularly used to provide VOIP services and is becoming the de facto standard protocol for this purpose.

sleep Of a computer, to run on low power with all activity suspended until required.

SLI (Scalable Link Interface) *Trademark*

A technology for linking two graphics cards (*see* expansion card) so they the act like one double-speed card. SLI was introduced in 2004 by NVIDIA Corporation.

SLIP /slip/ (**serial line Internet protocol**) The older protocol, now replaced by PPP, for the transmission of TCP/IP data packets over serial communication lines.

small computer systems interface *See* SCSI.

Smalltalk /**smawl**-tawk/ An object-oriented language that was developed by Xerox Palo Alto Research Center (PARC) in the 1970s. Many of the ideas used are now incorporated in other programming languages. *See also* object-oriented programming.

smart card A plastic credit card that contains a microprocessor and memory. It offers greater security and can be interrogated by automatic cash machines.

smart terminal *See* terminal.

SMDS (switched multimegabit data services) An extremely high-speed connectionless packet-switched data-transport service that connects LANs and WANs.

smiley /**smÿ**-lee/ *See* emoticon.

smoothing /**smooth**-ing/ The removal of JAGGIES from a curved line to make the line look smoother. *See also* anti-aliasing.

SMP (symmetric multiprocessing) A computer architecture that provides fast processing by using multiple CPUs sharing the same memory to complete individual processes in parallel (*multiprocessing*). An idle processor can be allocated any task and more CPUs can be added to improve performance. UNIX and Windows both support SMP.

SMT *See* surface-mount technology.

SMTP (simple mail transfer protocol) A protocol for sending e-mail messages between servers. An e-mail client can then retrieve the messages using POP or IMAP.

SNMP (**simple network management protocol**)   A TCP/IP protocol. Hardware or software SNMP agents can monitor network device activity and report to a workstation.

snowflake curve   *See* fractal.

SOAP   A protocol that uses XML and HTTP to send REMOTE PROCEDURE CALLS and other structured communications between programs running on different computers linked by the Internet. It was initially designed in 1998 with support from Microsoft and is now controlled by the W3C. SOAP initially stood for 'Simple Object Access Protocol', but this meaning has been dropped.

soft font   A FONT where the instructions that define the shape of each character are held as data, separate from the software and hardware in a VDU, printer, etc., and are used to interpret the font data and render the character. Soft fonts may be held in the ROM of a particular output device, but they are more often stored as data files in a computer's backing store. In the latter case the same data is used to create the font on each output device, ensuring compatible results. *See also* downloadable font.

soft keyboard   A KEYBOARD in which the meaning of each key, i.e. the code to be generated by each key, can be allocated and possibly changed by program control. Generally, not all the keys are 'soft'; some are permanently wired so that they have a fixed meaning.

soft return   A line break inserted in a document by a word processor or text editor to prevent a line exceeding the maximum length allowed. The positions of soft returns are adjusted automatically as the text is edited.

soft-sectored disk   *See* sector.

software /**sôft**-wair/   The programs that can be run on a computer system, as distinct from the physical components – the HARDWARE – of that computer system. There are two basic forms: SYSTEMS SOFTWARE and APPLICATIONS programs. Systems software, which includes the OPERATING SYSTEM, is an essential accompaniment to the hardware in providing an effective overall computer system; it is thus normally supplied by the computer manufacturer. The applications programs relate to the role that the computer system plays within a given organization. The DOCUMENTATION associated with all these programs is also considered part of the software.

software engineering   The entire range of activities involved in the design and development of software. The aim is to produce programs that are reliable and efficient by following standards of quality and adhering to specifications, as is done in other engineering disciplines.

software house   A commercial organization whose main business is to produce software or assist in the production of software. Software houses may offer a range of services, including consultancy, the hiring out of suitably qualified personnel to work within a client's team, and a complete system design and development service.

software interrupt   *See* interrupt.

software package   *See* application package.

software piracy   *See* piracy.

software tool   One of a set of programs that are used in the development, repair, or enhancement of other programs or of the computer's hardware. All the programs in the tool set have a common mode of use and employ files and other facilities in a well-defined standard way. A typical set of tools might consist of a TEXT EDITOR, COMPILER, LINK EDITOR, and some form of DEBUG TOOL. *See* program.

SOHO /**soh**-hoh/ (**small office home of-**

fice)   The market for computers and soft-ware composed of home computer users and small businesses.

solid-state memory   *See* semiconductor memory.

sorting   The process of rearranging in-formation into a desired order. This is achieved by means of *sort keys* that are as-sociated with each RECORD of information. Sorting is achieved by comparing sort keys and rearranging the records accordingly. The sort keys are derived from the values stored in the fields of the records in such a way that by ordering the sort keys in as-cending or descending sequence, the asso-ciated records will be placed in the desired order. For instance, if the sort key were to be derived from date of birth and surname, then the result of a sort would be to place the records in order of age, with those born on the same day in alphabetical order.

There are many methods of sorting. For example, in the *straight insertion sort* each sort key in a collection of records is checked in turn, and the record associated with the key is repositioned (if necessary) with respect to all previous sort keys. In a *bubble sort* a pair of records is inter-changed if their sort keys are not in the cor-rect order, different pairs being checked on different passes through the collection of records. This continues until no further in-terchanges are necessary.

sort key   *See* sorting.

SoundBlaster /**sownd**-blass-ter/   A range of sound cards manufactured by Creative Technology and its subsidiaries; it has be-come the de facto standard.

sound card   An expansion board giving the capability of playing sound from CDs, DVDs, MIDI files, or other sound files.

source code   *See* source language.

source language   The programming language in which a SOURCE PROGRAM is written. The program is in the form of *source code*.

source program   A program that is written by a programmer in a PROGRAM-MING LANGUAGE, but cannot be executed directly by a computer. It must first be con-verted into a suitable form by a specialized program. This program may be a COM-PILER, INTERPRETER, or ASSEMBLER. *See also* object module; high-level language.

space character   A member of a CHAR-ACTER SET that is usually represented by a blank site on a screen, printout, etc. It causes the print or display position to move forward by one position without produc-ing any mark. Within the computer a space is treated in exactly the same way as any other character.

spam /spam/   Junk mail on the Internet. The volume of spam has increased hugely in recent years. While such countermea-sures as KILL FILES used to be adequate, to-day's avalanche of spam requires more sophisticated solutions and anti-spam fil-ters are now a standard part of e-mail sys-tems.

special character   A symbol in a CHAR-ACTER SET that is a GRAPHIC CHARACTER but is not a letter, nor a digit, nor (usually) a SPACE CHARACTER. Punctuation marks, brackets, and arithmetic symbols are spe-cial characters. The special characters on the keyboard vary from one country to an-other, and would include, for instance, the national currency symbol. *Compare* al-phanumeric character; control character.

specification /spess-ă-fă-**kay**-shŏn/   *See* system specification.

speech generation device   A means of producing spoken messages in response to signals from a data processing or control system. The selection of messages are pro-duced by assembling speech sounds from a set of fundamental sounds that may be ar-tificial in origin or may have been extracted by processing human speech.

speech recognition   The process whereby a computer interprets spoken words in order to determine their data content. *See also* voice input device.

speech synthesis   The production of artificial speech by a computer, usually from textual input. The technique involves storing prerecorded words or sounds, or artificially generated versions of such sounds, and using software that analyzes text and joins the sounds together for output.

speech understanding   The process of using SPEECH RECOGNITION in order to perform some task making use of speech, such as voice input.

spider   *See* crawler.

spool   To transfer information intended for a slow peripheral device into an intermediate store on magnetic disk. This process of *spooling* allows information to be sent to the peripheral (such as a printer) at a more convenient time: programs that produce printed output can then run even when the printer is busy or unavailable. Without spooling, the program would have to slow down so that its speed matched that of the slow peripheral. Formerly this was undesirable because it wasted valuable processor time; today, in an age of cheap interactive computing, it would waste the user's time.

spreadsheet   /**spred**-sheet/   (spreadsheet program)   A program that manipulates tables consisting of rows and columns of *cells*, and displays them on a screen. The cells contain numerical or textual information or formulae. Each cell has a unique row and column identifier; different spreadsheets use different conventions so the top left-hand cell may be A1, 1A, or 1 1. The value in a cell is either typed in or is calculated from a formula in the cell; this formula can involve other cells. Each time the value of a cell is changed by typing in a new value from the keyboard, the value of all other cells whose values depend on this one are recalculated.

A common characteristic of all spreadsheets is the way in which the display screen of the computer acts as a WINDOW

on to the array of cells; if there are more rows and columns than will fit on the screen, then the spreadsheet can be scrolled horizontally or vertically to bring into view previously hidden rows or columns. To change a value it is only necessary to move the CURSOR into the required cell displayed on the screen and type in the new value.

Spreadsheets can be used for storing and amending accounts, 'what if?' financial projections, and many other applications involving tables with interdependent rows and columns. A spreadsheet is often a component of an INTEGRATED OFFICE SYSTEM.

sprite   /sprÿt/   A small animated image on a computer screen. Sprites are often used in computer games.

spyware   A piece of software that subverts the operation of a computer in some way, usually by sending unauthorized information to a third party over the Internet. This may be a record of Web sites visited, the user's personal details, or even a complete record of all keys typed; it is used for purposes ranging from marketing to fraud. Spyware is usually concealed in Web pages or software downloaded from the Internet. It has become a significant security risk to Internet users and several programs have been developed to detect and remove it. *See also* adware; malware.

SQL (structured query language)   A standardized programming language provided in a database management system to enable queries to be set up and processed. Most DBMSs implement a version of SQL. However, proprietary extensions to the standard are common, so a complex SQL program from one system might need extensive adjustment to run on another.

SQL Server   A relational DATABASE MANAGEMENT SYSTEM from Microsoft Corporation. Originally developed by Sybase, Inc., it was launched in 1989 for the OS/2 operating system as Ashton-Tate/Microsoft SQL Server. The current version, SQL 2000, is an exclusively Microsoft product and runs on Windows. (Sybase markets its own development of the earlier product called *Adaptive Server Enterprise*.)

SQR   *See* Basic.

SRAM /**ess**-ram/   *See* static RAM.

stack (**pushdown list**)   A LIST that is constructed and maintained in computer memory so that data items can only be inserted at or removed from one end of the list (called the *top*). The most recently inserted item is thus the first to be removed, i.e. the stack works on a last in first out (LIFO) basis. The operations *push* and *pop* refer respectively to the insertion and removal of items at the top of a stack. Stacks are frequently used in computing, especially in connection with subroutine CALLS and INTERRUPT handling. *See also* queue.

stand-alone system   A computer or a computer program whose operation is independent of any other device or program. A stand-alone microcomputer, for example, is capable of operation without being connected to any other piece of equipment or network. A diagnostic program that runs alone in a computer system without even the operating system is also termed stand-alone.

standard deviation   A measure of the amount of variation within a group of numerical values, such as measurements of a quantity. If the values are represented by
$$x_1, x_2, x_3, ..., x_n$$
so that there are n values in all, their standard deviation is given by the formula
$$\sqrt{[1/n\ ((x_1 - \overline{x})^2 + (x_2 - \overline{x})^2 + ... \\ (x_n - \overline{x})^2)]}$$
where $\overline{x}$ is the MEAN of the group. This formula is usually given in the form
$$\sqrt{[1/n\ \Sigma\ (x_i - \overline{x})^2]}$$
where i ranges from 1 to n and Σ is read as 'the sum of'. The square of the standard deviation (i.e. the above formulae without the square root) is called the *variance*.

standard generalized markup language   *See* SGML.

standard interface   *See* interface.

star network   *See* network.

startup disk /**start**-up/ (**system disk**)   A disk that contains an operating system and can be used to boot a computer.

statement   The basic building block of a programming language, HIGH-LEVEL or LOW-LEVEL, and used in a PROGRAM for a particular purpose or function. It may be an *executable statement* and thus specify some operation to be performed by a computer. Alternatively it may give information about the structure of the program, for example a DECLARATION, in which the names and types of variables are defined.

A program consists of a sequence of statements. It is translated into a sequence of MACHINE INSTRUCTIONS by a COMPILER, ASSEMBLER, or INTERPRETER. In the case of a high-level language, one executable statement is translated into many machine instructions, sometimes hundreds. An executable statement in an ASSEMBLY LANGUAGE is usually translated into one machine instruction.

static   1. Not changing or incapable of being changed over a period of time, usually while a system or device is in operation or a program is running. For example, a *static allocation* of a resource cannot be changed while processing is taking place, and *static RAM* is a type of semiconductor memory that retains its contents until written to.
2. Unable to take place during the execution of a program. For example, a *static dump* is usually taken at the end of a program run.
Compare dynamic.

static RAM (**SRAM**)   A type of semiconductor memory, using flip-flops and needing no refresh memory signals, normally seen in caches, where the information is retained as long as there is sufficient power. It is faster but more expensive than DRAM.

STOP   *See* Basic.

storage   1. (store; memory) A device or

medium in which data or program instructions can be held for subsequent use by a computer. The two basic types of storage in a computer are MAIN STORE (or main storage) and BACKING STORE (or backing storage). *See also* storage device; media.

2. The retention of data or program instructions in a storage device or storage medium.

3. The act of entering data in a storage device or storage medium.

storage allocation The allocation of specified areas of storage to the various processing tasks that are active in a computer. This controlled use of storage is necessary in a computer system in which several individual processes may be running. Each process is allocated sufficient WORKING SPACE in a way that does not interfere with another process's allocated working space. The allocation, and the amount of storage allocated, is controlled by the OPERATING SYSTEM of the computer; with some operating systems the amount can be changed while a program is running. Storage allocation is one form of RESOURCE ALLOCATION.

storage area network (**SAN**) A high-speed device connected to a network's servers, on which their data is stored rather than on local hard disks. It typically consists of a network of RAID disk arrays and supporting hardware. This consolidation of data storage and its separation from other server functions simplifies both network management and the addition of new storage capacity. A SAN is faster but less flexible than NETWORK-ATTACHED STORAGE and is usually found on large networks. The two technologies can be combined, with an NAS device using a SAN to store its data.

storage capacity *See* capacity.

storage cycle **1.** The sequence of events occurring when a unit of a storage device goes from one inactive state through a read and/or write phase, and back to an inactive state.

2. The minimum period of time required between successive accesses (read or write) to a storage device.

storage device A device that can receive data and retain it for subsequent use. Such devices cover a wide range of storage CAPACITIES and speeds of ACCESS. The semiconductor devices used as MAIN STORE have very fast speeds of access but the cost of storing each bit is relatively high. In comparison the devices – MAGNETIC DISKS, CDs, DVDs, etc. – used as BACKING STORE have a lower cost per bit and a greater storage capacity, but take much longer to retrieve data.

storage location (**store location**) *See* location.

storage medium *See* media.

storage protection Any of many facilities that limit and hence control access to a storage device or to one or more storage locations. The intention is to prevent inadvertent interference by users and/or to provide system SECURITY. It is achieved by prohibiting unauthorized reading or writing of data, or both. In storage devices that operate at high speed, protection is implemented by hardware (to maintain speed). For slower devices it may be done entirely by software.

store **1.** *See* storage.

2. To enter or retain data or instructions for subsequent use by a computer.

stored program A set of instructions that can be submitted as a unit to some computer system and stored in its memory, in advance of being processed. It is then possible for instructions to be extracted automatically from memory at the appropriate time. This is done, without human intervention, by the CONTROL UNIT of the computer.

The data to be manipulated by the computer, in accordance with the program instructions, is also stored in the computer memory. Instructions and data are both represented in binary form within the computer – i.e. as sequences of the binary digits 0 and 1 – and cannot be distinguished (*see* binary representation). As a result a program and the data handled by that program can be stored together, sharing the same storage space.

The idea of program and data sharing the same memory is fundamental to the great majority of modern digital computers. The concept of the stored program was documented in 1946 by John von Neumann, forming the basis of his design of EDVAC (*see* first generation computers). The concept was also considered by John W. Mauchly and J. Presper Eckert at about the same time. The first operational stored-program computers were the Manchester Mark I and EDSAC. All digital computers since then have been stored-program computers, and stored programs have long been called, simply, *programs*.

The storage of programs and data in READ/WRITE MEMORY allows the stored data to be modified by the operations performed on it, and if necessary the program itself to be modified by the program instructions. Modifying a running program is normally avoided since it is a major source of error. The opportunities for program modification in present-day computers are reduced by dividing the program into two parts: one part must not be modified, and is therefore either stored in a protected area of memory or is permanently resident in read-only memory (ROM, PROM, or EPROM); the other part may be modified, and is stored in an area of read/write memory.

See also program.

storm   A sudden excessive increase in traffic on a network.

straight insertion sort   *See* sorting.

streamer   *See* tape streamer.

streaming   The movement of data from one computer to another in such a way that it does not have to be completely transferred before the receiving computer can start using it. Often multimedia files are streaming files, meaning the user can watch a video or listen to a sound file while it is still being downloaded from, say, the Internet.

Stretch   *See* second generation computers.

string   One form in which a collection of data items can be held in computer memory. A string is a list of letters, digits, or other CHARACTERS, examples being

cat
735
Sue Jones
COMP

It is therefore an item of textual data. Strings are in fact used mainly for handling text. The *length* of a string, i.e. the number of characters in it, is not fixed. There may be only one character. An EMPTY STRING is also a possibility. Any sequence of characters within a string is called a *substring*; *bc* and *abc* are possible substrings of *abcde*.

A string is represented in memory as a line of adjacent LOCATIONS; either the first location holds a number that gives the string length and the remaining locations hold the characters in correct order, or the string is terminated by a location containing a special noncharacter code. A string is denoted in a program by various means, depending on the programming language. In Basic, for example, a dollar sign ($) as the final character of a VARIABLE name declares it to be a string. In many languages the characters making up a string are placed in quotation marks. An example in Basic would be

PET$ = 'HORSE'

Various operations can be performed on strings in addition to input/output and storage. They can be joined together (*see* concatenation); they can be compared to find out if they differ; they can be searched to see if they contain a given sequence of characters; a given substring can be extracted and replaced by another sequence of characters. The operations that can be performed are known collectively as *string manipulation*, the choice of operations depending on the programming language.

string manipulation   *See* string.

striping   A method of improving HARD DISK performance. Several disks are grouped together and act as one large disk, with the data being distributed across all the disks in a way designed to minimize access times. RAID systems use striping.

structured programming   An approach to producing an actual program in a high-level language in which only three structures need be used to govern the flow of control in the program. These three structures – CONTROL STRUCTURES – allow for sequential, conditional, and iterative flow of control. Arbitrary transfer of control – in the form of the GOTO STATEMENT – is expressly forbidden. As a result, for each statement or group of statements in a structured program there is precisely one entry point and one exit point. The program is thus (in theory) easy to follow, to debug, and to modify.

structured query language   *See* SQL.

StuffIt /stuff-it/   A Macintosh program for compressing and decompressing files. It is now commercially available for both Macintoshes and PCs.

subprogram /sub-proh-gram/   *See* program unit.

subroutine /sub-roo-teen/   **1.** A section of a program that carries out a well-defined operation on data. Subroutines function like PROCEDURES except that they cannot have PARAMETERS. They are found in ASSEMBLY LANGUAGE and a few HIGH-LEVEL LANGUAGES, particularly Basic. **2.** The term used for a procedure in such languages as Fortran.

subroutine library   A collection of SUBROUTINES, PROCEDURES, and FUNCTIONS kept in the form of object code in a single file on BACKING STORE. A subroutine library will often consist entirely of routines connected with a common subject area, such as graphics or numerical analysis. When a user program requiring library routines is combined with the library file by the LINK EDITOR, the library is searched for the required routines, and only they are extracted. If searching is not used, all the library routines would be combined with the user program whether they were required or not. *See also* program library.

subscript /sub-skript/ (index)   A value, usually an integer or integer expression, that is used in selecting a particular element in an ARRAY. In computing, subscripts cannot be put in a subscript position (e.g. x_i), as the name suggests, and some other notation must be used. This usually involves brackets, as in x(i) or x[i].

substring /sub-string/   *See* string.

suitcase   A Macintosh file containing fonts or desk accessories.

suite /sweet/   A set of PROGRAMS or program MODULES that is designed so as to meet some overall requirement, each program or module meeting some part of that requirement. For example, an accounting suite might consist of separate programs for stock control, inventory, payroll, etc.

Sun Microsystems   A US company founded in 1982 that builds computer hardware and software. It is best known for developing workstations and operating environments for UNIX operating systems and also for the development of JAVA.

supercomputer /soo-per-kŏm-**pyoo**-ter/   A computer that is capable of working at very great speed, and can thus process a very large amount of data within an acceptable time. Supercomputers are in fact now purpose-built to manipulate numbers, and are the most powerful form of NUMBER CRUNCHER. Advanced supercomputers can handle tens of billions of floating-point operations per second. These machines are used, for example, in meteorology, engineering, nuclear physics, and astronomy. The main manufacturers of supercomputers are Cray Inc., IBM, and NEC.

superuser   *See* administrator.

supervisor 1. (executive) The part of a large OPERATING SYSTEM that resides permanently in main store, as opposed to other sections of the operating system that are brought into main store when required. The supervisor is responsible for the supervision during some time period of many unconnected processing tasks running on the computer. Through the KERNEL, it controls the use by these processing tasks of the physical components of the computer system. The supervisor calls the routines it needs to perform a particular task, such as to display a dialog box or open a file. 2. *See* administrator.

surf To browse for information on the Internet.

surface-mount technology (SMT) A PCB manufacturing technology in which chips are attached directly to the surface of the board rather than being soldered into predrilled holes. Such boards are more compact and can accommodate densely packed interconnections on both sides of the board. *See also* DIP.

SVGA (super VGA) A video graphics standard introduced by VESA in 1989, which gives higher resolutions, such as 800 × 600 or 1280 × 1024, than VGA. Up to 16.7 million colors can also be achieved. At the larger resolutions SVGAs are slower and objects appear smaller. These resolutions appear better on larger monitors and are improved by the use of GRAPHICS ACCELERATORS.

swapping (roll-out roll-in) A method of handling MAIN STORE in which the contents of an area of main store are interchanged with the contents of an area of BACKING STORE. The contents are written to backing store during periods when they are not required by the processor, and are read back into main store when needed. Swapping is controlled automatically by the OPERATING SYSTEM of the computer and is used in systems where VIRTUAL STORAGE is available.

switching The use of a temporary (rather than a dedicated permanent) connection for a communications link. In a communications network messages are switched (routed) along a path of intermediate stations.

switching speed A measure of the rate at which an electronic LOGIC GATE or LOGIC CIRCUIT can change the state of its output from high to low or vice versa in response to changes at its inputs. It is extremely fast but does depend on the fabrication technology; TTL and ECL devices have much higher speeds than CMOS devices.

switch statement A conditional CONTROL STRUCTURE in certain programming languages (e.g. Algol, C) that allows a selection to be made between three or more choices. It is equivalent to the CASE STATEMENT that appears, for example, in Pascal.

symbol A string of one or more letters or characters used to represent an object, operation, quantity, relation, or function. See the appendix for a list of mathematical symbols.

symbolic addressing A method of addressing in which reference to an ADDRESS is made in an instruction by means of some convenient symbol chosen by the programmer. The symbol may be one or more letters or other characters (appropriate to the programming language), generally bearing some relationship to the meaning of the data expected to be located at that address. This *symbolic address* is replaced by some form of computed or computable address (such as the MACHINE ADDRESS) during the operation of an ASSEMBLER or COMPILER. *See also* addressing mode.

symbolic logic (formal logic) The branch of logic in which arguments, the terms used in them, the relationships between them, and the various operations that can be performed on them are all represented by symbols. The logical properties and implications of arguments can then be more easily studied strictly and formally, using algebraic techniques, proofs, and theorems in a mathematically rigorous way. It is sometimes called mathematical logic.

The simplest system of symbolic logic is *propositional logic* (sometimes called *propositional calculus*) in which letters, e.g. *P*, *Q*, *R*, etc., stand for propositions or statements, and various special symbols stand for relationships that can hold between them.

symbol table A list kept by a COMPILER or an ASSEMBLER of the IDENTIFIERS that have been used in a program. An identifier consists of a string of characters chosen by the programmer to identify a VARIABLE, an ARRAY, a FUNCTION, a PROCEDURE, or some other element in the program. The properties of these identifiers, such as their address, value, or size, form part of the table.

As the translation of the source program proceeds, the compiler or assembler recognizes occurrences of identifiers – or symbols – in the text and looks them up in the table. If a symbol is already in the table, then the associated value is used by the compiler or assembler instead of the symbol. If the symbol is not in the table, then it is entered and eventually, when a definition of the symbol is found, the correct information is also entered. The compilation or assembly will terminate with errors unless all symbols are defined.

symmetric digital subscriber line *See* SDSL.

symmetric multiprocessing *See* SMP.

synchronous /sink-rŏ-nŭs/ Involving or requiring a form of timing control in which sequential events or operations take place at fixed and predictable times, usually determined by an electronic signal generated by a CLOCK. This means that the length of time required by each set of events or operations must already be known.

Digital computers operate as synchronous machines: the start of every basic operation is under the control of the signal from an internal clock. The operations are therefore kept in synchrony – in step – with the clock signal (rather as the players in an orchestra are kept in time by the conduc-

tor's baton), and thus take place at regular and predictable times.

In *synchronous transmission* between two devices, the devices can operate continuously and each bit of data is transmitted at a fixed and predictable time. *Compare* asynchronous.

syntax /sin-taks/ Part of the description of a PROGRAMMING LANGUAGE, akin to the grammar of English or some other natural language. The syntax of a language is the set of rules defining which combinations of letters, digits, and other characters are permitted in that language. The rules allow a valid sequence of characters to be distinguished from an invalid, i.e. a meaningless combination. They determine the form of the various constructs – STATEMENTS, CONTROL STRUCTURES, etc. – but say nothing about the meaning of the constructs. The syntax of a language can be expressed precisely and unambiguously, allowing SYNTAX ERRORS to be detected. It is often expressed in a symbolic notation called *Backus-Naur form* (*BNF*) or some derivative of BNF. *See also* semantics.

syntax analysis (parsing) A phase in the COMPILATION of a program during which the program is checked for compliance with the SYNTAX of the programming language that has been used. If syntax errors are found then the compilation process is stopped and the errors can be corrected. Syntax analysis follows the process of LEXICAL ANALYSIS, and is achieved by means of a program known as a *syntax analyzer* or *parser*. The input to the syntax analyzer is a string of tokens from the lexical analyzer.

syntax error A programming ERROR in which the grammatical rules of the programming language are broken, i.e. the program fails to obey the SYNTAX of the programming language. Syntax errors are generally detected at COMPILE TIME. The compiler would then normally produce information indicating both the location of the error(s) in the program and the kind of error(s) involved – e.g. an unrecognized statement or an undeclared identifier.

sysop /**siss**-op/ The administrator of a BBS or forum. The word is a contraction of 'system operator'.

system A set of related components that can be regarded as a collective entity. In computing the word is most widely used to mean a related set of hardware units, or programs, or both. For example, the hardware units may make up a working computer installation, or one under design, as in the terms *computer system*, *system design*. The programs making up a system may, for instance, be those jointly controlling the operation of a computer, i.e. the OPERATING SYSTEM, or may be a group of programs used or required for a particular application. In the case of hardware units the word system may be broadened in meaning to include basic software, such as operating systems, associated with the hardware.

system crash *See* crash.

system design *See* systems analysis.

system disk *See* startup disk.

system file **1.** Any file that is part of a computer's operating system.
2. A Macintosh operating system resource file that contains such items as fonts, icons, default dialog boxes, etc.

system flowchart (**data flowchart**) *See* flowchart.

system library *See* program library.

systems analysis The analysis of the role of a computer system in, for example, fulfilling a particular job in a business, and the identification of a set of requirements that the system should meet. This can then be used as the starting point for an appropriate *system design*, from which a working version of the system is eventually produced.

In commercial programming, where the term systems analysis is most commonly used (with a variety of meanings), those involved in developing software can be de-

scribed as either *systems analysts* or *programmers*. Systems analysts are responsible for identifying a set of requirements and for producing a design that meets those requirements. The design is concerned with the main components of the system and their roles and interrelationships, and sometimes with the internal structure and operation of individual components. The procedures detailed in the design by the analysts are then encoded by the programmers using a suitable programming language. The programmers are thus responsible for the production of the working version of a system, i.e. its implementation.

systems analyst *See* systems analysis.

system specification A precise and detailed statement of what a computer system – hardware or software – is to do, generally without any commitment as to how the system is to do it. A good system specification defines only the externally observable behavior of the system, treating the system itself as a 'black box' whose internal makeup need not be considered. It is normally produced once the set of requirements to be met by the system has been identified. It is then used as a basis for the design of the system.

systems programming *See* systems software.

systems software The software required to produce a computer system useful to end-users, providing a software environment in which they can work effectively and on which they can build. It includes programs that are essential to the effective use of the system. OPERATING SYSTEMS, COMPILERS, UTILITY PROGRAMS, DATABASE MANAGEMENT SYSTEMS, and COMMUNICATION SYSTEMS are examples of systems software. Systems software is often bought together with the computer. It can also be produced or modified by competent users. The production, documentation, maintenance, and modification of systems software is known as *systems programming*.

tab key A key positioned toward the top left of the keyboard and marked with two opposing arrows. It is used for inserting a tab character into a document. Generally, the tab key will also move the FOCUS from place to place in a GUI system.

table A collection of data values, each one of which can be identified by some method. The values may be arranged in an ARRAY and a particular item is then identified by one or more subscript values. Alternatively the data may be stored in the form of RECORDS; one or more fields in each record form a unique KEY, and the key is the means of identification.

Values can be extracted from tables using the method of *table look-up (TLU)*. The subscript or key in the table is then the value to be looked up. The table may, for example, be a list of square roots. The 4th entry in the table would then be the square root of 4, i.e. 2. Again, the table may consist of prices of numbered items on a menu. Entering the number of a particular dish gives its price; the table can thus be used to calculate the bill for a particular selection of dishes.

table look-up (TLU) *See* table.

tag *See* SGML; XML.

tape *See* magnetic tape.

tape cartridge Usually, a casing containing MAGNETIC TAPE and from which the tape is not normally removed. The casing therefore protects the tape. The cartridge can be loaded in a suitable TAPE UNIT without the tape being handled. Cartridges are similar to CASSETTES but are faster and have larger storage capacities.

tape cassette *See* cassette.

tape deck *See* tape drive.

tape drive (tape transport; tape deck) A mechanism for moving MAGNETIC TAPE and controlling its movement. A tape drive together with magnetic read and write heads and associated electronics is known as a TAPE UNIT. The term tape drive is often used however as another name for the whole tape unit.

tape format *See* format; magnetic tape.

tape streamer (streamer) A type of TAPE UNIT specifically designed for the rapid BACKUP of magnetic disks using TAPE CARTRIDGES. The tape streamer is an efficient way of backing up the fixed disk systems.

tape transport *See* tape drive.

tape unit (magnetic tape unit, MTU) A peripheral device containing a mechanism for moving MAGNETIC TAPE and controlling its movement, together with recording and sensing HEADS (and associated electronics) that cause data to be written to and read from the tape. Magnetic tape is held in TAPE CARTRIDGES, CASSETTES, or, in older systems, on reels. The reel, cassette, or cartridge is mounted in the tape unit, and the tape is driven past the heads.

Data is recorded on the tape by the *write head*: the head receives an electrical signal coded with the data and converts it into patterns of magnetization on the tape. Data on the tape can be sensed by the *read head*: the patterns of magnetization induce a coded electrical signal in the read head, and this signal can be fed to the computer. Data is stored on and retrieved from magnetic tape by the method of SERIAL ACCESS, i.e. by first reading through all previous items of data on the tape until the required location is reached.

Tapes are nowadays no longer used as on-line backing store. However, they are still an important backup medium because they are relatively cheap and their main disadvantage – long access times – is unimportant.

tar A UNIX shell command that makes a single file (called an archive) from a set of files or extracts files from such an archive. The files are not compressed and the extension of the archive file is tar. Tar is a contraction of 'tape archive'.

target language The language into which source code is compiled.

task *See* process.

task bar A bar containing the start button, which appears at the foot of a Windows screen. Icons representing active applications or documents are also shown here and can be selected with the mouse.

T-carrier A long-distance digital communications line used for voice communication and Internet connection using pulse-code modulation and time-division multiplexing. The T-carrier service has several different capacity levels, from T1 to T4.

TCO *See* Total Cost of Ownership.

TCP/IP /tee-see-pee-ÿ-**pee**/ (**transmission control protocol/Internet protocol**) A set of protocols developed originally by the US Department of Defense, which has become the de facto standard for communications between computers and networks.

telecommuting /tel-ee-kŏ-**myoot**-ing/ The use of telecommunications and computer networks to work outside the normal office or workplace.

teleconferencing /tel-ee-**kon**-fĕ-rĕn-sing/ A computer-based system enabling its users to participate in a conference or some other joint activity despite being at different places and/or communicating at different times. Users typically have access to computers or terminals interconnected by communication lines. A contribution made by any member of the conference will be brought to the attention of each of the other members when they next use their computers or terminals. Sophisticated teleconferencing systems use audio and video equipment.

television monitor *See* monitor.

teleworking /tel-ĕ-wer-king/ The use of computers and communications technologies to interact with customers and business colleagues. *See also* SOHO.

Telnet /**tel**-net/ A high-level communications protocol that lets a user log on to a remote computer.

tera- (ter-a) A prefix indicating a multiple of a million million (i.e. 10^{12}) or, loosely, a multiple of 2^{40} (i.e. 1 099 511 627 776). In science and technology decimal notation is usually used, and powers of 10 are thus encountered – 10^3, 10^6, 10^9, etc. The symbol T is used for tera-, as in TV for teravolt. Binary notation is generally used in computing, and so the power of 2 nearest to 10^{12} has assumed the meaning of tera-.

The prefix is most frequently encountered in computing in the context of storage CAPACITY. With magnetic disks, magnetic tape, and main store, the capacity is normally reckoned in terms of the number of BYTES that can be stored, and the terabyte is used to mean 2^{40} bytes; terabyte is usually abbreviated to T byte, Tb, or just to T.

Tera- is part of a sequence
kilo-, mega-, giga-, tera-, peta-, ...
of increasing powers of 10^3 (or of 2^{10}).

teraflop /**te**-ră-flop/ *See* flop.

terminal An input/output device that is used for communication with a computer from a remote site. (Some terminals are restricted to the output of data or to the input of data.) The terminal may be linked by cable to the computer, which may be in the next room or a nearby building, or the two may be connected over a telephone line. In larger computer systems there may be many terminals linked to one computer (*see* multiaccess system).

The most common type of terminal is the VDU paired with a KEYBOARD: the user of the VDU types information to the computer and information from the computer is received on the VDU screen. Terminals can be designed for a particular application; examples include POINT-OF-SALE TERMINALS and cash dispensers at banks. If a terminal has a built-in capability to store and manipulate data – i.e. it contains a microcomputer – then it is classed as an *intelligent* (or *smart*) *terminal*; without this capability it is described as *dumb*.

termination The end of execution of a program or some other processing task. A processing task that reaches a successful conclusion terminates by returning control to the OPERATING SYSTEM; this is described as a *normal* termination. If the processing task reaches a point from which it cannot continue, it ABORTS itself, or is aborted by the operating system or the user; this is referred to as an *abnormal* termination.

terminator 1. A character or group of characters used to mark the end of a program STATEMENT, CONTROL STRUCTURE, or some other item of data. For example, in Fortran every statement is terminated by an end of line while in the language C every statement must have a semicolon at the end. *Compare* separator.
2. (rogue value) A value that is added at the end of a list of data values and that is recognized by the computer as a signal to terminate some operation on the data. It must not be possible for the terminator to appear in the list of data values. If the data values consist of, say, numbers from 0 to 50, then 99 could be selected as the terminator.

test data *See* testing.

testing Any activity that checks whether a system – software or hardware – behaves in the desired manner. It is achieved by means of a *test run* in which the system is supplied with input data, and the system's responses are recorded for analysis. The input data is referred to as *test data*. Testing can be performed on individual components of the system in isolation; when the components are brought together, further testing can be performed to check that the components operate together correctly. Alternatively the behavior of a system can be investigated without concern for individual components and their internal interfaces. Testing can never absolutely prove the correctness of a program. It can only show that the program works under certain circumstances. *See also* debugging.

test run *See* testing.

TeX /tek/ A large and complex typesetting system available as freeware for most operating systems, including Windows, UNIX, and Macintosh. It was created by Donald Knuth for the production of high-quality scientific, mathematical, or other technical documents. *See also* LaTeX.

texel /teks-ĕl/ The smallest graphical element in a two-dimensional texture mapping used in the rendition of a three-dimensional object. Pixels and texels are similar, but not the same, as although there may be a one-to-one correspondence in some parts of a rendition, they cannot usually be paired off in this way.

text box A type of CONTROL on a GUI FORM. The user types in content, which may be validated in some way but is not restricted to a set list of choices. *Compare* list box; *see also* combo box.

text editor A UTILITY PROGRAM used specifically for input and modification of information in textual form. For example, a document in English can be keyed into a computer and a text editor used to create a FILE. The same applies to a program in a high-level programming language. A text editor can also be used to inspect the contents of a file, and if necessary to modify them by inserting, deleting, or reordering data. Text editors are an essential means of communication between user and computer in an INTERACTIVE system. There is a considerable overlap between text editors and WORD PROCESSING systems.

A *screen editor* is a text editor that enables a particular portion of a file to be displayed through a WINDOW on a screen. The screen cursor can then be positioned at points where insertions, deletions, and other editing functions are to be performed. Since the window is movable, any portion of the file can be examined.

A *line editor* is a text editor that works with a single line at a time exclusively through keyboard commands. Line editors are normally used only when hardware or software limitations make a screen editor impractical.

text file   *See* plain text.

TFT display (**thin film transistor display**)   *See* active matrix display.

thermal printer   A type of nonimpact PRINTER – either a LINE PRINTER or a CHARACTER PRINTER – in which the printing mechanism contains tiny heating elements. Localized heating of special heat-sensitive paper by the printing mechanism causes visible marks to appear on the paper in the shapes of the required characters. The *thermal transfer printer* is a later development in which localized heating causes special 'thermoplastic' ink to be transferred to the paper from a ribbon or film, producing an image that is more permanent and of better quality; this printer is either a character or a page printer. Thermal and thermal transfer printers are quiet in operation compared with impact printers.

thin client (**network computer**)   A network device in a client/server system that can process information independently, but relies on the server for storage and applications and is centrally managed. *See* net PC.

third generation computers   Comput-

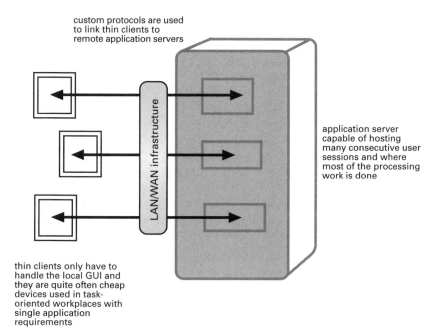

custom protocols are used to link thin clients to remote application servers

LAN/WAN infrastructure

application server capable of hosting many consecutive user sessions and where most of the processing work is done

thin clients only have to handle the local GUI and they are quite often cheap devices used in task-oriented workplaces with single application requirements

Thin client

ers designed in the 1960s or thereabouts. Since computer design is a continuous process by different groups in several countries, it is difficult to establish when a generation starts and finishes. One feature regarded as characterizing the third generation is the introduction of INTEGRATED CIRCUITS, although discrete transistors were still used in most of this generation. Main store was almost exclusively magnetic CORE STORE.

In general, SECOND GENERATION COMPUTERS were limited to what the engineers could put together and make work. Advances in electronic technology now made it possible to design a computer to suit the requirements of the tasks envisaged for the machine: the concept of computer ARCHITECTURE thus became a reality with this generation. Most manufacturers introduced at least three members of a computer family, architecturally similar but differing in price and performance. The IBM 360 series, first introduced in 1964, is an important example.

Comprehensive OPERATING SYSTEMS became, more or less, part of third generation machines. MULTITASKING was facilitated and much of the task of control of storage, input/output, and other resources became vested in the operating system or the machine itself. In addition new programming languages were introduced, such as COBOL, and later versions of existing languages, such as FORTRAN, came into use.

See also first, fourth, fifth generation computers.

thrashing 1. Excessive hard-disk use that occurs when a large number of short files scattered over the disk are being retrieved. Performance will be degraded. 2. The situation in which a VIRTUAL STORAGE system is continually swapping pages in and out of memory rather than running applications. Very little progress is made under these circumstances as memory or other resources have become exhausted or too limited to perform the necessary operations.

thread 1. A sequence of postings relating to an original newsgroup or forum message that continue on from each other. 2. *See* multitasking.

throughput /**throo**-pût/ A measure of the overall performance of a computer system, i.e. of the amount of work performed in a given period. It can, for example, be measured in terms of instructions per second.

thumbnail /**thum**-nayl/ A small version of an image or document page, which is much quicker to display on a screen or download from a Web site. Thumbnails make it quicker and easier to view or manage a group of larger images. They are often used in image catalogs.

thunk A code in an operating system that allows 16-bit code to call 32-bit code and vice versa. For example, Windows can run programs written in both 16-bit and 32-bit instruction sets.

TIFF /tiff/ (**tagged image file format**) A common format for scanning, storing, and interchanging raster graphics images. The files have the file extension tif.

time division multiplexing (TDM) *See* multiplexer.

timeout A condition that occurs when a process waiting for an external event reaches the end of a preset time interval before the external event has been detected. If, for example, the process has sent a message and no ACKNOWLEDGMENT has been detected at the end of the preset time period, then the process may take appropriate action, such as retransmitting the message.

timer *See* counter.

time sharing A technique, similar to TIME SLICING, whereby the time of a computer can be shared among several JOBS or users in such a way that each one appears to have sole use of the computer.

time slice *See* time slicing.

time slicing A technique whereby the time of a computer can be shared among several PROCESSES, a brief period being allocated (by the OPERATING SYSTEM) to each process in turn. During such a period – known as a *time slice* – the process is permitted to use the resources of the computer, i.e. the processor, main store, etc. Time slicing differs from TIME SHARING in that it disregards which jobs or users own the processes and so makes no effort to share the available time evenly between jobs or users. *See also* multitasking.

TLU (**table look-up**) *See* table.

token *See* lexical analysis.

token ring network A LAN in which the nodes are connected together in such a way that each node can only communicate with its neighbors on either side, leading to a logical ring. In practice star clusters of up to eight workstations can be connected up to a multiaccess unit that acts as a node on the ring. The single token is a series of bits sent around the ring from node to node, and it is used to arbitrate who has access to the ring at any one time. When a node wants to communicate it must grab the token when it is empty, and then it can send data to any other node; this data must be delivered successfully before any other node can grab the token. Originally developed by IBM and mainly used on IBM systems, the use of token ring networks has declined in recent years in favour of ETHERNET. *See also* network.

toolbar A block of icons showing in an application window, often in a row or column, that, when clicked on with the mouse, activate functions within the application. For example, a drawing toolbar could contain icons to draw a circle, pick up a paintbrush, shade an object, and so forth. Toolbars can often be customized by the user and moved and resized on the screen.

top-down design An approach to the design of a program, or a computer system, that starts with a statement of what is re-quired of the program or system. This is broken down into a succession of different levels that are progressively more detailed. At each level more basic elements are introduced to describe exactly what is required. These elements are then defined in the next stage by introducing even more basic elements. The process stops when there is sufficient detail for the elements to be written in the chosen programming language or to be brought together to form the desired system.

Total Cost of Ownership (TCO) A method of calculating the cost of software, hardware, etc., that takes into account not only the purchase price but also all projected subsequent costs over the item's lifetime: upgrades, repair, support, user training, etc. For example a low-price software application that is difficult to use and requires extensive user training and support might have a higher TCO that a rival high-price application that can be configured easily and used intuitively.

touchpad /**tuch**-pad/ *See* touch-sensitive device.

touch screen *See* touch-sensitive device.

touch-sensitive device A flat rectangular device that responds to the touch of, say, a finger by transmitting the coordinates of the touched point (i.e. its position) to a computer. The touch-sensitive area may be the VDU screen itself, in which case it is called a *touch screen*; alternatively it may be part of the keyboard or a separate unit that can be placed on a desk. In the first case the computer user can, for example, make a selection from a number of options displayed on the screen by touching one of them; in the latter two cases movement of the finger across the so-called *touchpad* can cause the CURSOR to move around the screen. *See also* mouse.

TP *See* transaction processing.

trace A report of the sequence of actions carried out during the execution of a program. It is produced by a *trace program*. A trace program records – or *traces* – the sequence in which the statements in the program are executed, and usually the results of executing the statements. It is thus possible to follow changes in data values at each stage of the program. There may be additional options offered by a trace program, for example it may trace changes to the value of a specific variable in the program. The trace information is stored in a *trace table*. A trace program can be used in DEBUGGING a program.

track   The portion of a magnetic storage medium along which data is stored. In MAGNETIC DISKS the tracks are concentric circles along which data is recorded as a stream of bits. There are several thousand tracks on a hard disk and 80 on a floppy disk. The tracks are subdivided into SECTORS for the purposes of reading and writing items of data.

trackerball /**trak**-er-bawl/ (**trackball**)   A device for generating signals that can cause the CURSOR or some other symbol to be moved about on a display. It consists of a ball supported on bearings so that it is free to rotate in any direction. The ball is held in a socket with less than half its surface exposed, and can be rotated by the operator's fingers. The direction of rotation produces a corresponding movement of the cursor. Trackerballs were used on laptop computers instead of mice, but have now been replaced by touchpads (*see* touch-sensitive device).

tractor feed   A technique for advancing CONTINUOUS STATIONERY through a PRINTER. The mechanism used to achieve this is called a *tractor* or *forms tractor*, and consists of a pair of loops that can be rotated in steps by the printer. The continuous stationery has a row of regularly spaced holes down each side, which engage with pegs or pintles on the loops and allow the paper to be accurately registered and advanced. *See also* cut sheet feed; friction feed.

train printer   *See* line printer.

transaction   1. A set of computer operations that must be treated as a unit; in particular, they must all be processed successfully or not processed at all. This means that, in the case of failure, any updates to data, etc., made by previous instructions in the transaction must be undone (*rolled back*). For example, making an entry in an accounting system typically involves crediting one account with an amount and debiting another account with the same amount; were the credit to be carried out but the debit to fail, or vice versa, the data files would be left in an inconsistent state. Grouping the credit and debit as a transaction ensures that this cannot happen. Transactions are used especially with databases.
2. A request for computer action that must be treated as such a transaction. Examples include making a booking or purchase online, getting cash from a cash dispenser, etc.

transaction file   A collection of TRANSACTIONS assembled prior to the automatic updating of a MASTER FILE or DATABASE. *See also* file updating.

transaction processing (**TP**)   A method of organizing a DATA PROCESSING system in which TRANSACTIONS are processed to completion as they arise, rather than being collected together for subsequent processing. An ON-LINE computer is therefore required. Transaction processing is used, for example, in travel agents to check the availability of airline seats, accommodation, etc.: if acceptable a booking can be made and the booking record updated on the spot. *See also* on-line transaction processing.

transfer   To send data from one point to another point. The process is known as a *transfer* or *data transfer*, and more specifically may be a FILE TRANSFER. The word transfer is used to describe the movement of data within a computer system – i.e. between storage LOCATIONS – or over a long-distance TRANSMISSION LINE.

transfer rate   *See* data transfer rate.

transistor /tran-**zis**-ter/   An electronic de-

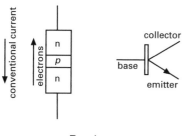

Transistor

vice constructed from semiconductor material and used for amplification and switching purposes. Various types can be produced with different characteristics. In computer hardware, transistors are used for their switching properties, especially in LOGIC CIRCUITS and SEMICONDUCTOR MEMORY. Invented in 1948, the transistor was used at first in discrete form. Compared with the electronic valve it was much smaller, consumed very little power, and was both cheap and reliable. From about 1959 it could be fabricated as part of an INTEGRATED CIRCUIT. In both forms it has made a great impact on computer development (*see* second, third, fourth generation computers).

transistor-transistor logic *See* TTL.

translation table A table of information that is stored within a processor or a peripheral device and is used to convert encoded information into another form of code with the same meaning. There are a wide variety of codes used in computing, and sometimes more than one code may be used within a single computer system. For example, the Macintosh uses a different character set to Windows. Appropriate translation tables allow documents created on one to be used on the other.

translator A program that converts a program written in one language into the equivalent program in another language. COMPILERS and ASSEMBLERS are examples: a compiler translates a program written in a high-level language, such as C++, into machine code; an assembler translates a pro-

gram written in assembly language into machine code.

transmission The sending of data from one place for reception elsewhere. The data is said to have been *transmitted*. The data may be analog or digital measurements, encoded characters, or information in general, and is carried as an encoded signal. The words transmission and transmit are usually restricted to use in telecommunications, describing the movement of data over, say, a telephone line.

transmission channel *See* communication channel.

transmission line (**communication line**) Any physical medium used to carry information between different locations. It may, for example, be a telephone line, an electric cable, an optical fiber, a radio beam, a microwave beam, or a laser beam. *Compare* data link.

transmit *See* transmission.

transport layer The fourth of the seven layers in the open systems interconnection (OSI) communications model. It ensures the reliable arrival of messages and provides error detection and correction, and data-flow controls.

tree One form in which a collection of data items can be held in computer memory. It is a nonlinear structure, similar in form to a family tree (see diagram). The set of locations at which data items are stored are called *nodes*. One particular item of data (at the *root node*) has LINKS to one or more other items, which in turn may possibly have links to one or more other items, and so on. There is thus a unique path from the root node in a tree to any other node. For instance, the data item C in the diagram can be accessed from the root node F via D and B.

The diagram shows an example of a *binary tree*. This is a tree in which each node has links, or arcs, to no more than two other nodes. Each node therefore consists of a data item and two links, left and right.

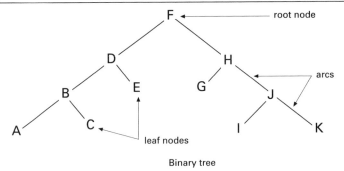

Binary tree

The *leaf nodes* at the bottom of the diagram have empty links. Note that trees are drawn upside down with the root at the top and the leaves at the bottom.

trojan /**troh**-jăn/ (**trojan horse**)   A small destructive program disguised as a game, application, or utility, which when activated causes damage although appearing to be of use.

true color   The specification of the color of a pixel on a display screen using a 24-bit value, 8 bits for each of the three primary RGB colors, giving the possibility of 16 777 216 colors.

TrueType /**troo**-tÿp/   A scalable font technology introduced first by Apple and then by Microsoft to enable high-grade fonts to be used in their operating systems. To view and print a TrueType font it is necessary to have the font file itself and TrueType raster graphics, which is inbuilt in Windows and Macintosh operating systems. TrueType fonts can be embedded in Web pages or word documents and viewed by users who do not have the font installed on their machines. Windows and Macintosh versions of TrueType fonts are not compatible. *See also* font.

truncation /trunk-**ay**-shŏn/   The process applied to a number whereby all digits after a specified number of significant digits are ignored. For example, truncation of the number 3.257 69 to 4 or to 5 digits yields 3.257 or 3.2576 respectively. In computing, truncation is usually required because the REGISTERS in a computer can hold only a fixed number of binary digits (bits). When the result of an arithmetic operation requires more than this fixed number of bits, either truncation or ROUNDING will take place. The truncated value is an approximation to the true value of the number, and leads to a *truncation error*. Repeated truncation can cause a build-up of such errors and the programmer should try to minimize this problem.

truth table   A table of values that describes a particular LOGIC OPERATION or LOGIC GATE. Truth tables were first used in connection with statements in logic that could be assigned one of two truth values – *true* or *false*. They can be used in computing to describe the logic operations, such as AND, OR, and NOT, that are performed for example in the ARITHMETIC AND LOGIC

P	Q	P AND Q	NOT (P AND Q)
0	0	0	1
0	1	0	1
1	0	0	1
1	1	1	0

P	Q	P OR Q	NOT (P OR Q)
0	0	0	1
0	1	1	0
1	0	1	0
1	1	1	0

Truth tables for AND, OR, and NOT operations

UNIT of a central processor. The truth table of a particular operation lists all possible combinations of the two truth values of the quantities on which the operation is to be performed (the operands), together with the truth value of the outcome of each of the possible combinations. If there are 2 operands, P and Q, there are 2^2, i.e. 4, possible outcomes. Truth tables for AND, OR, and NOT operations are shown in the diagram. The truth values are represented as 1 (for the value *true*) and 0 (for the value *false*).

Truth tables are also used to describe or determine the function of a logic gate. The truth table of a particular gate lists the output obtained for all the possible combinations of inputs. With 2 inputs, each of which can be at one of two voltages, there are 4 possible output voltages. With 3 inputs there are 8 possible outputs. The truth tables for AND and OR GATES with two inputs are identical to those shown in the table for P AND Q and P OR Q. The truth table for a NOT GATE (which always has one input) is identical to that for NOT P.

truth value Either of two values, *true* or *false*, that can be assigned to a logical expression or conclusion in BOOLEAN ALGEBRA. The values are symbolized by T and F. When used in computing these so-called *logical values* can be represented by and stored as a single bit (usually 1 for *true* and 0 for *false*). Larger units, such as a byte or word, are also used to store logical values since this is simpler and quicker, although less efficient in storage space. *See also* truth table.

TSR (**terminate and stay resident**) Describing MS-DOS programs that stay in memory at all times. Such programs can be instantly accessed from within other programs. TSRs are not needed in Windows but are still supported for BACKWARD COMPATIBILITY.

TTL (**transistor-transistor logic**) A family of LOGIC CIRCUITS that are all fabricated with a similar structure by the same INTEGRATED-CIRCUIT techniques. The circuits are bipolar in nature and are all character-ized by fairly high SWITCHING SPEEDS. *Schottky TTL* has higher switching speeds and lower power requirements than the standard TTL. Both versions have a lower PACKING DENSITY than CMOS. *See also* ECL.

tunneling /tun-ĕl-ing/ A method of transmitting data over connected networks using different protocols. The packet based on one protocol is wrapped in the packet of another protocol so that it can travel over the network using the second protocol. Tunneling is used in virtual private networks (VPNs).

Turing machine /tew-ring/ An abstract model of a computer, put forward by Alan Turing in 1936, that consists of a control or processing unit and an infinitely long tape divided into single squares along its length. At any given time each square of the tape is either blank or contains a single symbol $s_1, s_2, ..., s_n$. The Turing machine moves along the tape square by square and reads, erases, and prints symbols. At any given time the Turing machine is in one of a fixed finite number of states represented by $q_1, q_2, ..., q_n$. The 'program' for the machine is made up of a finite set of instructions of the form $q_i s_j s_k X q_j$, where X is either R (move to the right), L (move to the left), or N (stay in the same position). Here, q_i is the state of a machine reading s_j, which it changes to s_k, then moves left, right, or stays and completes the operation by going into state q_j.

The Turing machine is considered the theoretical model for digital computers. It is still important in computer science.

turnaround time 1. The time that elapses between the submission of a job to a computer facility and the return of the results. This is the only time that really matters to someone using a computer in this way.
2. The time taken to reverse the direction of transmission on a one-way (SIMPLEX) or a HALF-DUPLEX communication channel.

turnkey operation /tern-kee/ The delivery and installation of a complete computer system plus applications programs so that the system can be placed into immediate operational use. This operational system is called a *turnkey system*.

turtle graphics A method by which information from a computer can be turned into pictures or patterns. The original drawing device was a simple pen-plotter known as a *turtle*, a small motorized carriage carrying one or more pens and connected to the computer by a flexible cable. The carriage wheels could be precisely controlled, enabling the turtle to be steered by commands from the computer in any direction across a floor or other flat surface covered in paper or similar material; the pens could be raised or lowered by control signals to produce the drawing.

The action of the turtle can now be simulated by graphics on the VDU screen: the *screen turtle* is usually a triangular arrow that may or may not produce a line as it is made to move across the screen. *See also* Logo.

TWAIN /twayn/ A standard for interfacing code that allows graphics manipulation programs to obtain an image from a scanner or digital camera. Nearly all such devices contain a TWAIN driver, which hides the detailed operation of the device behind the standard interface. A TWAIN-aware application can thus use any device with a TWAIN driver.

tweening /**tween**-ing/ A technique used in computer animation that enables smoother animation to be achieved by the automatic insertion of extra frames between the originally created frames.

twisted pair An electric cable consisting of two similar wires that are each electrically insulated by a plastic sheath and are twisted around one another to improve the transmission of signals. Bundles of tens or hundreds of twisted pairs are commonly encased in an outer protective sheath to make large cables.

two's complement *See* complement.

type *See* data type.

typewriter terminal *See* teletypewriter.

U

UART /**yoo**-art/ (**universal asynchronous receiver–transmitter**) A module, usually consisting of only one circuit, that can both receive and transmit asynchronous communications. It is commonly used in modems. *See* USART.

UDP (**user datagram protocol**) A TCP/IP connectionless protocol that offers a limited amount of service when messages are exchanged using Internet protocol. UDP does not split data into packets nor reassemble at the other end; this has to be done by application software. Neither does UDP provide sequencing of the packets that the data arrives in. It is equivalent to the transport layer in the ISO/OSI reference model. It is less reliable, but more efficient, than TCP, the reliability depending on the applications generating and receiving the message; UDP is therefore suitable for applications dealing with small data units and for SNMP.

UI *See* user interface.

ULA *See* uncommitted logic array.

unary minus /**yoo**-nă-ree, -nair-ee/ The arithmetic operation that takes only one OPERAND (hence the word 'unary') and that makes a positive operand negative and a negative operand positive. The OPERATOR for unary minus is the minus sign, –. It is usually highest in the ORDER OF PRECEDENCE.

unary operator (**monadic operator**) *See* operator.

unbundling /un-**bun**-děl-ing/ The separation of SYSTEMS SOFTWARE charges from hardware charges in the marketing of computer systems. Originally systems software was included in the purchase without additional charges since it was minimal and thus represented a small part of the total cost of the system. Unbundling was brought about by hardware becoming less expensive while systems software became a larger proportion of the total cost. Many microcomputer systems, however, reverse the process by including a number of application programs, such as WORD PROCESSORS or SPREADSHEETS, in the purchase price as an incentive to buy. This can be attractive since the computer manufacturer buys many copies of the software at a large discount and passes much of the saving on to the customer. This practice is known as *bundling* and the included software is described as *bundled*.

UNC (**Universal Naming Convention**) A method of identifying the location of a resource, such as a shared printer or NETWORK DIRECTORY, on a network. For example a file called 'ComputingDictionary.xml' in network folder called 'Dictionary' on a Windows computer called 'EdmundXP' could be accessed from the rest of the network by specifying the UNC name 'EdmundXP\Dictionary\ComputingDictionary.xml'. Unix networks and the Internet also use UNC names, but with forward slashes rather than backslashes.

uncommitted logic array (**ULA**) A semiconductor CHIP on which sets of LOGIC GATES have been fabricated in a standard form but are not yet connected. The interconnections are made after manufacture so that a required circuit pattern can be achieved.

unconditional jump (unconditional branch) *See* jump.

undelete /**un**-di-leet/ To restore a deleted file. This is possible because the data in a deleted file is normally not erased immediately (*see* deletion).

underflow The condition arising when the result of an arithmetic operation in a computer is too small to be stored in the location allocated to it. For example, a fraction represented by 32 leading 0s before the significant bits could not be stored in a 32-bit word. If there were no means to detect underflow, results would be unreliable and might be incorrect. Facilities are therefore provided to detect underflow. It can then be corrected. Normally the offending number is replaced by 0. *See also* overflow.

undo To reverse an action and return to the previous state. For example, an editing action in word processing can be reversed. Many applications have several levels over which one can backtrack.

Unicode /**yoo**-nĭ-kohd/ An encoding standard developed by the Unicode Consortium since the late 1980s, in which one or more bytes are used to represent each character. This gives Unicode the capability of representing a huge number of characters, although the current limit is 1 114 112; over 90 000 have been defined, many of them being used for Chinese ideographs.

Uniform Resource Indicator *See* URI.

uninstall /un-in-**stawl**/ To completely remove software from a computer, including any references to files or components held on system files, such as the Windows registry. All applications should be supplied with a suitable uninstaller program.

uninterruptible power supply *See* UPS.

UNIVAC /**yoo**-nĭ-vak/ *See* first generation computers.

universal asynchronous receiver–transmitter *See* UART.

Universal Naming Convention *See* UNC.

UNIX A multiuser multitasking operating system developed by AT&T Bell laboratories in 1971 for use on minicomputers. Over its long evolution it has become a powerful and complex operating system. It has been adapted to run on many platforms, for example Linux on Intel and other chips and A/UX for the Macintosh, and is used extensively as a network operating system. There are many freeware versions available.

unpack *See* pack.

unrecoverable error An error that causes a program to terminate or a system to crash, so that continuation is impossible. It is often called a *fatal error*.

unzip To uncompress a file that has been previously compressed by PKZIP.

upgrade /**up**-grayd/ A newer version of an existing application, usually offering greater functionality, or a patch to an existing application.

uplink The transmission link from a ground station to a communications satellite. *Compare* downlink.

upload /**up**-lohd/ To transfer a copy of a file from a local to a remote computer, such as a larger central system or an FTP site. *Compare* download.

upper case The capital form of an alphabetic character, as distinct from lower case.

UPS (uninterruptible power supply) A battery-powered device connected between a computer or other electronic equipment and a power source that automatically provides backup power in the event of power failure of the source. A UPS is activated when it detects a loss of power from the source; it provides enough running time to cover short breaks and then, if necessary, for the system or device to be shut down in an orderly fashion so that data is not lost. The device also protects against sudden power surges.

uptime /**up**-tÿm/ The amount of time a system has been powered up and running.

upward compatibility (**forward compatibility**) The ability of a computer system or software to work beside or on more advanced equipment or versions. *See also* backward compatibility.

URI (**Uniform Resource Indicator**) An identifier for a physical or abstract object that follows the syntax laid down in RFC 3986 (2005). Internet URLS form one subset of URIs: in 'http://www.microsoft.com', the part before the colon ('http') is call the *scheme* and the remainder of the URI follows the syntax specified for the http scheme. Other schemes include the 'tel' scheme for telephone numbers: the URI 'tel:+1-800-123-4567' represents the US telephone number 800-1230-4567. The use of such standardized names makes their manipulation be computer easier.

URL /yoo-ar-**el**/ (**uniform resource locator**) The address of a resource on the Internet, which specifies the protocol to be used in accessing the resource, the name of the server where the resource is to be located, and if necessary the path to the resource. An example address is
http://www.microsoft.com
indicating a WWW page on the commercial server microsoft.com. *See also* URI.

USART /yoo-sart/ (**universal synchronous/asynchronous receiver–transmitter**) A module that can send and receive both asynchronous and synchronous communications. *See also* UART.

USB (**universal serial bus**) A serial bus used to connect up to 127 peripherals to a microcomputer through a single general port, by daisy-chaining peripherals together. It is a plug-and-play INTERFACE in which new peripherals can be automatically detected and configured without shutting the computer down and without the need for an adapter card. All modern microcomputers support USB.

Usenet /yooz-net/ A worldwide network of UNIX systems with thousands of subject-based newsgroups. It has now become part of the Internet and has set the standard for other bulletin-board systems.

user datagram protocol /**day**-tă-gram/ *See* UDP.

user-friendly Denoting a piece of hardware or software that a user should find easy to use, convenient, or helpful, i.e. that a user who is not a computer specialist should not find confusing or intimidating.

user group A group of people with an interest in a particular application or computer system. Some user groups are large and provide forums in which ideas can be exchanged, problems aired, and help provided.

user interface The means of communication between a person and a computer system, referring in particular to the use of input/output devices with supporting software. The operating system provides the basic facilities in the form of a GRAPHICAL USER INTERFACE and/or a COMMAND LINE INTERFACE, which individual applications adapt to suit their requirements. A good user interface is important to the success of an application.

username /**yoo**-zer-naym/ The name by which a user is recognized on a computer or network system. When logging on a user must enter a username and, usually, a valid password.

user port An electrical socket in a microcomputer that with associated circuitry allows the computer user to connect additional peripheral devices. The individual pins of the port may be set or tested from a user's program. The port can be used to control or acquire data from external devices.

user program   Usually, a program written by a user of a computer system for his or her own use. Such programs may be shared among, say, members of a workgroup, but are not intended for wide distribution or commercial sale. *See also* application.

UTF-8   (8-bit Unicode Transformation Format)   A method of encoding UNICODE characters so that they take up less space. Unicode currently allows for over one million different characters, so each character would theoretically need to take up three bytes in a computer file or memory; in practice, computer architectures dictate that this would be extended to four bytes. UTF-8 reduces this space requirement drastically by allowing ASCII characters – character numbers 0 to 127, which include nearly all the keys found on a computer keyboard – to represent themselves and occupy one byte, while all other characters are encoded using between two and four bytes. UTF-8 was invented in 1992 and is defined in RFC 3629 (2003).

There are other varieties of UTF encoding. *UTF-16* uses either two or four bytes and is more efficient then UTF-8 for text that uses, e.g., Chinese ideographs. *UTF-32* uses four bytes for all characters; it is thus the simplest representation but offers no compression. UTF-8 is the default encoding used by XML and all programs that process XML data must be able to handle files that use UTF-8 or UTF-16.

UTF-16   *See* UTF-8.

UTF-32   *See* UTF-8.

utility program   Any of a collection of programs that forms part of every computer system and provides a variety of generally useful functions. Examples of utility programs are LOADERS, TEXT EDITORS, LINK EDITORS, programs for copying FILES from one storage device to another, for file deletion and FILE MAINTENANCE, for SEARCHING and SORTING, and for DEBUGGING of programs.

UUCP   (UNIX-to-UNIX copy)   A group of software programs used to help the transmission of data between UNIX systems using serial data connections.

uudecode   /yoo-yoo-dee-**kohd**/   *See* uuencode.

uuencode   /yoo-yoo-en-**kohd**/   A program, originating on UNIX systems, that is used to translate an 8-bit binary file into a 7-bit ASCII text file without loss of data. It produces a file about 4/3 the original size. It is used mainly for sending images or executable code by e-mails or newsgroups over links that do not support the transmission of 8-bit data. The reverse process, conversion back to 8-bit, is effected by the program *uudecode*. Modern e-mail applications perform these operations automatically when required.

UUPC   The DOS, Windows, and OS/2 version of UUCP.

V

V.90 A standard for 56 K modems approved by the International Telecommunications Union in 1998. It resolved incompatibilities between rival technologies and was intended to be the final 56 K modem standard. However, the subsequent *V.92* standard has added a few new features.

V.92 *See* V.90.

vaccine /**vak**-seen/ A utility program to check for computer viruses. It is often resident in memory to perform different levels of virus checking according to program settings.

validation /val-ă-**day**-shŏn/ The checking of data for correctness, or for compliance with the restrictions imposed on it. The checks are known as *validity checks*. They are done by computer, and an error message will be produced if an error is detected. For example, a validity check is performed after the input of data to ensure that the value of each item of data lies within the acceptable range, that the item is of the right type, etc. If the values are meant to lie between, say, 0 and 500, the validity check would detect any number exceeding 500 or any negative number. It would also detect that a letter had been input when a digit was intended. A validity check ought to be written into a program by the programmer. *See also* data cleaning.

validity check *See* validation.

value-added network (VAN) A network that gives increased functionality, such as message routing and conversion facilities, to users communicating at different speeds and using different protocols.

VAN *See* value-added network.

variable /**vair**-ee-ă-băl/ A CHARACTER or group of characters that is used to denote some value stored in computer memory and that can be changed during the execution of a program. Each variable used in a program is associated with a particular LOCATION or group of locations in memory.

The variable can take any value from a set of specified values of the same kind. The values may, for example, be INTEGERS from the set 0 to 99, they may be the REAL NUMBERS that a computer is able to handle, they may be the logical values *true* or *false*, or they may even be colors from the set *red, green, blue*. The particular value taken by the variable can be changed by means of an ASSIGNMENT STATEMENT. The NAME selected by the programmer to identify the variable, together with the SCOPE of the variable and the type of data involved (e.g. integers, reals, etc.), is stated in the program in a DECLARATION.

variance /**vair**-ee-ăns/ *See* standard deviation.

VAX /vaks/ (**virtual address extension**) A range of 32-bit processor computers introduced in 1978 by DEC (Digital Equipment Corporation). Development ceased in the early 1990s, but many VAX systems are still in use.

VBScript /vee-bee-**skript**/ (**Visual Basic scripting**) A scripting language based on the Visual Basic language but much simpler, developed by Microsoft and supported by Internet Explorer. Similar in many ways to JavaScript, it enables authors to include controls, such as buttons and scrollbars, on their Web pages.

vCard /**vee**-kard/ A specification for creating an electronic business card for the exchange of information in e-mails and teleconferencing.

VDSL (**very high bit-rate digital subscriber line**) A variant of DSL technology that transfers data over an ordinary copper telephone wire at very high speeds, but only over short distances. Like ADSL it can share the wire with a telephone and transfers data at different speeds in the two directions: up to 52 Mbps *downstream* (i.e. when downloading) and 16 Mbps *upstream* (i.e. when uploading). It is expected that VDSL will become more widely used as copper wires are replaced by fiber-optic cables, which overcome the distance problem. *See also* SDSL.

VDU (**visual display unit**) A device that can display computer output temporarily on a screen. The information can be composed of letters, numbers, or other characters, or can be in the form of pictures, graphs, or charts. The information can be changed or erased under the control of a computer; this computer may be at a remote site or may be a microcomputer of which the VDU forms a part. The VDU is usually paired with a KEYBOARD, by which information can be fed into the computer. A VDU plus keyboard is now the most commonly used computer TERMINAL.
　　Modern GUI-based operating systems treat all output to VDUs as graphics. However, VDUs also have a facility for HARDWARE CHARACTER GENERATION that allows simple text-only output; this is normally used only for low-level maintenance or compatibility.

vector /**vek**-ter/ A one-dimensional ARRAY.

vector graphics A method of producing pictorial images in which the image is stored as mathematical instructions that define curves, lines, etc., rather than pixels. *See also* computer graphics. *Compare* raster graphics.

Venn diagram /ven/ A diagram used to show the relationships between sets. The universal set, *E*, is shown as a rectangle. Inside this, other sets are shown as circles. Intersecting or overlapping circles are intersecting sets. Separate circles are sets that have no intersection. A circle inside another is a subset. A group of elements, or a subset, defined by any of these relationships may be indicated by a shaded area in the diagram.

verification /ve-ră-fă-**kay**-shŏn / The checking of the accuracy of data after it has been transcribed. It is done most commonly when data has been encoded by an operator at a keyboard reading from a document. The document is subsequently rekeyed by another operator, and the two sets of input data compared by machine. Any differences are indicated and suitable action is taken.

Veronica /vĕ-**ron**-ă-kă/ (**very easy rodent-oriented network index to computerized archives**) A GOPHER searching service developed by the University of Nevada, similar to ARCHIE, which is used on FTP servers.

vertical application A specialized application designed for use in a particular area of business or by a particular type of user; for example, insurance, government, or hospitals.

very high bit-rate digital subscriber line *See* VDSL.

VESA /**vee**-să/ (**Video Electronics Standards Association**) An organization devoted to the improvement of standards for video and multimedia devices.

VESA local bus (**VL bus; VLB**) A local-bus architecture used in the early 1990s on computers with an Intel 80486 CPU. It was superseded on Pentium-based machines by the PCI local bus.

VFAT /vee-eff-ay-**tee**/ (**virtual file allocation table**) An extension of the FAT FILE SYSTEM, introduced in Windows 95, that allowed long file names.

VGA (**video graphics array**) A computer

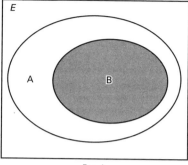

B ⊂ A

Venn diagram showing B as a subset of A

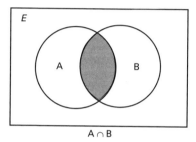

A ∩ B

Venn diagram showing the intersection of two sets

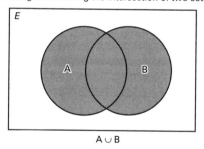

A ∪ B

Venn diagram showing the union of two sets

display system introduced in 1987 in which the maximum resolution depended on the number of colors displayed; for example, 16 colors at 640 × 480 pixels or 256 colors at 320 × 200 pixels. *See also* SVGA.

video adapter The components used to provide the output to a computer monitor. It is included on the motherboard or is on an expansion board, called a *display adapter* or *video display board*. Video

adapters usually contain a GRAPHICS ACCELERATOR.

video capture board An expansion board that is used to convert analog video signals to a digitized form from a plugged-in video cassette recorder. Some video capture boards can also perform the reverse action.

video conferencing A virtual meeting

between two or more participants in different locations achieved by using computer networks. Each participant has a video camera and speakers attached to his or her computer, and audio and video data is transmitted over the network. Participants can use a virtual conference room and appear as if they were sitting in a room next to each other. Hardware cost has made video conferencing too expensive for the small user in the past, but now inexpensive video conferencing software and digital cameras are available.

video display board   *See* video adapter.

video graphics board   A video adapter that generates the signals for displaying graphical images on a video screen. The term has fallen into disuse because nearly all modern video adapters have a graphics capability.

videophone   /**vid**-ee-oh-fohn/   A telephone that also has a camera and screen attached and is capable of transmitting and receiving video as well as audio signals. With conventional telephone lines, the only video that can be transmitted is freeze frame.

video RAM   (**VRAM**)   The RAM included on a video adapter board. All types of video RAM are special arrangements of dynamic RAM and act as a buffer between the CPU and the display. Most forms of video RAM are dual ported, which enables the processor to write a new image to the RAM while the display is reading from the RAM to refresh its current content.

virtual address   *See* virtual memory.

virtual memory (**virtual storage**)   The use of BACKING STORE, especially a hard disk, as if it were part of the MAIN STORE of a computer. In virtual memory systems, programs can be written requiring more main store than is physically present in the computer (or more than the OPERATING SYSTEM is prepared to allow them). The programs can address areas of main store that do not in fact exist using *virtual addresses*,

and during execution parts of the program not currently needed remain on backing store.

Data and instructions are divided by the operating system into blocks of fixed length; these blocks are called *pages*. When a reference to a virtual address is made, the system hardware detects whether or not the required address is physically present in main store. If it is not, an INTERRUPT is generated and the page containing the required location is transferred, by the operating system, from backing store to main store. A page is thus the unit of interchange between backing store and main store. Once in main store, special hardware or software maps the relevant virtual addresses onto the actual physical addresses. When a page held in main store is no longer required by the processor, it may be transferred back to backing store. The backing-store device is known as a *swapping device*.

virtual network   The connection of several LANs together so that they appear as one network or as a part of a network. For example, workstations on several sites can be grouped in the LAN definition in terms of department or type of user, rather than by geographical location, making the control and management much easier.

virtual reality   The use of a computer system to simulate a three-dimensional environment and allow the user to experience and interact with a realistic set of conditions, provided by sound, real-time motion effects, and sometimes with tactile feedback. It is used in such applications as flight simulation, the design of buildings, medical teaching, and computer games. *See* HMD.

virtual reality modeling language (**VRML**)   A language for creating interactive three-dimensional image sequences and possible user interaction with them. Links to VRML files can be embedded in HTML documents, but a VRML-enabled browser is needed to view such Web pages.

virtual storage   *See* virtual memory.

virus /vŷ-rŭs/   A program or macro code that attaches itself to other programs or files and can cause data corruption or hardware failure when activated, usually when the program is started or the file opened. Viruses can be transmitted as attachments to e-mails, by downloading an infected program, or by using a file from an infected portable disk. *See* vaccine; worm.

Visual Basic   A programming language and environment developed by Microsoft and launched in 1990. It is based on the Basic language and is particularly used for developing user interfaces. A programmer can produce code by dragging and dropping such controls as text boxes, buttons, etc. and then defining their characteristics. Visual Basic is sometimes called a 'rapid application development language' (RAD) because it enables programmers to develop prototype software quickly.

visual display unit   *See* VDU.

VLAN /vee-lan/ (**virtual local area network**)   A network of computers that behave as if they are connected to the same wire although they are physically located on different segments of the LAN. They are configured by software, which makes them very flexible. For example, when a computer is moved to a different location it can stay on the same VLAN without any hardware reconfiguration.

VLSI (**very large scale integration**)   *See* integrated circuit.

voice input device   A device in which speech is used to feed data or system commands directly into a computer system. Such equipment involves the use of SPEECH RECOGNITION processes, and can replace or supplement other input devices. Some voice input devices can recognize spoken words from a predefined vocabulary; other devices have to be trained for a particular speaker.

voice mail   A computer system that can store telephone messages in memory for later retrieval by a user. Each user has a mailbox and can copy, store, or forward the messages as appropriate.

voice output   The ability of a computer to produce spoken messages. Software can achieve this by either producing the sounds that make up words or assembling together prerecorded words. *See also* speech generation device.

Voice over IP   *See* VoIP.

voice recognition   The ability of a computer to understand speech, usually in the form of commands or data input from a user. Systems are limited to basic use at the moment because of the difficulty in recognizing accents, speech patterns, and differing syntaxes.

VoIP (**Voice over IP; Internet Voice**)   The transmission of verbal conversations over an IP-based network, especially the Internet. Sound is converted to digital form by an A/D CONVERTER, transmitted over the network, and converted back to sound at the destination. In its simplest form two computer users at machines equipped with microphones and loudspeakers can talk to each other as if on the telephone. Adaptors are available that connect telephone handsets to computers and emulate dialing tones, etc., as are special VoIP handsets that do not require a computer. For VoIP on the Internet, users must have a BROADBAND Internet connection and an account with a provider of VoIP services; they can then contact other subscribers to the same provider and often traditional telephones as well. VoIP has several advantages over the telephone network: it makes more efficient use of the available bandwidth; it is more flexible and VoIP can easily be integrated with other Internet services, such as CHAT and data transfer; and the same VoIP account can be used anywhere in the world. Above all it is currently significantly cheaper than the telephone, especially for international calls, and often completely free of charge. However, since VoIP is a comparatively new service there are still problems to be overcome; for example a user's Internet connection can run slowly and affect the reliability of VoIP, and it can be difficult to trace the geographical location of emergency calls. Nevertheless, it is expected that VoIP will be major growth area in the next few years. *See also* ENUM; SIP.

volatile memory /vol-ă-tăl/ A type of memory whose contents are lost as soon as the power supply to the memory is switched off. *Compare* nonvolatile memory.

voxel /**voks**-el/ A volume pixel, i.e. the smallest distinguishable box-shaped part of a three-dimensional image. Depth is added to an image using voxelization. The image is divided into two-dimensional slices each made up of pixels, and a volume is produced by stacking slices together in memory. Opacity transformation, which gives voxels different opacity values, is then used to produce a true three-dimensional image. Voxel imaging is used in magnetic resonance imaging to produce accurate models of the human body. It is also used in computer games.

VPN (virtual private network) The use of encryption techniques and tunneling protocols by a group of users on a public network, especially the Internet, to ensure that their messages are secure. The network nodes appear to be connected by private lines and so form a virtual private network. The effect is to give companies the same capabilities as on a dedicated communications system, but at much lower cost.

VRAM /**vee**-ram/ *See* video RAM.

VRML /**vee**-ar-em-**el**, **ver**-măl/ *See* virtual reality modeling language.

W

wafer /way-fer/ A thin flat piece of semiconductor crystal on which multiple copies of a circuit are formed, and then cut out, in the manufacture of integrated circuits.

wait state The state in which a program or system or part of a system is inactive while waiting for some event or process to finish. For example, a CPU has to enter a wait state if requested data cannot be transferred from main store quickly enough to match its processing speed.

wallpaper A pattern or picture, chosen by the user, that is displayed as a screen background.

WAN /wan/ *See* wide area network.

wand /wond/ A hand-held penlike device that is connected to a computer and is used to read printed BAR CODES or OCR characters. It is moved steadily over the surface of the printing, and an audible and/or visual signal is produced if the data has been sensed satisfactorily.

WAP /wap/ (**wireless application protocol**) A specification for the protocol used by wireless devices, such as cellular telephones and radio transceivers, to communicate with the Internet.

warm boot To restart a computer without turning the power off, usually by commanding the operating system to restart or by pressing the reset button combination.

warm start *See* restart.

WAV /wayv/ A Windows file format for sounds stored as waveforms. These sound files have the file extension wav.

W3C /dub-'l-yoo-three-**see**/ (**World Wide Web Consortium**) A group of industrial organizations seeking to promote standards for the evolution of the Web and interoperability of Web products. It is responsible for various standards crucial to the Web, e.g. HTML, XML, etc.

WDM (**wavelength division multiplexing**) A type of multiplexing that modulates each of several data streams onto a different part of the light spectrum. It was developed for use with optical fibers and is the optical equivalent of frequency division multiplexing.

Web *See* World Wide Web.

Web address *See* URL.

Web browser A program developed to allow users to browse documents on the WWW. A Web browser can also provide access to documents on a network, an intranet, or a local hard drive. Web browsers interpret HTML documents and can use hyperlinks to allow the user to jump from document to document in any order. They are also capable of downloading and transferring files, providing newsgroup access, playing audio and video clips associated with a document, displaying graphics files, and running small programs, such as ActiveX controls. *See also* Internet Explorer.

webcam A digital camera that provides live images to a WEB SERVER, from where they can be viewed by anybody browsing the Web. Webcams do not provide continuous video but rather a stream of still images taken at predefined intervals, usually a minimum of about a second. Many webcams remain in place for long periods showing panoramic views, for example of weather conditions on volcanoes in the Cascade mountain range or of Times Square, New York. Webcams are also used in VIDEO CONFERENCING.

Webcasting /**web**-kass-ting, -kahss-/ The broadcasting of information on the World Wide Web. It is received automatically by all subscribers to the relevant broadcast service.

Web feed (feed; channel) A notification of new or updated items on a Web site. Web feeds typically contain the item's title, a short summary of its content, and its URL. They are used particularly for Web sites that provide news, stock market information, BLOGS, podcasts (*see* podcasting), and other content that changes frequently. Web feeds are in one of several standard formats, such as RSS, and are read by special programs or extensions to Web browsers called *news aggregators*. Because Web feeds are small they are downloaded quickly; they are then reviewed as a list and users can follow the link to any item that interests them. Users can select which Web sites they wish to monitor and many news aggregators will notify them when a new item is available on these sites.

weblog *See* blog.

Web page A document on the World Wide Web, identified by its URL. It can contain text, images, sound, and links to other pages.

Web server A program that serves the files that form Web pages to Web users. Every computer on the Internet that contains Web pages must run a Web server. They are often included in a larger package of Internet and intranet programs.

Web site A group of Web pages that are usually connected by hyperlinks and have a home page as their starting point. A Web browser and an Internet connection are needed in order to access a Web site.

Web TV The provision of access to the WWW and the ability to send and receive e-mails on a television set by the use of a modem contained in a *set-top box*. A user must subscribe to a Web TV network and also have an Internet Service Provider (ISP).

Webzine /**web**-zeen/ An electronic magazine available at a Web site.

while loop (while do loop) *See* loop.

whiteboard An area on a display screen that can be written to, or drawn on, by several users. These are used in teleconferencing to provide visual as well as audio communication.

white space Any character that does not have a visual form and is displayed as a blank space on the screen or printer.

whois /**hoo**-iz/ A service provided at some Internet domains for finding e-mail addresses.

wide area network (WAN) A computer NETWORK that is distinguished from a LOCAL AREA NETWORK because of its longer-distance communications. The network may cover, say, a whole country or may include the sites of a large multinational organization. The communications between sites are usually provided by one or more national or international entities rather than by the operators of the WAN.

wideband /**wÿd**-band/ *See* broadband.

wiki (wee-kee) **1.** A WEB SERVER application that allows Web users to alter the contents of the WEB PAGES it manages and to create new ones. Wikis facilitate the collaborative authoring of large Web sites, for example *Wikipedia*, an online encyclopedia. The term 'wiki' derives from the Hawaiian 'wiki wiki' ('fast' or 'quick'). **2.** A WEB SITE created and managed by such an application. The first wiki was created in 1995 for the Portland Pattern Repository.

Wikipedia *Trademark See* wiki.

wildcard character /wȳld-kard/ A symbol used to represent any character or characters. The asterisk (*) is commonly used for this purpose for any number of characters, and the question mark (?) is used to represent a single character. Wildcard searches are available in many applications. For example, in word processing searching for d*g would find all words (of any length) starting with 'd' and ending with 'g'. Searching for d???g would find all such words with five letters.

WIMP /wimp/ (**windows, icons, mouse, and pointers**) A GUI (graphical user interface) as used in the Macintosh and Windows operating systems. It was first developed in the 1970s at PARC (Xerox Palo Alto Research Center).

Winchester disk drive /win-chess-ter/ A type of DISK DRIVE introduced by IBM in the early 1970s. It was adopted by many manufacturers and used in large and small computers. It uses one or more hard MAGNETIC DISKS – known as *Winchester disks* – that are mounted inside a hermetically sealed container together with the read/write heads and their supporting mechanism. All modern fixed hard disk units are Winchester disk drives.

window 1. A rectangular area on a display screen inside which a portion of a stored image or file can be displayed. The window can be of any size up to that of the screen; more than one window can usually be displayed at once. The process of making a portion of an image or file visible through the window is called *windowing*. The visible window formed by the process can be moved to allow examination of the different portions making up the entire stored image or file.
2. In a GUI, a similar area on a display screen used to display a FORM.

Windows *Trademark* A Microsoft GUI operating system introduced in 1983. Early versions were in fact MS-DOS applications that extended its capabilities and increasingly bypassed its limitations. *Windows 3.x* became the dominant PC operating system in the early 1990s, but suffered from problems of stability and performance. *Windows 95* and its successors, *Windows 98* and *Windows ME*, were true operating systems, but had to make design compromises for BACKWARD COMPATIBILITY with their MS-DOS and Windows 3.x heritage. A completely different line of Windows development, begun with *Windows NT* (Windows new technology) in 1993, was independent of MS-DOS and provided better stability, security, and multitasking; however, it required more expensive hardware and was too slow for some uses, especially computer games. It was succeeded by *Windows 2000* and then by the current version, *Windows XP*, which finally unified the two lines of development. *Windows CE* is a smaller version of Windows for use on hand-held and palm computers. See tables overleaf.

Windows 3.x *See* Windows.

Windows 95 *See* Windows.

Windows 98 *See* Windows.

Windows 2000 *See* Windows.

Windows CE *See* Windows.

Windows ME *See* Windows.

Windows NT *See* Windows.

Windows open services architecture (**WOSA**) A set of application programming interfaces in Microsoft Windows to enable Windows applications from a variety of sources to communicate with each other.

Windows XP *See* Windows.

Winsock /win-sok/ The programming interface in a Windows operating system that handles input and output requests for Internet applications. A comparable interface exists for Macintosh computers.

Wintel /win-tel/ An Intel-based PC run-

WINDOWS VERSIONS

Version	Date	Main technologies
Windows 1	November 1985	Windows dialogs, task switching, CGA and EGA graphics. A very basic mouse-driven GUI; resizable windows were not standard, windows could only be tiled not overlapped and most of the functionality was in drop-down menus. There were very few Windows programs.
Windows 2	November 1987	Data sharing for applications through dynamic data exchange, extended memory support, icons for programs and files, Word and Excel made their debut.
Windows/286	May 1988	Windows 2.1 giving extended memory support.
Windows/386	May 1988	Windows 286 with true multitasking, achieved by using the memory addressing capabilities of the Intel 80386 processor.
Windows 3	May 1990	More consistent and streamlined with support for VGA. Support for 16MB of RAM and virtual memory.
	1991	Support for CD-ROM drives and audio hardware.
Windows 3.1	April 1992	OLE drag and drop operations, support for SVGA and XGA graphics, and scalable True Type fonts.
Windows for Workgroups 3.1	June 1992	Peer-to-peer workgroups and domain-network support. Applications could be scheduled, user file and printer sharing and a simple e-mail client were implemented. Version 3.11(needed a 386/SX processor to run on) introduced security features and a remote access server.
Windows NT 3.1	May 1993	A 32-bit operating system providing more stablility and security with integrated networking and pre-emptive multitasking. It introduced the NTFS filing system and the Win 32 API.
Windows 95	August 1995	An updated GUI bringing in the start button, the taskbar and the Desktop. A central Registry was used to control operations. Various releases gave the FAT32 file system and USB support.

ning a Microsoft Windows operating system.

WinZip A Windows program equivalent to the MS-DOS-based PKZIP and PKUNZIP, which compresses and uncompresses zip files.

wireframe model /wȳ-er-fraym/ The representation of a three-dimensional object by a set of lines resembling a wire frame. For example, a sphere can be represented as a polygon made up of a large number of triangular faces. *See also* rendering.

wireless LAN (WLAN) A LAN that communicates using radio signals, infrared optical signals, or microwaves.

wizard /wiz-erd/ An interactive help utility that provides step by step help to install software or to perform a particular task.

WLAN /dub-'l-yoo-lan/ *See* wireless LAN.

WINDOWS VERSIONS

Version	Date	Main technologies
Windows NT4	August 1996	This had a Windows 95 interface but with much improved network and security handling. The server version provided the capability for good TCP/IP and network management.
Windows 98	June 1998	This integrated Internet Explorer with the operating system, gave improved USB support and introduced ACTIVE desktop.
Windows 2000 Professional	February 2000	Support for dynamic plug-and-play, IEEE-1394, Direct X 6. Proper ACPI power management with hibernation and suspend modes.
Windows ME	September 2000	Easier for users to maintain with system restore and auto update. A fully functional media player and scanner and camera wizards.
Windows XP	October 2001	Professional and home versions. A unified NT codebase with a new start panel that changed to reflect the applications being used. A good Windows for mobile users with 802.1b Wi-Fi support and better offline folder synchronization. A 64-bit version was also released.
Windows Media Center Edition	September 2002	The living room PC. Windows XP with a different GUI, giving a remote-control-driven interface for playing music, showing DVDs and watching/recording live TV.
Windows XP Tablet PC Edition	November 2002	A pen-based 'ink' input system, with voice recognition and handwriting analysis. A 2005 edition included several enhancements making it easier to use the 'ink' interface with standard Windows applications.
Windows Vista	Autumn 2006?	A media-friendly interface with virtual folders, translucent Windows, XML metadata emphasis and a new Start menu.

word (machine word; computer word) A collection of bits that is treated as a single unit by the hardware of a computer. The number of bits in a word is called the *word length*. The word length is fixed in most computers, and in present-day machines is usually 32 bits. Main store is divided into either words or BYTES. In the former case, each storage location holds one word, and the word is thus the smallest addressable unit in the computer system, i.e. the smallest unit to be identified by an ADDRESS.

What the contents of a word signify is determined entirely by its context. A given bit sequence could represent a MACHINE INSTRUCTION in one part of main store, while in another part it might represent an INTEGER number, a REAL NUMBER, or perhaps one or more CHARACTERS, depending on word length. A 32-bit word, for example, could hold integers between –2 147 483 648 and +2 147 483 647, or up to four characters, or a FLOATING-POINT NUMBER with a PRECISION of about seven digits, or a machine instruction.

Word (**Microsoft Word**) A WORD-PRO-CESSING application developed by Microsoft. First released for MS-DOS systems in 1983, versions followed for the Macintosh in 1985 and Windows in 1989. It has been upgraded several times and the current versions are Word 2003 (Windows) and Word 2004 (Macintosh). The most widely used word processor in the world, it forms part of Microsoft OFFICE.

word length *See* word.

word processing (**wp, WP**) Using a computer program to compose letters, reports, and other documents and to edit, reformat, store, and print them. The document is automatically displayed on the screen as it is typed on the keyboard, and words can be corrected, inserted, and deleted. The program allows the user to specify left and right margins between which text is automatically justified and hyphenated (if desired) as it is typed in. It is possible to reposition blocks of text in the document, to search for and replace strings of text in the document, to add text from other files, to specify FONTS, typesize, boldface, italics, and other typographic attributes, and to include tables, pictures or other graphics, etc. Spelling and grammar can usually be checked. The document can be stored for future use and/or printed any number of times. Variable information such as names, addresses, or prices can be substituted when printing the document for easy production of form letters, contracts, etc.

word processor (**wp, WP**) A computer program that provides a WORD PROCESSING facility.

work file A FILE held very briefly on backing store and used to overcome space limitations in main store or to carry information between one job and the next.

workgroup computing The computing that occurs when all the individuals in a workgroup have computers connected to a network that gives them such functionality as sharing files, sending e-mail to each other, scheduling meetings, and so on. In sophisticated workgroup systems workflows can be defined so that information is automatically forwarded to the appropriate people at the appropriate time.

working space (**workspace; working store; working area; scratchpad**) A portion of MAIN STORE that is used for the temporary storage of data during processing, including intermediate results needed during a calculation. The contents of the working space will not be preserved after the program has finished. It is the responsibility of the programmer to ensure that the working space does not overlap the program code, although in many high-level languages and operating systems this is made difficult or impossible.

workspace /**werk**-spayss/ *See* working space.

workstation /**werk**-stay-shŏn/ **1.** A position in, say, an office or factory where an operator has access to all the facilities required to perform a particular task that involves a computer in some way.
2. A powerful desktop computer, usually part of a network.

World Wide Web (**Web; WWW**) A part of the INTERNET consisting of a very large number of documents (Web pages) linked by hyperlinks.

World Wide Web Consortium *See* W3C.

worm A program that is introduced into a host and replicates itself to other computers across a NETWORK. Like a VIRUS, a worm can carry a malicious PAYLOAD.

WORM /werm/ (**write once read many times**) An optical disk that, once recorded, can be read over and over again but cannot be written to again. *See* CD; DVD.

WOSA /**wo**-să/ *See* Windows open services architecture.

wp (WP) *See* word processing; word processor.

wrap-around A feature of a VDU or printer, allowing it to output lines of text that would otherwise be too long to be displayed or printed in their entirety. The line appears as two or more successive lines on the screen or paper.

write To record data in some form of storage, usually either in a magnetic medium or in semiconductor RAM.

write head *See* tape unit.

write instruction A MACHINE INSTRUCTION that causes an item of data to be recorded on some form of storage.

write protection A system for the prevention of recording (writing) information usually on a disk.

write-protect notch A small opening in the jacket of a FLOPPY DISK that is used to prevent the disk from being written to. There is a tab in a hole in the corner of the disk that, when removed, means the disk is write-protected.

WWW *See* World Wide Web.

WYSIWYG /**wiz**-ă-wig/ (**what you see is what you get**) What is seen on a screen is an accurate representation of how text, graphics, etc., will appear when printed.

X

X25 A network protocol that allows computers on different public networks to communicate over long distances by means of an intermediary computer at the network level layer.

x86 The generic name for the Intel series of processors that includes the 8086, the 80286, the 386, the 486, and PENTIUM families.

X500 A standard way to develop an electronic directory service of employees in an organization. Many companies have created an X500 directory and these are organized into a single global directory available to anyone on the Internet.

x and y coordinates The horizontal and vertical address of any pixel on a computer screen.

xDSL A term referring collectively to all types of digital subscriber lines (DSLs).

Xeon /**zee**-on/ *See* Pentium.

XHTML *See* HTML.

XML (**extensible markup language**) A simpler and easier to use version of SGML. Introduced for Web development, it allows customized tags to be created and gives greater flexibility than is possible with older HTML document coding systems.

XML Query (**XQuery**) A language for querying and extracting data from XML documents. XML Query uses XPATH expressions to isolate the required parts of the document(s) and provides facilities for manipulating the retrieved data. It thus performs a similar function to that of SQL

on RELATIONAL DATABASES. XML Query was released by the W3C as a working draft in 2003 and is being regularly updated.

Xmodem /**eks**-moh-dem/ An error-correcting modem protocol developed in 1978, which has become the de facto standard. Data is sent in 128-byte blocks. *See also* Ymodem; Zmodem.

XON/XOFF /eks-**on** eks-**off**/ *See* flow control.

XOR gate /eks-**or**/ (**exclusive-OR gate**) *See* nonequivalence gate.

XPath A language for addressing and selecting parts of an XML document. For example, when applied to an XML file containing a list of books the XPath expression 'Book[Title['Facts On File Dictionary of Computing']]' might select the Book ELEMENT describing this dictionary. XPath is used in other XML technologies, such as XML QUERY and XSL. The XPath specification was released as a recommendation in 1999 and is maintained by the W3C.

XQuery *See* XML Query.

X series A set of recommendations adopted by the International Telecommunications Union Telecommunication Standardization Sector (ITU-T) and ISO. These standardize equipment and protocols for use in public-access and private-computer networks.

XSL (**extensible stylesheet language**) A set of languages for transforming and formatting XML documents. XSL consists of three parts. *XSLT* (*XSL Transformations*) is used to specify rules that manipulate and restructure all or selected parts of an XML document. The output can be an XML document, but can also be PLAIN TEXT, HTML, etc.: a common use of XSLT is to change XML to HTML so it can be download over the Web. *XSL-FO* (*XSL Formatting Objects*) is used with XSLT to specify presentational details (typeface, type size, color, page layout, etc.) for the various parts of an XML document. XSL-FO processors typically use these specifications to produce output in a format that can be viewed on screen or printed, such as POST-SCRIPT or PDF (*see* Acrobat). The third part of XSL is the XPATH language, which is used to select which parts of the input document are subject to each transformation or formatting rule. The specification of all the rules necessary to achieve a particular transformation and/or presentation is contained in a separate file called an *XSL stylesheet*, which is itself a valid XML document. XSL was first proposed in 1997 and was released as a W3C recommendation in 2001. XSLT was quickly adopted because its transformation capabilities met a need among XML developers. XSL-FO, however, was taken up much more slowly and is still not widely used because were already other methods of formatting XML, e.g. CASCADING STYLE SHEETS.

Yahoo! A Web PORTAL that began as a hierarchical directory of Web sites organized by category. Yahoo! links to several other popular Web search sites if a search is unsuccessful on Yahoo!.

Y2K *See* millennium bug.

Ymodem /wÿ-moh-dem/ A modem protocol based on XMODEM, which sends data in 1024-byte blocks.

Z

zap /zap/ **1.** In programming, a correction for a computer code problem. Incorrect compiled code can be overlaid with a fix, which is called a zap.
2. To erase or remove something. For example a Macintosh PROM can be zapped, i.e. erased, so that the system can rebuild its contents.

zero wait state The situation in which a processor is used with no WAIT STATE pauses.

ZIF socket /ziff/ (**zero insertion forces socket**) A chip socket in which a chip can be inserted and removed without using special tools. The ZIF socket contains a lever that opens and closes, or a screw, thus securing the microprocessor in place. ZIF sockets take up more space and are more expensive than conventional sockets.

zip The file extension of a file that has been compressed using PKZIP.

Zip drive *Trademark* A small portable disk drive supplied by Iomega Corporation. Disks come in three sizes, 100 Mb, 250 Mb, and 750 Mb. A Zip drive is needed to read and write 3.5-inch Zip disks.

Zmodem /**zee**-moh-dem/ A modem protocol based on XMODEM, which sends data in 512-byte blocks.

zoom /zoom/ An enlarged or diminished view of an image or part of an image. Many applications include a zoom feature to enable editing at a more detailed level.

APPENDIXES

Symbols and Notation

Arithmetic and algebra

equal to	$=$
not equal to	$\neq$
identity	$\equiv$
approximately equal to	$\simeq$
approaches	$\rightarrow$
proportional to	$\propto$
less than	$<$
greater than	$>$
less than or equal to	$\leq$
greater than or equal to	$\geq$
much less than	$\ll$
much greater than	$\gg$
plus, positive	$+$
minus, negative	$-$
plus or minus	$\pm$
multiplication	$a \times b$
	$a.b$
division	$a \div b$
	a/b
magnitude of a	$\lvert a \rvert$
factorial a	$a!$
logarithm (to base b)	$\log_b a$
common logarithm	$\log_{10} a$
natural logarithm	$\log_e a$ or $\ln a$
summation	Σ
continued product	Π

Symbols and Notation (continued)

Geometry and trigonometry

angle	∠
triangle	△
square	□
circle	○
parallel to	‖
perpendicular to	⊥
congruent to	≡
similar to	~
sine	sin
cosine	cos
tangent	tan
cotangent	cot, ctn
secant	sec
cosecant	cosec, csc
inverse sine	$\sin^{-1}$, arc sin
inverse cosine	$\cos^{-1}$, arc cos
inverse tangent	$\tan^{-1}$, arc tan
etc.	etc.
Cartesian coordinates	(x, y, z)
spherical coordinates	(r, θ, ϕ)
cylindrical coordinates	(r, θ, z)
direction numbers or cosines	(l, m, n)

Symbols and Notation (continued)

Sets and logic

implies that	$\Rightarrow$
is implied by	$\Leftarrow$
implies and is implied by (if and only if)	$\Leftrightarrow$
set a, b, c,...	$\{a, b, c, ...\}$
is an element of	$\in$
is not an element of	$\notin$
such that	$:$
number of elements in set S	$n(S)$
universal set	E or $\mathcal{E}$
empty set	$\varnothing$
complement of S	S'
union	$\cup$
intersection	$\cap$
is a subset of	$\subset$
corresponds one-to-one with	$\leftrightarrow$
x is mapped onto y	$x \rightarrow y$
conjunction	$\wedge$
disjunction	$\vee$
negation (of p)	$\sim p$ or $\neg p$
implication	$\rightarrow$ or $\supset$
biconditional (equivalence)	$\equiv$ or $\leftrightarrow$
the set of natural numbers	$\mathbb{N}$
the set of integers	$\mathbb{Z}$
the set of rational numbers	$\mathbb{Q}$
the set of real numbers	$\mathbb{R}$
the set of complex numbers	$\mathbb{C}$

Symbols and Notation (continued)

Calculus

increment of x	$\triangle x$ or δx
limit of function of x as x approaches a	$\underset{x \to a}{\text{Lim}}\ f(x)$
derivative of $f(x)$	$df(x)/dx$ or $f'(x)$
second derivative of $f(x)$ etc.	$d^2f(x)/dx^2$ or $f''(x)$ etc.
indefinite integral of $f(x)$ with respect to x	$\int f(x)dx$
definite integral with limits a and b	$\int_{b}^{a} f(x)dx$
partial derivative of function $f(x,y)$ with respect to x	$\partial f(x,y)/\partial x$

File Extensions

There are a large number of different file extensions in use. Some of the more common extension names are listed below.

.7z	7-zip compressed archive file	.cmd	command file in Windows, MS-DOS, and OS/2
.afm	PostScript font metrics file	.cmf	Corel Metafile format graphics file
.ai	Adobe Illustrator graphics file	.com	command or program file
.arc	compressed archive file	.cpp	C++ source file
.asc	ASCII text file	.cs	C# source file
.asp	ASP web page	.css	cascading style sheets file
.atm	Adobe Type Manager file	.csv	comma separated values file
.au	audio file	.ct	Paint Shop Pro graphics file
.avi	audio visual interleaved multimedia file	.dat	data file
.bac	backup file	.dbf	dBase and FoxPro database file
.bak	backup file	.dcr	Macromedia Shockwave multimedia file
.bas	Basic source file	.dll	dynamic link library file
.bat	batch program file in MS-DOS and Windows	.doc	document file in Microsoft Word (and some other word-processing applications)
.bin	compressed MacBinary file	.dos	MS-DOS file
.bmp	bitmapped graphics file	.dot	Microsoft Word document template
.c	C source file	.drv	device driver file
.c++	C++ source file	.dtd	document type definition file for SGML and XML
.cab	Cabinet compressed archive file	.dtp	Microsoft publisher document file
.cbl	Cobol source file	.dv	video file
.cdr	CorelDraw graphics file	.eps	encapsulated PostScript file
.cgi	Common Gateway Interface script file	.exe	executable program file
.cgm	Computer Graphics metafile format file	.fax	fax file
.chk	Windows unidentified data saved by ScanDisk	.fla	Macromedia Flash movie file
.class	Java class file		
.clp	Windows clipboard file		

File Extensions (continued)

.fon	Windows system font file	.msi	Windows installer file
.for	Fortran source file	.obj	compiled program module file
.gid	Windows index file	.ole	Microsoft OLE object
.gif	Gif format graphics file	.p	Pascal source file
.gtar	compressed Unix file	.pas	Pascal source file
.gz	compressed Unix file	.pcd	Kodak Photo-CD file
.h	C or C++ header file	.pcx	Paintbrush bitmap graphics file
.hex	Macintosh BinHex encoded file	.pdd	Photoshop file
.hlp	Windows help file	.pdf	Adobe Acrobat portable document format file
.htm	HTML file	.pfb	PostScript font file
.html	HTML file	.pfm	PostScript font metrics file
.ico	Windows icon file	.pic	PC Paint graphics file
.ini	initialization file in MS-DOS	.pict	Macintosh PICT graphics file
.iso	CD image file	.png	PNG format graphics file
.jar	Java archive file	.pps	Microsoft PowerPoint slide show file
.java	Java source file	.ppt	Microsoft PowerPoint presentation file
.jpe	JPEG format graphics file	.prn	Windows print file
.jpeg	JPEG format graphics file	.ps	PostScript print file
.jpg	JPEG format graphics file	.psd	Photoshop file
.js	JavaScript source file	.pub	document file in certain DTP programs
.lb	library file	.qti	Apple QuickTime image file
.lzh	LZH archive file	.qtm	Apple QuickTime movie file
.mac	MacPaint graphics file	.qxd	QuarkXpress document file
.mdb	Microsoft Access database file	.qxl	QuarkXpress library file
.mid	MIDI format audio file	.ra	RealAudio sound file
.midi	MIDI format audio file	.ras	Sun raster image bitmap file
.mime	MIME format file	.rm	RealAudio movie file
.mov	Apple QuickTime video file	.rtf	rich text format file
.mp3	MPEG compressed audio file	.sea	Macintosh self-extracting file compressed using StuffIt
.mpeg	MPEG video file		
.mpg	MPEG video file		
.msg	Microsoft Outlook file		

File Extensions (continued)

.sgm	SGML file		.uud	file produced in ASCII by uuencode
.sgml	SGML file			
.shtml	HTML file with SSI		.uue	file produced in binary by uudecode
.sit	Macintosh file compressed with StuffIt		.vb	Visual Basic source file
.spl	Macromedia Shockwave Flash file		.vbs	VBScript source file
			.vp	Ventura Publisher document file
.stm	HTML file with SSI		.wav	WAV format audio file
.stml	HTML file with SSI		.wp	Corel WordPerfect document file
.sun	Sun graphics file			
.svg	SVG format graphics file		.wpg	Corel WordPerfect graphics file
.sw	Macromedia Shockwave audio file		.wps	Microsoft Works document file
			.wri	Microsoft Write document file
.sys	system file		.xht	XHTML file
.tar	Unix archive file		.xhtm	XHTML file
.taz	Unix archive file		.xhtml	XHTML file
.tga	Targa format bitmap graphics file		xls	Microsoft Excel spreadsheet file
			xlm	XML file
.tif	TIFF format graphics file		xsd	XML schema file
.tiff	TIFF format graphics file		xsl	XSL file
.tmp	temporary file		xslt	XSL-T file
.ttf	TrueType font file		.xtg	QuarkXpress tagged format file
.txt	ASCII text file		.zip	ZIP compressed archive file

Organizational Top-Level Domain Names

Internet top-level domain names are of two types: organizational names and geographic names. Organizational domain names indicate the type of Internet site and are used as follows:

.aero	air transport industry
.biz	businesses
.cat	relating to Catalan language or culture
.com	any purpose (originally for commercial organizations)
.coop	cooperative associations
.edu	colleges and universities (mainly US)
.gov	US government (nonmilitary)
.info	any purpose
.int	international organizations established by treaty
.jobs	businesses advertising jobs
.mil	US military; this has three subdomains:

	.af.mil	Air Force
	.army.mil	Army
	.navy.mil	Navy

.mobi	mobile devices and related services
.museum	museums
.name	individuals
.net	any purpose (originally for network-related organizations)
.org	any purpose, but especially nonprofit-making organizations
.pro	professionals (doctors, lawyers, accountants)
.travel	travel industry

The following proposed new top-level domains are under consideration:

.asia	Asia and the Pacific region
.post	postal-service industry
.tel	telephone-Internet communications services
.xxx	sexually explicit material

Geographic Top-Level Domain Names

Geographic names indicate the country. When a geographic name is the top-level domain, the second-level subdomain in the US may indicate an organization or a state (see page 264). For other countries, it may indicate an organization. For example

.co.uk	a company in the United Kingdom
.ed.uk	an educational establishment in the United Kingdom

Geographic names are as shown below

.ac	Ascension Island	.bg	Bulgaria
.ad	Andorra	.bh	Bahrain
.ae	United Arab Emirates	.bi	Burundi
.af	Afghanistan	.bj	Benin
.ag	Antigua and Barbuda	.bm	Bermuda
.ai	Anguilla	.bn	Brunei Darussalam
.al	Albania	.bo	Bolivia
.am	Armenia	.br	Brazil
.an	Netherlands Antilles	.bs	Bahamas
.ao	Angola	.bt	Bhutan
.aq	Antarctica	.bv	Bouvet Island
.ar	Argentina	.bw	Botswana
.as	American Samoa	.by	Belarus
.at	Austria	.bz	Belize
.au	Australia	.ca	Canada
.aw	Aruba	.cc	Cocos Islands
.ax	Aland Islands	.cd	Congo
.az	Azerbaijan	.cf	Central Africa Republic
.ba	Bosnia-Herzegovina	.cg	Congo-Brazzaville
.bb	Barbados	.ch	Switzerland
.bd	Bangladesh	.ci	Cote D'Ivoire
.be	Belgium	.ck	Cook Islands
.bf	Burkina Faso	.cl	Chile

Geographic Top-Level Domain Names (continued)

.cm	Cameroon	.gb	United Kingdom (also .uk)
.cn	China	.gd	Grenada
.co	Colombia	.ge	Georgia (Republic of)
.cr	Costa Rica	.gf	French Guiana
.cs	Serbia and Montenegro (formerly Czechoslovakia)	.gg	Guernsey
		.gh	Ghana
.cu	Cuba	.gi	Gibraltar
.cv	Cape Verde	.gl	Greenland
.cx	Christmas Island	.gm	Gambia
.cy	Cyprus	.gn	Guinea
.cz	Czech Republic	.gp	Guadaloupe
.de	Germany	.gq	Equatorial Guinea
.dj	Djibouti	.gr	Greece
.dk	Denmark	.gs	South Georgia and South Sandwich Islands
.dm	Dominica		
.do	Dominican Republic	.gt	Guatemala
.dz	Algeria	.gu	Guam
.ec	Ecuador	.gw	Guinea-Bissau
.ee	Estonia	.gy	Guyana
.eg	Egypt	.hk	Hong Kong
.eh	Western Sahara	.hm	Heard Island and McDonald Islands
.er	Eritrea		
.es	Spain	.hn	Honduras
.et	Ethiopia	.hr	Croatia (Hrvatska)
.eu	European Union	.ht	Haiti
.fi	Finland	.hu	Hungary
.fj	Fiji	.id	Indonesia
.fk	Falkland Islands	.ie	Ireland
.fm	Micronesia (Federated States of)	.il	Israel
.fo	Faeroe Islands	.im	Isle of Man
.fr	France	.in	India
.ga	Gabon	.io	British Indian Ocean Territory

Geographic Top-Level Domain Names (continued)

.iq	Iraq	.mc	Monaco
.ir	Iran	.md	Moldova
.is	Iceland	.mg	Madagascar
.it	Italy	.mh	Marshall Islands
.je	Jersey	.mk	Macedonia
.jm	Jamaica	.ml	Mali
.jo	Jordan	.mm	Myanmar
.jp	Japan	.mn	Mongolia
.ke	Kenya	.mo	Macau
.kg	Kyrgyzstan	.mp	Northern Mariana Islands
.kh	Cambodia	.mq	Martinique
.ki	Kiribati	.mr	Mauritania
.km	Comoros	.ms	Montserrat
.kn	St. Kitts-Nevis	.mt	Malta
.kp	Korea (Democratic People's Republic)	.mu	Mauritius
		.mv	Maldives
.kr	Korea (Republic of)	.mw	Malawi
.kw	Kuwait	.mx	Mexico
.ky	Cayman Islands	.my	Malaysia
.kz	Kazakhstan	.mz	Mozambique
.la	Laos	.na	Namibia
.lb	Lebanon	.nc	New Caledonia
.lc	Saint Lucia	.ne	Niger
.li	Liechtenstein	.nf	Norfolk Island
.lk	Sri Lanka	.ng	Nigeria
.lr	Liberia	.ni	Nicaragua
.ls	Lesotho	.nl	Netherlands
.lt	Lithuania	.no	Norway
.lu	Luxembourg	.np	Nepal
.lv	Latvia	.nr	Nauru
.ly	Libya	.nu	Niue
.ma	Morocco	.nz	New Zealand

Geographic Top-Level Domain Names (continued)

.om	Oman	.sn	Senegal
.pa	Panama	.so	Somalia
.pe	Peru	.sr	Suriname
.pf	French Polynesia	.st	São Tomé and Principe
.pg	Papua New Guinea	.sv	El Salvador
.ph	Philippines	.sy	Syrian Arab Republic
.pk	Pakistan	.sz	Swaziland
.pl	Poland	.tc	Turks and Caicos Islands
.pm	St. Pierre and Miquelon	.td	Chad
.pn	Pitcairn Island	.tf	French Southern Territories
.pr	Puerto Rico	.tg	Togo
.ps	Palestinian Territories	.th	Thailand
.pt	Portugal	.tj	Tajikistan
.pw	Palau	.tk	Tokelau
.py	Paraguay	.tl	Timor-Leste
.qu	Qatar	.tm	Turkmenistan
.re	Reunion Island	.tn	Tunisia
.ro	Romania	.to	Tonga
.ru	Russian Federation	.tp	East Timor
.rw	Rwanda	.tr	Turkey
.sa	Saudi Arabia	.tt	Trinidad and Tobago
.sb	Solomon Islands	.tv	Tuvalu
.sc	Seychelles	.tw	Taiwan
.sd	Sudan	.tz	Tanzania
.se	Sweden	.ua	Ukraine
.sg	Singapore	.ug	Uganda
.sh	Saint Helena	.uk	United Kingdom (also .gb)
.si	Slovenia	.um	United States Minor Outlying Islands
.sj	Svalbard and Jan Mayen Islands		
.sk	Slovakia	.us	United States
.sl	Sierra Leone	.uy	Uruguay
.sm	San Marino	.uz	Uzbekistan

Geographic Top-Level Domain Names (continued)

.va	Vatican City State	.ws	Samoa (formerly Western Samoa)
.vc	Saint Vincent and the Grenadines	.ye	Yemen
.ve	Venezuela	.yt	Mayotte
.vg	British Virgin Islands	.yu	Yugoslavia
.vi	United States Virgin Islands	.za	South Africa
.vn	Vietnam	.zm	Zambia
.vu	Vanuatu	.zw	Zimbabwe
.wf	Wallis and Futuna Islands		

United States Subdomain Names

There are various subdomains in use with the top-level domain name .us. A few of these refer to organizations, for example:

.lib.us	United States library
.state.us	state government

Most common subdomain names are used for states, for example:

.ak.us	Alaska
.ar.us	Arkansas
.ca.us	California
.fl.us	Florida
.ga.us	Georgia
.il.us	Illinois
.in.us	Indiana
.la.us	Louisiana
.ma.us	Massachusetts
.md.us	Maryland
.mi.us	Michigan
.mn.us	Minnesota
.ms.us	Mississippi
.nc.us	North Carolina
.ne.us	Nebraska
.nh.us	New Hampshire
.ny.us	New York (state)
.oh.us	Ohio
.ok.us	Oklahoma
.or.us	Oregon
.va.us	Virginia
.vt.us	Vermont
.wv.us	West Virginia
.wy.us	Wyoming

A third-level subdomain may be used for a town or city, as in:

.atl.ga.us	Atlanta, Georgia
.nyc.ny.us	New York City, New York
.sf.ca.us	San Francisco, California

Number Conversions

Decimal	Binary	Octal	Hex
1	1	1	1
2	10	2	2
3	11	3	3
4	100	4	4
5	101	5	5
6	110	6	6
7	111	7	7
8	1000	10	8
9	1001	11	9
10	1010	12	A
11	1011	13	B
12	1100	14	C
13	1101	15	D
14	1110	16	E
15	1111	17	F
16	10000	20	10
17	10001	21	11
18	10010	22	12
19	10011	23	13
20	10100	24	14
21	10101	25	15
22	10110	26	16
23	10111	27	17
24	11000	30	18
25	11001	31	19
26	11010	32	1A
27	11011	33	1B
28	11100	34	1C
29	11101	35	1D
30	11110	36	1E
31	11111	37	1F
32	100000	40	20
33	100001	41	21
34	100010	42	22
35	100011	43	23
36	100100	44	24
37	100101	45	25
38	100110	46	26
39	100111	47	27
40	101000	50	28

Number Conversions (continued)

Decimal	Binary	Octal	Hex
41	101001	51	29
42	101010	52	2A
43	101011	53	2B
44	101100	54	2C
45	101101	55	2D
46	101110	56	2E
47	101111	57	2F
48	110000	60	30
49	110001	61	31
50	110010	62	32
51	110011	63	33
52	110100	64	34
53	110101	65	35
54	110110	66	36
55	110111	67	37
56	111000	70	38
57	111001	71	39
58	111010	72	3A
59	111011	73	3B
60	111100	74	3C
61	111101	75	3D
62	111110	76	3E
63	111111	77	3F
64	1000000	100	40
65	1000001	101	41
66	1000010	102	42
67	1000011	103	43
68	1000100	104	44
69	1000101	105	45
70	1000110	106	46
71	1000111	107	47
72	1001000	110	48
73	1001001	111	49
74	1001010	112	4A
75	1001011	113	4B
76	1001100	114	4C
77	1001101	115	4D
78	1001110	116	4E
79	1001111	117	4F
80	1010000	120	50
81	1010001	121	51

Number Conversions (continued)

Decimal	Binary	Octal	Hex
82	1010010	122	52
83	1010011	123	53
84	1010100	124	54
85	1010101	125	55
86	1010110	126	56
87	1010111	127	57
88	1011000	130	58
89	1011001	131	59
90	1011010	132	5A
91	1011011	133	5B
92	1011100	134	5C
93	1011101	135	5D
94	1011110	136	5E
95	1011111	137	5F
96	1100000	140	60
97	1100001	141	61
98	1100010	142	62
99	1100011	143	63
100	1100100	144	64
101	1100101	145	65
102	1100110	146	66
103	1100111	147	67
104	1101000	150	68
105	1101001	151	69
106	1101010	152	6A
107	1101011	153	6B
108	1101100	154	6C
109	1101101	155	6D
110	1101110	156	6E
111	1101111	157	6F
112	1110000	160	70
113	1110001	161	71
114	1110010	162	72
115	1110011	163	73
116	1110100	164	74
117	1110101	165	75
118	1110110	166	76
119	1110111	167	77
120	1111000	170	78
121	1111001	171	79
122	1111010	172	7A

Number Conversions (continued)

Decimal	Binary	Octal	Hex
123	1111011	173	7B
124	1111100	174	7C
125	1111101	175	7D
126	1111110	176	7E
127	1111111	177	7F
128	10000000	200	80
129	10000001	201	81
130	10000010	202	82
131	10000011	203	83
132	10000100	204	84
133	10000101	205	85
134	10000110	206	86
135	10000111	207	87
136	10001000	210	88
137	10001001	211	89
138	10001010	212	8A
139	10001011	213	8B
140	10001100	214	8C
141	10001101	215	8D
142	10001110	216	8E
143	10001111	217	8F
144	10010000	220	90
145	10010001	221	91
146	10010010	222	92
147	10010011	223	93
148	10010100	224	94
149	10010101	225	95
150	10010110	226	96
151	10010111	227	97
152	10011000	230	98
153	10011001	231	99
154	10011010	232	9A
155	10011011	233	9B
156	10011100	234	9C
157	10011101	235	9D
158	10011110	236	9E
159	10011111	237	9F
160	10100000	240	A0
161	10100001	241	A1
162	10100010	242	A2
163	10100011	243	A3

Number Conversions (continued)

Decimal	Binary	Octal	Hex
164	10100100	244	A4
165	10100101	245	A5
166	10100110	246	A6
167	10100111	247	A7
168	10101000	250	A8
169	10101001	251	A9
170	10101010	252	AA
171	10101011	253	AB
172	10101100	254	AC
173	10101101	255	AD
174	10101110	256	AE
175	10101111	257	AF
176	10110000	260	B0
177	10110001	261	B1
178	10110010	262	B2
179	10110011	263	B3
180	10110100	264	B4
181	10110101	265	B5
182	10110110	266	B6
183	10110111	267	B7
184	10111000	270	B8
185	10111001	271	B9
186	10111010	272	BA
187	10111011	273	BB
188	10111100	274	BC
189	10111101	275	BD
190	10111110	276	BE
191	10111111	277	BF
192	11000000	300	C0
193	11000001	301	C1
194	11000010	302	C2
195	11000011	303	C3
196	11000100	304	C4
197	11000101	305	C5
198	11000110	306	C6
199	11000111	307	C7
200	11001000	310	C8
201	11001001	311	C9
202	11001010	312	CA
203	11001011	313	CB
204	11001100	314	CC

Number Conversions (continued)

Decimal	Binary	Octal	Hex
205	11001101	315	CD
206	11001110	316	CE
207	11001111	317	CF
208	11010000	320	D0
209	11010001	321	D1
210	11010010	322	D2
211	11010011	323	D3
212	11010100	324	D4
213	11010101	325	D5
214	11010110	326	D6
215	11010111	327	D7
216	11011000	330	D8
217	11011001	331	D9
218	11011010	332	DA
219	11011011	333	DB
220	11011100	334	DC
221	11011101	335	DD
222	11011110	336	DE
223	11011111	337	DF
224	11100000	340	E0
225	11100001	341	E1
226	11100010	342	E2
227	11100011	343	E3
228	11100100	344	E4
229	11100101	345	E5
230	11100110	346	E6
231	11100111	347	E7
232	11101000	350	E8
233	11101001	351	E9
234	11101010	352	EA
235	11101011	353	EB
236	11101100	354	EC
237	11101101	355	ED
238	11101110	356	EE
239	11101111	357	EF
240	11110000	360	F0
241	11110001	361	F1
242	11110010	362	F2
243	11110011	363	F3
244	11110100	364	F4
245	11110101	365	F5

Number Conversions (continued)

Decimal	Binary	Octal	Hex
246	11110110	366	F6
247	11110111	367	F7
248	11111000	370	F8
249	11111001	371	F9
250	11111010	372	FA
251	11111011	373	FB
252	11111100	374	FC
253	11111101	375	FD
254	11111110	376	FE
255	11111111	377	FF

Web Sites

Apple Computer Inc	www.apple.com
British Computer Society	www.bcs.org
Cnet Magazine	www.cnet.com
Institute of Electrical and Electronics Engineers	www.ieee.org
International Standards	www.iso.org/iso/en/ CatalogueListPage.CatalogueList
Linux Online	www.linux.org
Microsoft Corporation	www.microsoft.com
Sun Developer Network, for Java	http://java.sun.com
The Open Group, for UNIX	www.unix.org
The Unicode Consortium	www.unicode.org
University at Albany (links)	http://library.albany.edu/subject/csci.htm
World Wide Web Consortium	www.w3.org
ZDNet	www.zdnet.com

Bibliography

Bacon, J. & Harris, T. *Operating Systems*. Boston: Addison-Wesley, 2003.

Broy, M. & Denert, E. (eds) *Software Pioneers: contributions to software engineering*. New York: Springer-Verlag, 2002.

Collins, H. & Pinch, T. *The Golem at large: what you should know about technology*. Cambridge, U.K.: Cambridge University Press, 1998.

Cormen, T.H., Leiserson, C.D., Rivest, R.L., & Stein, C. *Introduction to Algorithms*. Cambridge, MA: MIT Press, 2001.

Knuth, D.E. *The Art of Computer Programming*, Boston: Addison-Wesley, 1997-98.

Devlin, K. *Sets, Functions, and Logic: an introduction to abstract mathematics*. London: Chapman and Hall, 2003.

Nissanke, N. *Introductory Logic and Sets for Computer Scientists*. Boston: Addison-Wesley, 1999.

Patterson, D. & Hennessy, J. *Computer Organisation and Design*. San Francisco, CA: Morgan Kaufmann, 1998.

Pressman, R.S. *Software engineering*. New York: McGraw-Hill, 2001.

Silberschatz, A., Peterson, J.L., & Galvin, P.C. *Operating Systems Concepts*. New York: Addison-Wesley, 1998.

Simon, H.A. *The Sciences of the Artificial*. Cambridge, MA: MIT Press, 1996.

Tanenbaum, A.S. *Structured Computer Organisation*. London: Prentice-Hall, 1990.